COLLAPSE OF THE WAVE FUNCTION
Models, Ontology, Origin, and Implications

This is the first single volume about the collapse theories of quantum mechanics, which is becoming a very active field of research in both physics and philosophy. In standard quantum mechanics, it is postulated that when a quantum system is measured, its wave function no longer follows the Schrödinger equation but instantaneously and randomly collapses to one of the wave functions that correspond to definite measurement results. However, why and how a definite measurement result appears is unknown. A promising solution to this problem is collapse theories, in which the collapse of the wave function is spontaneous and dynamical. Chapters written by distinguished physicists and philosophers of physics discuss the origin and implications of wave-function collapse, the controversies around collapse models and their ontologies, and new arguments for the reality of wave-function collapse. This is an invaluable resource for students and researchers interested in the philosophy of physics and foundations of quantum mechanics.

SHAN GAO is Professor of Philosophy at the Research Center for Philosophy of Science and Technology at Shanxi University. He is the founder and managing editor of the *International Journal of Quantum Foundations* and is the author of several books, including the recent monograph *The Meaning of the Wave Function: In Search of the Ontology of Quantum Mechanics* (Cambridge University Press, 2017). His research focuses on the philosophy of physics and foundations of quantum mechanics. He is also interested in the philosophy of mind and the philosophy of science.

COLLAPSE OF THE WAVE FUNCTION

Models, Ontology, Origin, and Implications

Edited by

SHAN GAO
Shanxi University

CAMBRIDGE
UNIVERSITY PRESS

University Printing House, Cambridge CB2 8BS, United Kingdom

One Liberty Plaza, 20th Floor, New York, NY 10006, USA

477 Williamstown Road, Port Melbourne, VIC 3207, Australia

314–321, 3rd Floor, Plot 3, Splendor Forum, Jasola District Centre, New Delhi – 110025, India

79 Anson Road, #06–04/06, Singapore 079906

Cambridge University Press is part of the University of Cambridge.

It furthers the University's mission by disseminating knowledge in the pursuit of education, learning, and research at the highest international levels of excellence.

www.cambridge.org
Information on this title: www.cambridge.org/9781108428989
DOI: 10.1017/9781316995457

© Cambridge University Press 2018

This publication is in copyright. Subject to statutory exception and to the provisions of relevant collective licensing agreements, no reproduction of any part may take place without the written permission of Cambridge University Press.

First published 2018

Printed in the United Kingdom by Clays, St Ives plc

A catalogue record for this publication is available from the British Library.

ISBN 978-1-108-42898-9 Hardback

Cambridge University Press has no responsibility for the persistence or accuracy of URLs for external or third-party internet websites referred to in this publication and does not guarantee that any content on such websites is, or will remain, accurate or appropriate.

Contents

List of Contributors		*page* vii
Preface		ix
Part I Models		1
1	How to Teach and Think About Spontaneous Wave Function Collapse Theories: Not Like Before LAJOS DIÓSI	3
2	What Really Matters in Hilbert-Space Stochastic Processes GIANCARLO GHIRARDI, ORESTE NICROSINI, AND ALBERTO RIMINI	12
3	Dynamical Collapse for Photons PHILIP PEARLE	23
4	Quantum State Reduction DORJE C. BRODY AND LANE P. HUGHSTON	47
5	Collapse Models and Space–time Symmetries DANIEL J. BEDINGHAM	74
Part II Ontology		95
6	Ontology for Collapse Theories WAYNE C. MYRVOLD	97
7	Properties and the Born Rule in GRW Theory ROMAN FRIGG	124
8	Paradoxes and Primitive Ontology in Collapse Theories of Quantum Mechanics RODERICH TUMULKA	134

9	On the Status of Primitive Ontology PETER J. LEWIS	154
10	Collapse or No Collapse? What Is the Best Ontology of Quantum Mechanics in the Primitive Ontology Framework? MICHAEL ESFELD	167

Part III Origin — 185

11	Quantum State Reduction via Gravity, and Possible Tests Using Bose–Einstein Condensates IVETTE FUENTES AND ROGER PENROSE	187
12	Collapse. What Else? NICOLAS GISIN	207
13	Three Arguments for the Reality of Wave-Function Collapse SHAN GAO	225
14	Could Inelastic Interactions Induce Quantum Probabilistic Transitions? NICHOLAS MAXWELL	257
15	How the Schrödinger Equation Would Predict Collapse: An Explicit Mechanism ROLAND OMNÈS	274

Part IV Implications — 293

16	Wave-Function Collapse, Non-locality, and Space–time Structure TEJINDER P. SINGH	295
17	The Weight of Collapse: Dynamical Reduction Models in General Relativistic Contexts ELIAS OKON AND DANIEL SUDARSKY	312

Index — 346

Contributors

Daniel J. Bedingham
Royal Holloway, University of London

Dorje C. Brody
Brunel University London and St Petersburg National Research University of Information Technologies, Mechanics and Optics

Lajos Diósi
Wigner Research Center for Physics

Michael Esfeld
University of Lausanne

Roman Frigg
London School of Economics and Political Science

Ivette Fuentes
University of Nottingham and University of Vienna

Shan Gao
Shanxi University

Nicolas Gisin
University of Geneva

Lane P. Hughston
Brunel University London

GianCarlo Ghirardi
University of Trieste and Abdus Salam International Centre for Theoretical Physics

Peter J. Lewis
Dartmouth College

Nicholas Maxwell
University College London

Wayne C. Myrvold
University of Western Ontario

Oreste Nicrosini
INFN, Pavia Unit

Elias Okon
National Autonomous University of Mexico

Roland Omnès
University of Paris XI

Philip Pearle
Hamilton College

Roger Penrose
University of Oxford

Alberto Rimini (deceased)
University of Pavia

Tejinder P. Singh
Tata Institute of Fundamental Research

Daniel Sudarsky
National Autonomous University of Mexico

Roderich Tumulka
Eberhard-Karls University Tübingen

Preface

In standard quantum mechanics, it is postulated that when a quantum system is measured by a measuring device, its wave function no longer follows the linear Schrödinger equation but instantaneously and randomly collapses to one of the wave functions that correspond to definite measurement results. However, this collapse postulate is ad hoc, and the theory does not tell us why and how a definite measurement result appears. A promising solution to this measurement problem is collapse theories, in which the collapse of the wave function is spontaneous and dynamical, and it is integrated with the Schrödinger evolution into a unified dynamics.

The origin of collapse theories can be traced back to as early as 1952. Shortly after David Bohm suggested his hidden-variable interpretation of quantum mechanics (Bohm, 1952), Lajos Jánossy published a paper titled "The physical aspects of the wave-particle problem," in which he argued for the reality of spontaneous wave-function collapse and also tried to alter quantum mechanics by introducing the so-called damping term into the Schrödinger equation (Jánossy, 1952). Jánossy's collapse model is incomplete in many aspects but it does illustrate the possibility of a dynamical collapse theory. Although Jánossy's work has gone unnoticed by later researchers of collapse theories, it was referred to and discussed by Werner Heisenberg, who regarded it as the "most careful attempt in this direction" (Heisenberg, 1958).

Today, collapse theories, together with Bohm's theory and Everett's theory, are widely considered as three main realistic alternatives to standard quantum mechanics. This book is the first single volume about collapse theories. It is intended not mainly as a comprehensive review of collapse theories[1] but more as an anthology reflecting the latest thoughts of leading experts on the subject. The

---

[1] See Bassi and Ghirardi (2003), Diósi (2005), Pearle (2007, 2009), Bassi et al. (2013), and Ghirardi (2016) for helpful reviews of collapse theories.

book is accessible to graduate students in physics. It will be of value to students and researchers with an interest in the philosophy of physics and especially to physicists and philosophers working on the foundations of quantum mechanics.

This book is composed of four parts. The first part is about collapse models. There are (at least) two key issues for collapse models. The first one is the preferred basis problem. It is still unclear what the preferred bases should be in collapse theories. If the preferred bases are smeared position eigenstates, as is commonly believed, then it seems that the theories will inevitably violate the law of conservation of energy. But if the preferred bases are energy eigenstates, then the violation can be readily avoided.[2] In addition, it is still unknown whether a relativistically invariant collapse model can be formulated in a satisfactory way, although the recent progress seems to suggest a positive answer. These two issues, among others, will be addressed in the first part of this book.

The second part of this book concerns the ontology of collapse theories. Since collapse theories are a realistic alternative to standard quantum mechanics, it must have an ontology. Indeed, several ontologies of collapse theories have been proposed, such as mass density ontology, flash ontology, and the recently suggested ontology of RDM (random discontinuous motion) of particles (Gao, 2017). The question, then, is which one the right ontology of collapse theories is. Another issue is whether the ontology for collapse theories may contain only the wave function. Proponents of the primitive ontology approach will give a negative answer to this question, while others may disagree. The key is to understand the physical meaning of the wave function. For example, if the wave function indeed represents a property of particles in three-dimensional space, which is itself a primitive ontology, then the answer to the question may be positive even if assuming the primitive ontology approach. These concerns about the ontology of collapse theories will be addressed and debated in the second part of this book.

The third part of this book is about the physical origin of wave-function collapse. In fact, the first question collapse theories must answer is just this: What causes the (spontaneous) collapse of the wave function? There have been a few suggested answers to this question, a well-known one of which is Penrose's gravity-induced collapse conjecture. In this part, several authors, including Penrose himself, will give further new analysis of the origin of wave-function collapse. The last part of this book concerns the possible implications of collapse theories for our understandings of space–time and cosmology. This part may be more speculative. But if collapse theories are indeed in the right direction to solve the measurement problem, then it will be quite reasonable or even imperative to consider the applications

---

[2] It has been shown that a collapse model with the preferred bases being energy eigenstates can still be consistent with experiments and our experience (Gao, 2013, 2017).

of these theories, which may help solve the deep puzzles in other fields of modern physics, such as quantum gravity and cosmology. The last two parts of this book also contain a concise review of experimental tests of collapse theories (see Chapter 16) and a new test suggestion (see Chapter 11).

Different from other quantum theories, collapse theories are becoming a lively field of research in both philosophy and physics. I hope this anthology will arouse more researchers' interest in these promising and testable quantum theories. I thank all contributors for taking the time to write these new essays in the anthology. I am particularly grateful to Steve Adler, Lajos Diósi, Roman Frigg, GianCarlo Ghirardi, Nicolas Gisin, Philip Pearle, and Roger Penrose for their help and support. I thank Simon Capelin of Cambridge University Press for his kind support as I worked on this project, and the referees who gave helpful suggestions on how the work could best serve its targeted audience. Finally, I am deeply indebted to my parents, QingFeng Gao and LiHua Zhao, my wife Huixia, and my daughter Ruiqi for their unflagging love and support.

References

Bassi, A. and Ghirardi, G. C. (2003). Dynamical reduction models. *Phys. Rep.* 379, 257.
Bassi, A., Lochan, K., Satin, S., Singh, T. P., and Ulbricht, H. (2013). Models of wavefunction collapse, underlying theories, and experimental tests. *Rev. Mod. Phys.* 85, 471.
Bohm, D. (1952). A suggested interpretation of quantum theory in terms of "hidden" variables, I and II. *Physical Review* 85, 166.
Diósi, L. (2005). Intrinsic time-uncertainties and decoherence: comparison of 4 models. *Brazilian Journal of Physics* 35, 260.
Gao, S. (2013). A discrete model of energy-conserved wavefunction collapse, *Proceedings of the Royal Society* A 469, 20120526.
Gao, S. (2017). *The Meaning of the Wave Function: In Search of the Ontology of Quantum Mechanics*. Cambridge: Cambridge University Press.
Ghirardi, G (2016). Collapse Theories. *The Stanford Encyclopedia of Philosophy* (Spring 2016 Edition), Edward N. Zalta (ed.), https://plato.stanford.edu/archives/spr2016/entries/qm-collapse.
Heisenberg, W. (1958). *Physics and Philosophy: The Revolution in Modern Science*. New York: Harper and Row. p.113.
Jánossy, L. (1952). The physical aspects of the wave-particle problem. *Acta Physica Academiae Scientiarum Hungaricae* 1(4), 423.
Pearle, P. (2007). How stands collapse I. *Journal of Physics A: Mathematical and General* 40, 3189.
Pearle, P. (2009). How stands collapse II. in Myrvold, W. C. and Christian, J. eds., *Quantum Reality, Relativistic Causality, and Closing the Epistemic Circle: Essays in Honour of Abner Shimony*. The University of Western Ontario Series in Philosophy of Science, 73(IV), 257.

Part I

Models

1
How to Teach and Think About Spontaneous Wave Function Collapse Theories: Not Like Before

LAJOS DIÓSI

A simple and natural introduction to the concept and formalism of spontaneous wave function collapse can and should be based on textbook knowledge of standard quantum state collapse and monitoring. This approach explains the origin of noise driving the paradigmatic stochastic Schrödinger equations of spontaneous localization of the wave function Ψ. It reveals, on the other hand, that these equations are empirically redundant and the master equations of the noise-averaged state $\hat{\rho}$ are the only empirically testable dynamics in current spontaneous collapse theories.

1.1 Introduction

"We are being captured in the old castle of standard quantum mechanics. Sometimes we think that we have walked into a new wing. It belongs to the old one, however." [1]

The year 1986 marked the birth of two theories, prototypes of what we call the theory of spontaneous wave function collapse. Both the GRW paper published in *Physical Review D* [2] followed by Bell's insightful work [3] and the author's thesis [4] constructed strict stochastic jump equations to explain unconditional emergence of classical behavior in large quantum systems. Subsequently, both theories obtained their time-continuous versions, driven by white noise rather than by stochastic jumps. The corresponding refinement of the GRW proposal [5, 6, 7] is the Continuous Spontaneous Localization (CSL) theory. The author's gravity-related spontaneous collapse theory [4, 8, 9] used to be called DP theory after Penrose concluded to the same equation for the characteristic time of spontaneous collapse in large bodies [10]. These theories modified the standard theory of quantum mechanics in order to describe the irreversible process of wave function collapse. The mathematical structure of modification surprised the proponents themselves, and it looked strange and original for many of the interested as well. In fact, these theories were considered new physics with new mathematical structures to replace standard equations like Schrödinger's. The predicted effects of spontaneous collapses are extremely small and have thus remained untestable for the lack of experimental technique. After three decades, fortunately, tests on nanomasses are now becoming gradually available. Theories like GRW, CSL, and DP have not changed over the decades apart from their parameter ambiguities; see reviews by Bassi *et al.* [11, 12]. But our understanding and teaching of spontaneous collapse should be revised radically.

Personally, I knew that GRW's random jumps looked like unsharp measurements, but, in the late 1980s, I believed that unsharp measurements were phenomenological modifications of von Neumann standard ones. My belief extended also for the time-continuous limit of unsharp measurements [13] that DP collapse equations [9] were based on. Finally in the 1990s I got rid of my ignorance and learned that unsharp measurements and my time-continuous measurement (monitoring) could have equally been derived from standard quantum theory [14, 15].

That was disappointing [1]. Excitement about the radical novelty of our modified quantum mechanics evaporated. Novelty got reduced to the concept that tiny collapses which get amplified for bulk degrees of freedom happen everywhere and without measurement devices. That's why we call them spontaneous. But they are standard collapses otherwise. I have accordingly stressed their revised interpretation recently [16], and the present work is arguing further toward such demand.

1.2 How to Teach GRW Spontaneous Collapse?

We should build as much as possible on standard knowledge, using standard concepts, equations, and terminology. The key notion is unsharp generalized measurement, which has been standard ever since von Neumann showed how inserting an ancilla between object and measuring device will control measurement unsharpness [17]. Hence we are in the best pedagogical position to explain GRW theory to educated physicists. No doubt, for old generations measurement means the projective (sharp) one, but this has changed recently due to the boom in quantum information science. For younger scientists, generalized measurements are the standard ones, while projective measurements are the specific case [18, 19]. For the new generation, there is a natural way to get acquainted with spontaneous collapse. The correct and efficient teaching goes like this.

GRW theory assumes that independent position measurements of unsharpness (precision) $r_C/\sqrt{2}$, with GRW choice $r_C = 10^{-5}$ cm, are happening randomly at average frequency $\lambda = 10^{-19}$ Hz on each (non-relativistic) particle in the Universe. The two parameters r_C, λ are considered new universal constants of Nature. The mathematical model of unsharp measurements is exactly the same as for independent von Neumann detectors [17] where the Gaussian ancilla wave function has the width $\sigma = r_C/\sqrt{2}$ [1]. The difference from standard von Neumann detection is the concept of being spontaneous: GRW are measurements supposed to happen *without* the presence of detectors.

The merit of GRW is wave function *localization* in bulk degrees of freedom, e.g., the center of mass (c.o.m.) of large objects. Quantum theory allows for arbitrary large quantum fluctuations of macroscopic degrees of freedom in large quantized systems. The extreme example is a Schrödinger cat state in which two macroscopically different wave functions would be superposed. In GRW theory such macroscopic superpositions or fluctuations become suppressed by GRW spontaneous measurements but the superpositions of microscopic degrees of freedom will invariably survive. These complementary features are guaranteed by the chosen values of parameters σ and λ. Due to the extreme low rate of measurements, individual particles are almost never measured. But among an Avogadro number (A) of constituents some $N = A\lambda \sim 10^4$ become spontaneously measured in each second, meaning that their collective variables, e.g. center of mass, are measured each second with a precision of $\sigma/\sqrt{N} \sim 10^{-7}$ cm, leading to extreme sharp c.o.m. localization on the long run. That's what we expect of spontaneous localization theories.

The *mathematical model* is the following. We model the Universe or part of it by a quantized N-body system satisfying the Schrödinger equation

$$\frac{d|\Psi\rangle}{dt} = -\frac{i}{\hbar}\hat{H}|\Psi\rangle \tag{1.1}$$

apart from instances of spontaneous position measurements that happen randomly and independently at rate λ on every constituent. Spontaneous position measurements are standard generalized measurements. Accordingly, when the kth coordinate $\hat{\mathbf{x}}_k$ endures a measurement, the quantum state undergoes the following collapse:

$$|\Psi\rangle \Longrightarrow \frac{\sqrt{G(\mathbf{x}_k - \hat{\mathbf{x}}_k)}|\Psi\rangle}{\|\sqrt{G(\mathbf{x}_k - \hat{\mathbf{x}}_k)}|\Psi\rangle\|}. \tag{1.2}$$

The effects of unsharp position measurement take the Gaussian form:

$$G(\mathbf{x}_k - \hat{\mathbf{x}}_k) = \frac{1}{(2\pi\sigma^2)^{3/2}} \exp\left(-\frac{(\mathbf{x}_k - \hat{\mathbf{x}}_k)^2}{2\sigma^2}\right), \tag{1.3}$$

where $\mathbf{x}_k$ is the random outcome of the unsharp position measurement on $\hat{\mathbf{x}}_k$, and σ sets the scale of unsharpness (precision). The probability of the outcomes $\mathbf{x}_k$ is defined by the standard rule:

$$p(\mathbf{x}_k) = \|\sqrt{G(\mathbf{x}_k - \hat{\mathbf{x}}_k)}|\Psi\rangle\|^2. \tag{1.4}$$

We have thus specified the mathematical model of GRW in terms of standard unsharp position measurements targeting every constituent at rate λ and precision σ. These measurements are *selective* measurements if we assume that the measurement outcomes $\mathbf{x}_k$ are accessible. If they are not, we talk about *non-selective* measurements, and the jump equation (1.2) should be averaged over the outcomes, according to the probability distribution (1.4). The mathematical model of the GRW theory reduces to the following master equation for the density matrix $\hat{\rho}$:

$$\begin{aligned}\frac{d\hat{\rho}}{dt} &= -\frac{i}{\hbar}[\hat{H}, \hat{\rho}] + \lambda \sum_k \left(\int d\mathbf{x}_k \sqrt{G(\mathbf{x}_k - \hat{\mathbf{x}}_k)} \hat{\rho} \sqrt{G(\mathbf{x}_k - \hat{\mathbf{x}}_k)} \right) - \lambda \hat{\rho} \\ &= -\frac{i}{\hbar}[\hat{H}, \hat{\rho}] + \lambda \sum_k \mathcal{D}[\hat{\mathbf{x}}_k]\hat{\rho}. \end{aligned} \tag{1.5}$$

The decoherence superoperator is defined by

$$\mathcal{D}[\hat{\mathbf{x}}]\hat{\rho} = \int d\mathbf{x} \sqrt{G(\mathbf{x} - \hat{\mathbf{x}})} \hat{\rho} \sqrt{G(\mathbf{x} - \hat{\mathbf{x}})} - \hat{\rho}. \tag{1.6}$$

We can analytically calculate it in coordinate representation $\rho(\mathbf{x}, \mathbf{x}')$ of the density matrix. Its contribution on the rhs of the master equation (1.5) shows spatial decoherence, saturating for large separations:

$$\frac{d\rho(\mathbf{x}, \mathbf{x}')}{dt} = \cdots - \lambda \sum_k \left(1 - \exp\left(-\frac{(\mathbf{x}_k - \mathbf{x}'_k)^2}{8\sigma^2}\right) \right) \rho(\mathbf{x}, \mathbf{x}'), \tag{1.7}$$

where ellipsis stands for the Hamiltonian part.

The *amplification mechanism* is best illustrated in c.o.m. dynamics. As we said, for the individual particles the decoherence term remains negligible, whereas for bulk degrees of freedom, e.g. the c.o.m., it becomes crucial to damp Schrödinger cats, as we desired. Assume, for simplicity, free spatial motion of a many-body object. Then the non-selective GRW equation (1.5) yields the following autonomous equation for the reduced c.o.m. density matrix $\hat{\rho}_{cm}$:

$$\frac{d\hat{\rho}_{cm}}{dt} = -\frac{i}{\hbar}[\hat{H}_{cm}, \hat{\rho}_{cm}] + N\lambda \mathcal{D}[\hat{\mathbf{x}}_{cm}]\hat{\rho}_{cm}. \tag{1.8}$$

As we see, the decoherence term concerning the c.o.m. coordinate has been amplified by the number N of the constituents [2], ensuring the desired fast decay of macroscopic superpositions:

$$\frac{d\rho_{cm}(\mathbf{x}_{cm}, \mathbf{x}'_{cm})}{dt} = \cdots - N\lambda \left(1 - \exp\left(-\frac{(\mathbf{x}_{cm} - \mathbf{x}'_{cm})^2}{8\sigma^2}\right) \right) \rho(\mathbf{x}_{cm}, \mathbf{x}'_{cm}). \tag{1.9}$$

In the selective evolution, the individual GRW measurements (1.2) entangle the c.o.m., rotation and internal degrees of freedom; hence $|\Psi_{cm}\rangle$ does not exist in general. It does exist in a limiting case of rigid many-body motion when the unitary evolution of $|\Psi_{cm}\rangle$ is interrupted by spontaneous σ-precision measurements of the c.o.m. coordinate $\hat{x}_{cm}$ similar to (1.2), just the average rate of the measurements becomes $N\lambda$ [20] instead of λ.

1.3 Localization Is Not Testable, but Decoherence Is

The standard concept of selective measurement implies that we have access to the measurement outcomes, which are the values $\mathbf{x}_k$ in GRW. If they are accessible variables, then the stochastic jump process of the GRW state vector $|\Psi\rangle$ is testable; otherwise it is not. If not, then the same spontaneous measurement is called nonselective, and what is testable is the density operator $\hat{\rho}$. The stochastic jump process (1.1–1.4) becomes *illusory*, and the master equation (1.5) contains the whole GRW physics.

This latter sentence holds in GRW where, as a matter of fact, the $\mathbf{x}_k$s remain inaccessible. Consider the conservative preparation-detection scenario. Assume we prepare a well-defined pure initial state $\hat{\rho}_0 = |\Psi_0\rangle\langle\Psi_0|$ and after time t we desire to test it for the presence of GRW collapses (1.2), but we perform no test prior to this one. As a matter of fact, the relevant state is $\hat{\rho}_t$, being the solution of the master equation (1.5), which does not know about GRW collapses but about GRW decoherence. This is equally valid in the particular case of the macroscopic Schrödinger cat initial state, i.e., a superposition of c.o.m. at two distant locations. The c.o.m. GRW master equation (1.8) will exhaustively predict the results of all subsequent tests on the c.o.m. (including the results and statistics of possible naked-eye observations).

Obviously, inference on stochastic collapse assumes our access to the measurement outcomes. In real laboratory quantum measurements it is the detector design and operation that determine if we have full (or partial) access to the measurement outcomes or we have no access at all. In the case of GRW collapse, accessibility of outcomes is not a matter of postulation. It is useless to postulate that $\mathbf{x}_k$s are accessible without a prescription of how to access them.

1.4 Digression: Random Unitary Process Indistinguishable From GRW

Let us consider an alternative to GRW random process in which the stochastic nonlinear GRW jumps (1.2) are replaced by the following stochastic unitary jumps:

$$|\Psi\rangle \Longrightarrow e^{i\mathbf{k}\hat{\mathbf{x}}_k}|\Psi\rangle, \qquad (1.10)$$

corresponding to the transfer of momentum $\hbar\mathbf{k}$ to the kth constituent. The probability distribution of momentum transfer is universal, independent of the particle and of the state:

$$p(\mathbf{k}) = \frac{1}{(2\pi\sigma^{-2})^{3/2}} \exp\left(-\frac{\mathbf{k}^2}{2\sigma^{-2}}\right). \qquad (1.11)$$

The decoherence superoperator acts as

$$\mathcal{D}[\hat{\mathbf{x}}]\hat{\rho} = \int d\mathbf{k}\, p(\mathbf{k}) e^{i\mathbf{k}\hat{\mathbf{x}}} \hat{\rho} e^{-i\mathbf{k}\hat{\mathbf{x}}} - \hat{\rho}, \qquad (1.12)$$

which looks completely different from the GRW structure (1.6) but coincides with it! Hence the master equation for the Schrödinger (1.1) dynamics with the averaged unitary jumps (1.10) will be the the master equation (1.5) derived earlier for the GRW theory. As we argued in Sec. 1.3, the GRW theory can only be tested at the level of the density operator; no experiment could tell us whether the underlying stochastic process of $|\Psi\rangle$ was the GRW stochastic localizing process (1.1–1.4) or the stochastic unitary process.

1.5 How to Think About CSL?

We could repeat what we said concerning correct and efficient teaching of GRW in Section 1.2. This time the standard discipline of modern physics, relevant to CSL, is time-continuous quantum measurement (monitoring), which is just the time-continuous limit of unsharp sequential measurements similar to those underlying GRW in Section 1.2. Quantum monitoring theory was not yet conceived in 1986 (GRW). It was born in 1988, and it became widely known in the 1990s as the standard theory of quantum monitoring in the laboratory [14, 15]. It played an instrumental role for semiclassical gravity's consistent introduction to spontaneous collapse theories [21, 22]. In what follows, I utilize the summary of standard Markovian quantum monitoring theory from [21].

So, how should we interpret CSL? It derives from GRW. The discrete sequence of spontaneous unsharp position measurements is replaced by spontaneous monitoring of the spatial number distribution of particles [6] (or, in a later version, of the spatial mass distribution of particles [7]). Accordingly, CSL introduces the smeared mass distribution

$$\hat{n}(\mathbf{x}) = \sum_k G(\mathbf{x} - \hat{\mathbf{x}}_k), \qquad (1.13)$$

where, this time, the width of the Gaussian is r_C. Monitoring yields the measured signal in the form

$$n_t(\mathbf{x}) = \langle\Psi_t|\hat{n}(\mathbf{x})|\Psi_t\rangle + \delta n_t(\mathbf{x}), \qquad (1.14)$$

where $\delta n_t(\mathbf{x})$ is the signal white noise, still depending on the spatial resolution/correlation of monitoring. The CSL signal noise is a spatially uncorrelated white noise:

$$\mathbb{E}\delta n_t(\mathbf{x})\delta n_s(\mathbf{y}) = \frac{1}{4\gamma}\delta(\mathbf{x}-\mathbf{y})\delta(t-s). \tag{1.15}$$

Just like in the case of GRW sequential spontaneous measurements, the conditional quantum state evolves stochastically, this time according to the following stochastic Schrödinger equation, *driven by the signal noise* in the Ito-sense:

$$\frac{d|\Psi\rangle}{dt} = \left\{-\frac{i}{\hbar}\hat{H} - \frac{\gamma}{2}\int d\mathbf{x}\left(\hat{n}(\mathbf{x})-\langle\hat{n}(\mathbf{x})\rangle\right)^2 + 4\gamma\int d\mathbf{x}\left(\hat{n}(\mathbf{x})-\langle\hat{n}(\mathbf{x})\rangle\right)\delta n(\mathbf{x})\right\}|\Psi\rangle. \tag{1.16}$$

So far we have introduced the equations of selective spontaneous monitoring, assuming that the signal (1.14) is accessible, which won't be the case, similar to GRW. In non-selective monitoring, the CSL physics reduces to the signal-averaged evolution of the conditional state, i.e., to the CSL master equation:

$$\frac{d\hat{\rho}}{dt} = -\frac{i}{\hbar}[\hat{H},\hat{\rho}] - \frac{\gamma}{2}\int d\mathbf{x}[\hat{n}(\mathbf{x}),[\hat{n}(\mathbf{x}),\hat{\rho}]]. \tag{1.17}$$

(Note $\gamma = (4\pi r_C^2)^{3/2}\lambda$ would ensure the coincidence with GRW's spatial decoherence rate at the single-particle level, although CSL defined a slightly different γ [6]).

The traditional CSL teaching differs in a single major point: it does not mention the theory of monitoring. Hence it does not use the notion of signal $n_t(\mathbf{x})$, and the equation (1.14) is not part of it. Instead, CSL's traditional definition postulates the stochastic Schrödinger equation:

$$\frac{d|\Psi\rangle}{dt} = \left\{-\frac{i}{\hbar}\hat{H} - \frac{\gamma}{2}\int d\mathbf{x}\left(\hat{n}(\mathbf{x})-\langle\hat{n}(\mathbf{x})\rangle\right)^2 + \sqrt{\gamma}\int d\mathbf{x}\left(\hat{n}(\mathbf{x})-\langle\hat{n}(\mathbf{x})\rangle\right)w(\mathbf{x})\right\}|\Psi\rangle, \tag{1.18}$$

which would correspond to the replacement $\delta n_t(\mathbf{x}) = 2\sqrt{\gamma}w_t(\mathbf{x})$ had CSL derived it from our (1.16). The traditional CSL dynamics is driven by the spatially uncorrelated standard white noise, satisfying

$$\mathbb{E}\delta w_t(\mathbf{x})\delta w_s(\mathbf{y}) = \delta(\mathbf{x}-\mathbf{y})\delta(t-s). \tag{1.19}$$

In CSL narrative (e.g. [12]) the origin of the noise field as well its anti-Hermitian coupling to density $\hat{n}(\mathbf{x})$ are mentioned among theory elements yet to be justified, still without reference to the spontaneous monitoring interpretation available already for a long enough time.

From arguments of Section 1.3, it follows that all testable predictions follow from the CSL master equation (1.17), the stochastic Schrödinger equation (1.18) is *empirically redundant*, collapse in the claimed quantitative sense is an illusion.

1.6 Final Remarks

Disregarding that spontaneous collapse theories are rooted in standard quantum mechanical collapse theories with hidden detectors has had too many drawbacks.

The principle one is the illusion that the quantitative models of spontaneous collapse (localization) in their current forms are relevant empirically like master equations of spontaneous decoherence are, which have already been under empiric tests due to recent breakthroughs in technology. This illusion is surviving despite no proposals having been ever made for a future experiment to test underlying localization effects of $|\Psi\rangle$ beyond decoherence of $\hat{\rho}$; all proposals have so far concerned the dynamical features (e.g. spontaneous decoherence) of the averaged state $\hat{\rho}$.

Secondary drawbacks concern illusions that teaching and interpretation of spontaneous collapse necessitate radical departure from standard quantum theory both conceptually and mathematically. This may have kept philosophers excited and may have prevented students from learning the subject faster and physicists from going deeper into their fundational investigations.

Physics research will gradually adapt itself to the option that spontaneous collapse fits better to standard quantum knowledge than we thought of it before. Monitoring theory roots were revealed for DP spontaneous collapse from the beginning and have been detailed and exploited for CSL, too, recently in [21, 22].

References

[1] L. Diósi, Emergence of classicality: From collapse phenomenologies to hybrid dynamics, *Lect. Notes Phys.* **538**, 243–250 (2000).
[2] G. C. Ghirardi, A. Rimini, and T. Weber, Unified dynamics for microscopic and macroscopic systems, *Phys. Rev.* **D 34**, 470–491 (1986).
[3] J. S. Bell, Are there quantum jumps? In *Schrödinger Centenary of a polymath* (Cambridge, Cambridge University Press, 1987).
[4] L. Diósi, A quantum-stochastic gravitation model and the reduction of the wavefunction. Thesis, 1986. Available from: http://wigner.mta.hu/~diosi/prints/thesis1986.pdf.
[5] N. Gisin, Stochastic quantum dynamics and relativity, *Helv. Phys. Acta* **62**, 363–371 (1989).
[6] G. C. Ghirardi, P. Pearle, and A. Rimini, Markov processes in Hilbert space and continuous spontaneous localization of systems of identical particles, *Phys. Rev.* **A 42**, 78–89 (1990).
[7] G. C. Ghirardi, R. Grassi, and F. Benatti, Describing the macroscopic world: Closing the circle within the dynamical reduction program, *Found. Phys.* **25**, 5–38 (1995).

[8] L. Diósi, A universal master equation for the gravitational violation of the quantum mechanics, *Phys. Lett.* **120A**, 377–381 (1987).

[9] L. Diósi, Models for universal reduction of macroscopic quantum fluctuations, *Phys. Rev.* **A 40**, 1165–1174 (1989).

[10] R. Penrose, On gravity's role in quantum state reduction. *Gen. Rel. Grav.* **28**, 581 (1996).

[11] A. Bassi and G. C. Ghirardi, Dynamical reduction models, *Phys. Rep.* **379**, 257–426 (2003).

[12] A. Bassi, K. Lochan, S. Satin, T. P. Singh, and H. Ulbricht, Models of wave-function collapse, underlying theories, and experimental tests, *Rev. Mod. Phys.* **85**, 471–527 (2013).

[13] L. Diósi, Continuous quantum measurement and Ito-formalism, *Phys. Lett.* **129A**, 419–423 (1988).

[14] H. Carmichael, *An open system approach to quantum optics* (Berlin, Springer, 1993).

[15] H. M. Wiseman and G. J. Milburn, *Quantum measurement and control* (Cambridge, Cambridge University Press, 2010).

[16] L. Diósi, Is spontaneous wave function collapse testable at all? *J. Phys. Conf. Ser.* **626**, 012008-(5) (2015).

[17] J. von Neumann, *Matematische Grundlagen der Quantenmechanik* (Berlin, Julius Springer, 1932).

[18] M. A. Nielsen and I. L. Chuang, *Quantum computing and quantum information* (Cambridge, Cambridge University Press, 2000).

[19] L. Diósi: *Short course in quantum information theory* (Berlin, Springer, 2011).

[20] L. Diósi, On the motion of solids in modified quantum mechanics, *Eurohys. Lett.* **8**, 285–290 (1988).

[21] A. Tilloy and L. Diósi, Sourcing semiclassical gravity from spontaneously localized quantum matter, *Phys. Rev.* **D 93**, 024026-(12) (2016).

[22] A. Tilloy and L. Diósi, Principle of least decoherence for Newtonian semi-classical gravity, *Phys.Rev.* **D96**, 104045-(6) (2017).

2

What Really Matters in Hilbert-Space Stochastic Processes

GIANCARLO GHIRARDI, ORESTE NICROSINI, AND ALBERTO RIMINI

The relationship between discontinuous and continuous stochastic processes in Hilbert space is investigated. It is shown that for any continuos process, there is a parent discontinuous process that becomes the continuous one in the proper infinite frequency limit. From the point of view of solving the quantum measurement problem, what really matters is the choice of the set of operators whose value distributions are made sharp. In particular, the key role of position sharpening is emphasized.

2.1 Introduction

In attempting to overcome the difficulties connected with the quantum measurement problem, two types of stochastic processes in Hilbert space have been considered, the so-called hitting or discontinuous processes and the continuous ones. All proposed processes aim at describing reduction as a physical process. A common feature is non-linearity, which originates from the need of maintaining normalization while reducing the state vector. One finds sometimes the statement that the continuous processes allow us to deal with systems which cannot be treated with the hitting processes. The aim of the present contribution is to discuss the relationship between discontinuous and continuous processes. Actually, it will be shown that all continuous processes have a physically equivalent *parent* discontinuous one, applicable to the same class of physical systems.

In all examples of stochastic process in Hilbert space a set of *physical quantities* (observables) appears, represented by the corresponding set of self-adjoint operators. The process acts by inducing the *sharpening* of the distribution of values of those quantities around a stochastically chosen centre. What really matters from the physical point of view is the choice of the set of quantities rather than the details of the sharpening procedure.

Our dear colleague and friend Alberto Rimini sadly passed away after the completion of this work (13 April 2017). He will live in our hearts forever.

Both approaches have their benefits. In the discontinuous processes the meaning of the parameters appearing in their definition is transparent and the physical consequences are easy to grasp, while the continuous processes are undoubtedly more elegant from the mathematical point of view. The continuous processes depend on one parameter less than the corresponding discontinuous ones. But this difference is illusory since, as discussed in what follows, the effectiveness of the discontinuous process depends on two of its parameters only through their product.

In the next section a simplified heuristic derivation of a generic continuous process as the proper infinite frequency limit of the parent hitting process is given. In the last section the three physically most relevant examples of discontinuous and corresponding continuous processes are presented.

2.2 Infinite Frequency Limit

In the present section we show how a generic hitting process becomes a continuous stochastic process in the Hilbert space under a suitable infinite frequency limit.

A hitting process is characterized by a choice of the quantities whose distribution is to be made sharp by the mean frequency of the Poisson distribution of hitting times and by the accuracy of the sharpening of the chosen quantities. A probabilistic rule is assumed for the distribution of the hitting centres. The effectiveness of the hitting process actually depends on the products of the mean frequency times the accuracy parameters, so that a given effectiveness can be obtained by increasing the hitting frequency and, at the same time, appropriately decreasing the accuracy parameters. Taking the infinite frequency limit with this prescription, one gets the corresponding continuous process.

This feature of the effectiveness of a hitting process was first noticed by [1] in a different conceptual context. Later, the same feature was highlighted by [2], and the infinite frequency limit was considered by [3] with reference to the time evolution of the statistical operator. Last, [4] in a particular case and [5] in the general case considered the same limit for the time evolution of the state vector, obtaining the continuous stochastic process.

In what follows, for the sake of simplicity we use evenly spaced hitting times instead of random ones, and, at least initially, we ignore the Schrödinger evolution. In the present section the sharpened quantities are assumed, for simplicity, to have a purely discrete spectrum.

Let the set of compatible quantities characterizing the considered discontinuous stochastic process be

$$\hat{A} \equiv \{\hat{A}_p; p = 1, 2, \ldots, K\}, \qquad [\hat{A}_p, \hat{A}_q] = 0, \qquad \hat{A}_p^\dagger = \hat{A}_p, \qquad (2.1)$$

and the sharpening action be given by the operator

$$S_i = \left(\frac{\beta}{\pi}\right)^{K/4} \exp(R_i), \qquad R_i = -\tfrac{1}{2}\beta\,(\hat{A} - a_i)^2. \tag{2.2}$$

The parameter β rules the accuracy of the sharpening and a_i is the centre of the hitting i. It is assumed that the hittings occur with frequency μ.

The sharpening operator for the ith hitting S_i acts on the normalized state vector $|\psi_t\rangle$ giving the normalized state vector $|\psi'_{i,t}\rangle$ according to

$$|\psi'_{i,t}\rangle = \frac{|\chi_{i,t}\rangle}{\||\chi_{i,t}\rangle\|}, \qquad |\chi_{i,t}\rangle = S_i|\psi_t\rangle. \tag{2.3}$$

The probability that the hitting takes place around a_i is

$$\mathscr{P}(\psi_t|a_i) = \|\chi_{i,t}\|^2. \tag{2.4}$$

When considering n evenly spaced hittings in $(t, t + dt]$, so that the time interval between two adjacent hittings is $\tau = dt/n$, the hitting frequency is given by $\mu = 1/\tau$. In this case, assumptions (2.3) and (2.4) become

$$|\psi_{t+i\tau}\rangle = \frac{|\chi_{t+i\tau}\rangle}{\||\chi_{t+i\tau}\rangle\|}, \qquad |\chi_{t+i\tau}\rangle = S_i|\psi_{t+(i-1)\tau}\rangle, \tag{2.5}$$

with probability

$$\mathscr{P}(\psi_{t+(i-1)\tau}|a_i) = \mathscr{P}(a_1, a_2, \ldots, a_{i-1}|a_i) = \|\chi_{t+i\tau}\|^2. \tag{2.6}$$

The continuous process based on the same quantities $\hat{A} \equiv \{\hat{A}_p; p = 1, 2, \ldots, K\}$ is ruled by the Itô stochastic differential equation

$$d|\psi\rangle = \left[\sqrt{\gamma}\,(\hat{A} - \langle\hat{A}\rangle_{\psi_t}) \cdot d\mathbf{B} - \tfrac{1}{2}\gamma\,(\hat{A} - \langle\hat{A}\rangle_{\psi_t})^2 dt\right]|\psi\rangle, \tag{2.7}$$

where

$$\langle\hat{A}\rangle_{\psi_t} = \langle\psi_t|\hat{A}|\psi_t\rangle \tag{2.8}$$

and

$$d\mathbf{B} \equiv \{dB_p; p = 1, 2, \ldots, K\}, \qquad \overline{d\mathbf{B}} = 0, \qquad \overline{dB_p\, dB_q} = \delta_{pq}\,dt. \tag{2.9}$$

The parameter γ sets the effectiveness of the process.

We shall show that taking the infinite frequency limit of the discontinuous process (2.5) and (2.6) with the prescription

$$\beta\mu = \text{constant} = 2\gamma, \tag{2.10}$$

one gets the continuous process (2.7). As a consequence it becomes apparent that, for $t \to \infty$, the continuous process drives the state vector to a common eigenvector of the operators $\hat{A}$, the probability of a particular eigenvector $|a_r\rangle$ being $|\langle a_r|\psi_0\rangle|^2$, for the state vector $|\psi_0\rangle$ at a given arbitrary initial time.

In what follows for any set of stochastic variables v we use the notation $\overline{v_p}$ for the mean value and $\overline{v_p v_q} = \overline{(v_p - \overline{v_p})(v_q - \overline{v_q})}$ for the variances and covariances.

The effect of n hitting processes in the time interval $(t, t + dt]$ is described by

$$|\chi_{t+dt}\rangle = \left(\frac{\beta}{\pi}\right)^{nK/4} \exp(R_n) \ldots \exp(R_2) \exp(R_1) |\psi_t\rangle, \quad |\psi_{t+dt}\rangle = \frac{|\chi_{t+dt}\rangle}{\|\chi_{t+dt}\|}. \quad (2.11)$$

By using the properties of the exponential function, the final non-normalized state vector is then given by

$$|\chi_{t+dt}\rangle = \left(\frac{\beta}{\pi}\right)^{nK/4} \exp\left(-\tfrac{1}{2}\beta \sum_{i=1}^{n} a_i^2\right) \exp\left\{\sum_{i=1}^{n}\left[-\tfrac{1}{2}\beta\left(\hat{A}^2 - 2\hat{A} \cdot a_i\right)\right]\right\} |\psi_t\rangle$$

$$= F \exp\left\{-\tfrac{1}{2}\beta\left[n\hat{A}^2 - 2n\hat{A} \cdot \tfrac{1}{n}\sum_{i=1}^{n} a_i\right]\right\} |\psi_t\rangle$$

$$= F \exp\left\{-\tfrac{1}{2}\beta n\left[\hat{A}^2 - 2\hat{A} \cdot \tfrac{1}{n}\sum_{i=1}^{n}(a_i - \langle\hat{A}\rangle_{\psi_t}) - 2\hat{A} \cdot \langle\hat{A}\rangle_{\psi_t}\right]\right\} |\psi_t\rangle, \quad (2.12)$$

with joint probability

$$\mathscr{P}(a_1, \cdots, a_n) = \|\chi_{t+dt}\|^2. \quad (2.13)$$

Actually, in the limit $\beta \to 0$ the joint probability factorizes as

$$\mathscr{P}(a_1, \cdots, a_n) = \mathscr{P}(a_n) \cdots \mathscr{P}(a_1) \quad (2.14)$$

where

$$\mathscr{P}(a_i) = \left(\frac{\beta}{\pi}\right)^{K/2} \langle\psi_t| \exp\left(-\beta(\hat{A} - a_i)^2\right) |\psi_t\rangle \quad (2.15)$$

as illustrated in the following two-hitting example. The joint probability for two hittings is given by

$$\mathscr{P}(a_1, a_2) = \mathscr{P}(a_1|a_2)\mathscr{P}(a_1), \quad (2.16)$$

where $\mathscr{P}(a_1|a_2)$ is the conditional probability of a_2 given a_1. In turn,

$$\mathscr{P}(a_1|a_2) = \|\chi_{t+2\tau}\|^2$$

$$= \left(\frac{\beta}{\pi}\right)^{K/2} \langle\psi_{t+\tau}| \exp\left(-\beta(\hat{A} - a_2)^2\right)|\psi_{t+\tau}\rangle$$

$$= \left(\frac{\beta}{\pi}\right)^{K/2} \left[\left(\frac{\beta}{\pi}\right)^{-K/2} \langle\psi_t| \exp\left(-\beta(\hat{A} - a_1)^2\right)|\psi_t\rangle^{-1}\right]$$

$$\langle\chi_{t+\tau}| \exp\left(-\beta(\hat{A} - a_2)^2\right)|\chi_{t+\tau}\rangle$$

$$= \left(\frac{\beta}{\pi}\right)^{K/2} \left[\left(\frac{\beta}{\pi}\right)^{-K/2} \langle\psi_t|\exp\left(-\beta(\hat{A}-a_1)^2|\psi_t\rangle^{-1}\right]\right.$$

$$\left(\frac{\beta}{\pi}\right)^{K/2} \langle\psi_t|\exp\left(-\beta(\hat{A}-a_2)^2\exp\left(-\beta(\hat{A}-a_1)^2|\psi_t\rangle\right.\right.$$

$$\xrightarrow[\beta\to 0]{} \left(\frac{\beta}{\pi}\right)^{K/2} \langle\psi_t|\exp\left(-\beta(\hat{A}-a_2)^2|\psi_t\rangle = \mathscr{P}(a_2) \quad (2.17)$$

We now compute the statistical properties of the variables a_i. The average value is given by

$$\overline{a_i} \xrightarrow[\beta\to 0]{} \int da_i\, a_i\, \mathscr{P}(a_i) = \int da\, a\, \mathscr{P}(a)$$

$$= \int da\, a \left(\frac{\beta}{\pi}\right)^{K/2} \langle\psi_t|\exp\left(-\beta(\hat{A}-a)^2\right)|\psi_t\rangle. \quad (2.18)$$

By inserting the expansion of the identity in terms of the common eigenvectors of the operators A, satisfying $\hat{A}_p|\alpha_k\rangle = \alpha_{kp}|\alpha_k\rangle$, one finds

$$\overline{a_i} \xrightarrow[\beta\to 0]{} = \sum_k \int da\, a \left(\frac{\beta}{\pi}\right)^{K/2} \langle\psi_t|\exp\left(-\beta(\alpha_k-a)^2\right)|\alpha_k\rangle\langle\alpha_k|\psi_t\rangle$$

$$= \sum_k |\langle\alpha_k|\psi_t\rangle|^2 \int da\, a \left(\frac{\beta}{\pi}\right)^{K/2} \exp\left(-\beta(\alpha_k-a)^2\right)$$

$$= \sum_k \mathscr{P}_{\psi_t}(\alpha_k)\,\alpha_k = \langle\hat{A}\rangle_{\psi_t}. \quad (2.19)$$

Similarly, for the variances one gets

$$\overline{a_{ip}^2} \xrightarrow[\beta\to 0]{} \int da_i\, a_{ip}^2\, \mathscr{P}(a_i) - \overline{a_{ip}}^2$$

$$= \int da\, a_p^2\, \mathscr{P}(a) - \overline{a_p}^2$$

$$= \int da_p\, a_p^2 \left(\frac{\beta}{\pi}\right)^{1/2} \langle\psi_t|\exp\left(-\beta(\hat{A}_p-a_p)^2\right)|\psi_t\rangle - \overline{a_p}^2$$

$$= \sum_k \int da_p\, a_p^2 \left(\frac{\beta}{\pi}\right)^{1/2} \langle\psi_t|\exp\left(-\beta(\alpha_{kp}-a_p)^2\right)|\alpha_k\rangle\langle\alpha_k|\psi_t\rangle - \overline{a_p}^2. \quad (2.20)$$

By properly shifting the integration variable one then finds

$$\overline{\overline{a_{ip}^2}} \xrightarrow[\beta \to 0]{} \sum_k |\langle \alpha_k | \psi_t \rangle|^2 \int db\,(b+\alpha_{kp})^2 \left(\frac{\beta}{\pi}\right)^{1/2} \exp\left(-\beta b^2\right) - \overline{a_p}^2$$

$$= \sum_k \mathcal{P}_{\psi_t}(\alpha_k) \left(\frac{1}{2\beta} + \alpha_{kp}^2\right) - \overline{a_p}^2$$

$$= \frac{1}{2\beta} + \langle \hat{A}_p^2 \rangle_{\psi_t} - \langle \hat{A}_p \rangle_{\psi_t}^2 \xrightarrow[\beta \to 0]{} \frac{1}{2\beta}. \tag{2.21}$$

Last, for the covariances, without the need of shifting the integration variables one can write

$$\overline{a_{ip}a_{iq}} \xrightarrow[\beta \to 0]{} \int da_i\, a_{ip}\, a_{iq}\, \mathcal{P}(a_i) - \overline{a_{ip}}\,\overline{a_{iq}} = \int da\, a_p\, a_q\, \mathcal{P}(a) - \overline{a_p}\,\overline{a_q}$$

$$= \sum_k \mathcal{P}_{\psi_t}(\alpha_k)\, \alpha_{kp}\, \alpha_{kq} - \overline{a_p}\,\overline{a_q}$$

$$= \langle \hat{A}_p \hat{A}_q \rangle_{\psi_t} - \langle \hat{A}_p \rangle_{\psi_t} \langle \hat{A}_q \rangle_{\psi_t}. \tag{2.22}$$

Let us define the set of stochastic variables

$$d\mathbf{B} = \sqrt{\frac{2\beta}{\mu}} \sum_{i=1}^n \left(\mathbf{a}_i - \langle \hat{\mathbf{A}} \rangle_{\psi_t}\right). \tag{2.23}$$

Taking the limit $\beta \to 0$ according to the prescription (2.10), μ and n go to infinity in the same way, the conditions of the central limit theorem are satisfied so that the variables $d\mathbf{B}$ are Gaussian with the properties

$$\overline{d\mathbf{B}} = 0,$$

$$\overline{dB_p^2} = \frac{2\beta}{\mu}\frac{n}{2\beta} = \frac{n}{\mu} = n\tau = dt,$$

$$\overline{dB_p\, dB_q} = \frac{2\beta}{\mu} n\left(\langle \hat{A}_p \hat{A}_q \rangle_{\psi_t} - \langle \hat{A}_p \rangle_{\psi_t}\langle \hat{A}_q \rangle_{\psi_t}\right)$$

$$= 2\beta\, dt \left(\langle \hat{A}_p \hat{A}_q \rangle_{\psi_t} - \langle \hat{A}_p \rangle_{\psi_t}\langle \hat{A}_q \rangle_{\psi_t}\right) \xrightarrow[\beta \to 0]{} 0. \tag{2.24}$$

Inserting the definition (2.23) into equation (2.12) one gets

$$|\chi_{t+dt}\rangle = F \exp\left\{-\gamma\left(\hat{\mathbf{A}}^2 - 2\hat{\mathbf{A}}\cdot\langle \hat{\mathbf{A}} \rangle_{\psi_t}\right)dt + \sqrt{\gamma}\,\hat{\mathbf{A}}\cdot d\mathbf{B}\right\}|\psi_t\rangle. \tag{2.25}$$

By expanding the exponential and using the rules of Itô calculus one eventually obtains

$$|\chi_{t+dt}\rangle = F\left[1 - \tfrac{1}{2}\gamma\left(\hat{\mathbf{A}}^2 - 4\hat{\mathbf{A}}\cdot\langle \hat{\mathbf{A}} \rangle_{\psi_t}\right)dt + \sqrt{\gamma}\,\hat{\mathbf{A}}\cdot d\mathbf{B}\right]|\psi_t\rangle,$$

$$\|\chi_{t+dt}\|^{-1} = F^{-1}\left[1 - \tfrac{1}{2}\gamma\langle \hat{\mathbf{A}} \rangle_{\psi_t}^2\, dt - \sqrt{\gamma}\,\langle \hat{\mathbf{A}} \rangle_{\psi_t}\cdot d\mathbf{B}\right],$$

so that

$$d|\psi_t\rangle = |\psi_{t+dt}\rangle - |\psi_t\rangle = \|\phi_{t+dt}\|^{-1}|\phi_{t+dt}\rangle - |\psi_t\rangle$$

$$= \left[\sqrt{\gamma}\left(\hat{A} - \langle\hat{A}\rangle_{\psi_t}\right)\cdot d\mathbf{B} - \tfrac{1}{2}\gamma\left(\hat{A} - \langle\hat{A}\rangle_{\psi_t}\right)^2 dt\right]|\psi_t\rangle. \quad (2.26)$$

By assuming that both the Schrödinger evolution and the stochastic process are there, and taking into account that the two terms in the stochastic differential equation (2.26) are of the order $\sqrt{dt}$ and dt, respectively, one can write on the whole

$$d|\psi_t\rangle = \left[-\frac{i}{\hbar}\hat{H}dt + \sqrt{\gamma}\left(\hat{A} - \langle\hat{A}\rangle_{\psi_t}\right)\cdot d\mathbf{B} - \tfrac{1}{2}\gamma\left(\hat{A} - \langle\hat{A}\rangle_{\psi_t}\right)^2 dt\right]|\psi_t\rangle. \quad (2.27)$$

This is the form of the evolution equation normally assumed for continuous stochastic processes in Hilbert space, corresponding to eq. (2.7) with the addition of the term describing the Schrödinger dynamics.

This argument is worked out with reference to a case in which the quantity label runs over a finite numerable set. There are relevant situations in which the quantity label runs over a measurable continuous set. Two such cases will be examined in the following section, together with all the necessary changes.

2.3 Three Relevant Implementations

In the present section we present three physically most relevant implementations of discontinuous stochastic processes and the corresponding continuous evolution equations.

As discussed in Section 2.2, both the discontinuous and the equivalent continuous processes are characterized by the choice of the sharpened quantities $\hat{A}_p$, $p \in \{1,\ldots,K\}$.

The discontinuous process is further specified by a sharpening frequency μ and a sharpening accuracy β. The probability distribution of the hitting centres $a_{p,i}$ for the ith hitting is assumed to be

$$\mathscr{P}(\psi_t|\mathbf{a}_i) = \left(\frac{\beta}{\pi}\right)^{K/2}\langle\psi_t|\exp\left[-\beta\sum_{p=1}^{K}\left(\hat{A}_p - a_{p,i}\right)^2\right]|\psi_t\rangle. \quad (2.28)$$

The continuous process is ruled by eq. (2.27) specified by the strength parameter γ and by the properties of the Gaussian random variables

$$\overline{d\mathbf{B}} = 0,$$
$$\overline{dB_p^2} = dt,$$
$$\overline{dB_p\,dB_q} = 0. \quad (2.29)$$

For equivalence of the two processes, the parameter γ must be given by $\gamma = \beta\mu/2$.

Distinguishable particles

For N distinguishable particles the sharpened quantities are the three-dimensional positions $\hat{x}_l$, $l \in \{1, \ldots, N\}$.

The discontinuous process ([2]) is defined by the localization frequency λ_l for particle l and by the localization accuracy α. The probability distribution of localization centres $\bar{x}_{l,i}$ for the ith hitting on particle l is

$$\mathscr{P}\left(\psi_t | \bar{x}_{l,i}\right) = \left(\frac{\beta}{\pi}\right)^{3/2} \langle \psi_t | \exp\left[-\alpha \left(\hat{x}_l - \bar{x}_{l,i}\right)^2\right] | \psi_t \rangle. \tag{2.30}$$

The corresponding continuous process is ruled by the stochastic differential equation

$$d|\psi_t\rangle = \left[-\frac{i}{\hbar}\hat{H}dt + \sum_{l=1}^{N}\sqrt{\gamma_l}(\hat{x}_l - \langle\hat{x}_l\rangle_{\psi_t})\cdot d\boldsymbol{B}_l - \tfrac{1}{2}\sum_{l=1}^{N}\gamma_l(\hat{x}_l - \langle\hat{x}_l\rangle_{\psi_t})^2 dt\right]|\psi_t\rangle, \tag{2.31}$$

where the stochastic variables $d\boldsymbol{B}_l$ are N independent three-dimensional Gaussian variables whose statistical properties are described in eqs. (2.29). For equivalence, the strength parameters γ_l must be given by $\gamma_l = \alpha\lambda_l/2$.

Identical particles

In this case the localization effect is obtained by sharpening the particle density $\hat{N}(x)$ around each point x in physical space.[1] The particle densities can be defined in the second quantization language as

$$\hat{N}(x) = \left(\frac{\alpha}{2\pi}\right)^{3/2} \sum_s \int dx' \exp\left(-\tfrac{1}{2}\alpha\left(x'-x\right)^2\right) a^\dagger(x',s)a(x',s), \tag{2.32}$$

$a^\dagger(x,s)$ and $a(x,s)$ being the creation and annihilation operators of a particle at point x with spin component s. The *smooth* volume used to define the particle density has linear dimensions of the order of $1/\sqrt{\alpha}$.

For the discontinuous process the sharpening frequency and the sharpening accuracy of the density $\hat{N}(x)$ are μ and β, respectively. It is to be noted that, because of the nature of the domain of the quantity label x, the "centre" of the sharpening for the ith hitting is now a number density profile $n_i(x)$ and its probability density (in the functional space of number density profiles) is given by

$$\mathscr{P}[n_i] = |C|^2 \langle\psi_t|\exp\left[-\beta\int dx \left(\hat{N}(x) - n_i(x)\right)^2\right]|\psi_t\rangle. \tag{2.33}$$

---

[1] J.S. Bell, private comunication, 1987.

The coefficient C is given by the normalization condition

$$\int \mathscr{D}n \, \mathscr{P}[n] = 1. \tag{2.34}$$

The corresponding continuous process is ruled by the equation ([6] and [3])

$$d|\psi_t\rangle = \left[-\frac{i}{\hbar}\hat{H}dt + \sqrt{\gamma}\int d\boldsymbol{x}\left(\hat{N}(\boldsymbol{x}) - \langle\hat{N}(\boldsymbol{x})\rangle_{\psi_t}\right)dB(\boldsymbol{x})\right.$$

$$\left. -\frac{1}{2}\gamma\int d\boldsymbol{x}\left(\hat{N}(\boldsymbol{x}) - \langle\hat{N}(\boldsymbol{x})\rangle_{\psi_t}\right)^2 dt \right]|\psi_t\rangle, \tag{2.35}$$

where the Gaussian random variables $dB(\boldsymbol{x})$ have the properties

$$\overline{dB(\boldsymbol{x})} = 0, \qquad \overline{dB(\boldsymbol{x})dB(\boldsymbol{x}')} = \delta(\boldsymbol{x}-\boldsymbol{x}')dt. \tag{2.36}$$

For equivalence, the strength parameter γ must be given by $\gamma = \beta\mu/2$.

Several kinds of identical particles

In the case of several kinds of identical particles, the most established formulation sharpens the mass density around each point in physical space by using a universal stochastic field $dB(\boldsymbol{x})$. The particle density operators $\hat{N}(\boldsymbol{x})$ are then replaced, in both the discontinuous and the continuous processes, by the mass densities $\hat{M}(\boldsymbol{x})$ where

$$\hat{M}(\boldsymbol{x}) = \sum_k m_k \hat{N}_k(\boldsymbol{x}), \tag{2.37}$$

m_k being the mass of the particle of kind k.

The continuous process ([7]) is ruled by eq. (2.35), with $\hat{N}(\boldsymbol{x})$ replaced by $\hat{M}(\boldsymbol{x})$.

2.4 Final Considerations

Some final comments are in order. From the discussion of Sections 2.2 and 2.3 it is apparent that the discontinuous processes bear the same generality as the continuous ones as far as their applicability to physical systems is concerned. In particular, contrary to what has been sometimes stated in the literature, discontinuous processes can be formulated for systems of identical particles or of several kinds of identical particles. To deal with such physical systems resorting to continuous formulations is not necessary.

As explicitly shown in Section 2.2, discontinuous processes give rise, in a proper infinite frequency limit, to corresponding continuous ones, thus showing the physical equivalence of the two formulations for sufficiently high hitting frequencies. Stated differently, for any continuous process there is a discrete process which turns out to induce a dynamics as near as wanted to the corresponding continuous one and which becomes identical to it when the infinite frequency limit is taken.

One could ask what really means "sufficiently high" frequencies. From the purely formal point of view, the equivalence of discontinuous and continuous processes requires that, in the time interval $(t, t + dt)$, one has a large enough number of hittings, so that the central limit theorem can be applied. Having said that, the effectiveness of the hitting process depends on the product $\beta\mu$, so that for fixed effectiveness one can still maintain a finite frequency, provided that a sufficient number of hittings occur on the time scale relevant to the solution of the measurement process. For the sake of simplicity, in Section 2.2 it has been assumed that the hittings occur at evenly spaced times; in order to preserve time translation invariance ([2]) one should use random times with a certain mean frequency. Then the mean frequency has to be sufficiently large to guarantee that reduction takes place in the time interval of interest. There remains, however, a small probability that no reduction takes place. The same thing happens in the continuous process that leads certainly to a common eigenstate of the considered quantities only when $t \to \infty$.

As a last comment we stress that the continuous processes, on one hand, are undoubtedly mathematically more elegant, while, on the other hand, the discontinuous processes show immediately the physical effect of reduction, so that, taken for granted the infinite frequency limit, they show the reduction properties of the continuous ones too.

What really matters is the choice of the quantities induced to have a sharp distribution. We think it is important to stress the role of positions as the quantities that allow the strengthening of the process in going from microscopic to macroscopic degrees of freedom. In the case of distinguishable particles the variables undergoing the process are directly the positions of individual particles. In the case of identical particles or several kinds of identical particles the variables undergoing the process are the number or mass densities around the running point in physical space, that play the role of positions respecting the identity of particles. The final effect is again to make definite the position in space of macroscopic objects, thus providing a viable and conceptually simple solution to the measurement problem.

References

[1] A. Barchielli, L. Lanz, and G. M. Prosperi. A model for macroscopic description and continous observations in quantum mechanics. *Nuovo Cimento*, 72B:79, 1982.

[2] G. C. Ghirardi, A. Rimini, and T. Weber. Unified dynamics for microscopic and macroscopic systems. *Phys. Rev.*, D34:470, 1986.

[3] G. C. Ghirardi, P. Pearle, and A. Rimini. Markov processes in Hilbert space and continuous spontaneous localization of systems of identical particles. *Phys. Rev.*, A42:78–79, 1990.

[4] L. Diosi. Continuous quantum measurement and Itô formalism. *Phys. Lett.*, A129:419–423, 1988.
[5] O. Nicrosini and A. Rimini. On the relationship between continuous and discontinuous stochastic processes in Hilbert space. *Found. Phys.*, 20:1317–1327, 1990.
[6] P. Pearle. Combining stochastic dynamical state vector reduction with spontaneous localization. *Phys. Rev.*, A39:2277–2289, 1989.
[7] G. C. Ghirardi, R. Grassi, and F. Benatti. Describing the macroscopic world – Closing the circle within the dynamical reduction program. *Found. Phys.*, 25:5–38, 1995.

3
Dynamical Collapse for Photons

PHILIP PEARLE

I suggest a simple alteration of my CSL (Continuous Spontaneous Localization) theory, replacing the mass density collapse-generating operators by relativistic energy density operators. Some consequences of the density matrix evolution equation are explored. First, the expression for the mean energy increase of free particles is calculated (which, in the non-relativistic limit, agrees with the usual result). Then the density matrix evolution is applied to photons. The mean rate of loss of photon number from a laser beam pulse, the momentum distribution of the photons "excited" out of the laser beam pulse, and the alteration of the cosmic blackbody spectrum are all treated to first order in the collapse rate parameter λ. Associated possible experimental limits on λ are discussed.

3.1 Introduction

Some time ago, I proposed the idea of a stochastic dynamical collapse theory [1], where a term which depends upon a randomly fluctuating quantity is added to Schrödinger's equation. As a result, a superposition of states (in a particularly chosen basis) is continuously driven toward one such state, with (neglecting the usual Hamiltonian evolution) the Born probability.

In the CSL (Continuous Spontaneous Localization) theory [2, 3, 4], the randomly fluctuating quantity is a classical scalar field, and the term added to the Schrödinger equation depends as well upon a "collapse-generating operator." Initially, I chose this to be the particle number density operator [2] but later [5] replaced it by the mass density operator so that the collapse is toward a mass density eigenstate.

In addition to this modified Schrödinger equation, CSL is completed with the specification of the "probability rule," that the probability of a given fluctuating field is proportional to the squared norm of the state vector which evolved under that field.

An important aspect of CSL collapse behavior is that the collapse is very slow for micro-objects but fast for macro-objects, behavior which was first embodied in the thereby justly celebrated Spontaneous Localization (SL) theory of Ghirardi, Rimini and Weber [6] (where, however, the evolution is discontinuous: also, fermion or boson wave function symmetry is destroyed in SL, but a version which removes that flaw exists [7]). As a result, particle behavior is scarcely affected but, since we see macro-objects, "what you see is what you get" from the theory.

Events are common physical occurrences. Standard quantum theory predicts the probabilities of events but does not describe their occurrence: like Moses, it indicates the promised land but does not go there. Standard quantum theory may therefore be justifiably regarded as incomplete: CSL may be regarded as providing a completion.

Since photons do not have mass, in the present non-relativistic CSL theory (which has been, and is currently, the object of experimental scrutiny), photons do not contribute to collapse dynamics. It does not seem that there is a physical reason why this should be so. Since photons are relativistic particles, perhaps that has been waiting on the construction of a convincing, viable relativistic version of CSL [8]. Until that happens, I propose the following. In non-relativistic CSL, replace the mass density operators $\xi^\dagger(\mathbf{x})\xi(\mathbf{x})$ with the energy density operators $[K^{1/2}\xi^\dagger(\mathbf{x})][K^{1/2}\xi(\mathbf{x})]$, where $K^2 \equiv -\nabla^2 + M^2$, and $\xi(\mathbf{x})$ is the annihilation operator of a particle of mass M at location $\mathbf{x}$. Setting $M = 0$ then gives collapse dynamics for (one polarized species of) photons.

Another obvious possible choice of energy density operators is $\frac{1}{2}\xi^\dagger[K\xi(\mathbf{x})] + \frac{1}{2}[K\xi^\dagger(\mathbf{x})]\xi(\mathbf{x})$. Both operators are Hermitian, and their integral over all space gives the free particle relativistic Hamiltonian, the two basic requirements for the energy density operator. I have chosen to work with the one-term expression rather than the two-term expression simply because its square (which appears in the density matrix evolution equation) is one term, while the square of the other is more cumbersome, four terms. Whether there is a physical reason for preferring one over the other I do not know, nor have I looked to see how the other choice might affect the calculations in this chapter.

Of course, this is not a relativistically invariant theory, although collapse caused by differences in relativistic energy density does capture aspects of what one could expect in such a theory. One might consider the proposal here as representing collapse in the preferred, co-moving frame.

An important difference between the usual mass density operators and the energy density operators is that the commutator of the former operators at any two spatial points vanishes, while this is not so for the latter operators. In the former case, this allows one to make use of a theorem that, if all collapse-generating operators mutually commute, there is collapse toward the mutual joint eigenstates of these operators (neglecting the Hamiltonian evolution). Thus, one is assured of collapse toward mass density eigenstates in the former case.

For the latter case, one does not have that easy assurance. However, at least for massive particles, the commutator is quite small[1], although not vanishing, $\to -[(2\pi)^3 \lambda_M^3 \ |\mathbf{x} - \mathbf{x}'|^5]^{-1/2} e^{-|\mathbf{x}-\mathbf{x}'|/\lambda_M}$, for $|\mathbf{x} - \mathbf{x}'|/\lambda_M \gg 1$ ($\lambda_M \equiv \hbar/Mc$ is the reduced Compton wavelength of the particle). This suggests looking for an extension of the theorem to "almost" commuting operators, which shall not be pursued here.

Instead, one may look at examples to see how collapse dynamics evolves. The basic requirement of a collapse theory is that, when one considers a superposed state of many particles in two different places, there is collapse toward all particles being in one or the other place. An example is given in Appendix A, where the particles are moving, so that relativistic behavior may come into play. There, for the state $|\psi, 0\rangle = \frac{1}{\sqrt{2}}[|L\rangle + |R\rangle]$, the two spatially displaced states $|L\rangle, |R\rangle$ each consist of N particles, each particle in the same state occupying a volume $\sim \sigma^3$, each particle moving with well-defined momentum $\mathbf{k}_0$ in a direction orthogonal to their displacement vector $\mathbf{x}_L - \mathbf{x}_R$. In this example, the resulting density matrix behavior describing energy density generated collapse turns out to be identical to that when there is mass density generated collapse, except that the collapse rate factor $\sim M^2$ is replaced by $\omega^2(k_0) \equiv k_0^2 + M^2$.

We shall not review here the CSL dynamical equation for the state vector and how one derives the Lindblad equation for the density matrix from it and the probability rule [4, 9]. We shall just start with that density matrix evolution equation, with the given substitution:

$$\frac{\partial}{\partial t}\rho(t) = -i[H, \rho(t)] - \frac{\lambda}{2M_N^2} \int d\mathbf{x} \int d\mathbf{x}' e^{-(\mathbf{x}-\mathbf{x}')^2/4a^2} [K^{1/2}\xi^\dagger(\mathbf{x})K^{1/2}\xi(\mathbf{x}),$$

$$[K^{1/2}\xi^\dagger(\mathbf{x}')K^{1/2}\xi(\mathbf{x}'), \rho(t)]], \quad (3.1)$$

and proceed from there. (Here, M_N is the mass of the neutron, λ is the collapse rate and a the collapse range, typically chosen as the SL suggested values $\lambda \approx 10^{-16}$s and $a \approx 10^{-5}$cm, but limits on these phenomenological constants are being experimentally pursued.)

Applying (3.1) to the collapse example mentioned, the off-diagonal density matrix element between the two states, to order λ, is given by Eq. (A.44):

---

[1] $[K^{1/2}\xi^\dagger(\mathbf{x})K^{1/2}\xi(\mathbf{x}), K'^{1/2}\xi^\dagger(\mathbf{x}')K'^{1/2}\xi(\mathbf{x}')] = \left(K^{1/2}\xi^\dagger(\mathbf{x})K'^{1/2}\xi(\mathbf{x}') - K'^{1/2}\xi^\dagger(\mathbf{x}')K^{1/2}\xi(\mathbf{x})\right)K^{1/2}K'^{1/2}\delta(\mathbf{x} - \mathbf{x}')$ is the commutator, and $K^{1/2}K'^{1/2}\delta(\mathbf{x} - \mathbf{x}') = \frac{1}{(2\pi)^3}\int d\mathbf{k}\sqrt{k^2 + M^2}e^{i\mathbf{k}\cdot(\mathbf{x}-\mathbf{x}')} = \frac{1}{2\pi^2|\mathbf{x}-\mathbf{x}'|}\int_0^\infty k dk\sqrt{k^2 + M^2}\sin k|\mathbf{x} - \mathbf{x}'| = -\frac{M^2}{2\pi^2|\mathbf{x}-\mathbf{x}'|^2}\left[K_0(M|\mathbf{x} - \mathbf{x}'|) + \frac{1}{M|\mathbf{x}-\mathbf{x}'|}K_1(M|\mathbf{x} - \mathbf{x}'|)\right]$. The result follows from the large argument approximation of the Bessel function K_0.

$$\langle L|\rho(t)|R\rangle \approx \frac{1}{2} - \frac{N\lambda t\omega^2(k_0)}{2M_N^2}\left[N\left(\frac{a}{\sigma}\right)^3\left(1 - e^{-(\mathbf{x}_L-\mathbf{x}_R)^2/4\sigma^2}\right) + 1\right]$$

This clearly describes the decay of the matrix element, for any values of the parameters consistent with the assumptions underlying (A.44), $k_0\sigma \gg k_0 a \gg 1$.

In this chapter, we shall discuss the explicit collapse behavior no further than this example calculated in Appendix A. For we are particularly interested in the "anomalous" excitation of photons ($M = 0$) which is a byproduct of the collapse dynamics. (By "anomalous" is always meant behavior not predicted by standard quantum theory, and therefore open to experimental test.)

Because collapse narrows wave functions, the momentum and therefore the energy of particles is "anomalously" increased. In the non-relativistic theory based upon mass density-generated collapse, the rate of energy increase of N identical non-relativistic particles is [5]

$$\frac{d}{dt}\bar{H} = \lambda\frac{3\hbar^2}{4Ma^2}\frac{M^2}{M_N^2}N. \tag{3.2}$$

In Section 3.2, the comparable relativistic expression shall be obtained from Eq. (3.1) (with (3.2) as the non-relativistic limit).

In Sections 3.3, 3.4, we consider the effect of collapse on a beam or pulse of laser light. The state vector is a coherent state, a superposition of states of various numbers of photons of almost identical momentum, where the number of photons obeys Poisson statistics. These states have different energy densities. Insofar as the collapse dynamics tries to evolve the state vector toward one of these states, while this changes the statistics for a single beam, it doesn't affect the Poisson statistics for the ensemble of beams because the collapse dynamics respects the Born rule.

But these states are expected to be modified since the collapse mechanism also imparts energy to photons, which removes them from a coherent beam. That *will* affect the statistics of the ensemble and decrease the mean number of photons in the beam. In Section 3.3, we calculate the loss in the ensemble-mean number of photons from the laser beam to first order in the collapse rate parameter λ. We apply the result to an experimentally achieved intense laser beam pulse, which is in the infrared, and also to an experimentally achieved x-ray laser beam pulse to see what upper limits on λ could be implied.

Photon number is conserved. The photons lost from the beam are made more energetic by the collapse process. In Section 3.4 we calculate the momentum distribution of these "anomalous" photons. We consider how these photons are ejected from an experimentally achieved intense CW laser beam, again suggesting a limit on λ.

Section 3.5 is motivated by the consideration that the longer the collapse process acts, the more photons are excited. Therefore we discuss the cosmic blackbody photons as they are affected by collapse over the time interval since recombination sent them freely on their way, almost over the age of the universe. There is an ensuing distortion of the blackbody spectrum, but the resulting effect is small. This is partly because there are so few photons involved, $\approx 16\pi(kT/hc)^3 \approx 400$ photons/cc, and partly because their energy $\approx 2.5 \times 10^{-6}$eV to 2.5×10^{-2}eV is so small (photons with wavelength 50cm to .05cm).

3.2 Energy Increase

The mean energy of a collection of identical particles of mass M described by the density matrix $\rho(t)$ is $\bar{H}(t) \equiv TrH\rho(t)$, where $H \equiv \int d\mathbf{x}\xi^\dagger(\mathbf{x})K\xi(\mathbf{x})$, and Tr is the trace operation. Then by Eq. (3.1),

$$\frac{\partial}{\partial t}\bar{H}(t) = -\frac{\lambda}{2M_N^2}Tr\rho(t)\int d\mathbf{x}\int d\mathbf{x}'\int d\mathbf{x}''e^{-(\mathbf{x}-\mathbf{x}')^2/4a^2}$$
$$[K^{1/2}\xi^\dagger(\mathbf{x})K^{1/2}\xi(\mathbf{x}), [K^{1/2}\xi^\dagger(\mathbf{x}')K^{1/2}\xi(\mathbf{x}'), \xi^\dagger(\mathbf{x}'')K\xi(\mathbf{x}'')]]. \quad (3.3)$$

Writing $\xi(\mathbf{x}) \equiv \frac{1}{(2\pi)^{3/2}}\int d\mathbf{k}a(\mathbf{k})e^{i\mathbf{k}\cdot\mathbf{x}}$ (i.e., $a(\mathbf{k})$ is the annihilation operator of a particle of momentum $\mathbf{k}$), Eq. (3.3) becomes, with $\omega(k) \equiv \sqrt{k^2 + M^2}$:

$$\frac{\partial}{\partial t}\bar{H}(t) = -\frac{\lambda}{2M_N^2}\frac{1}{(2\pi)^6}Tr\rho(t)\int d\mathbf{x}\int d\mathbf{x}'$$
$$\int d\mathbf{k}_1 d\mathbf{k}_2 d\mathbf{k}_3 d\mathbf{k}_4 d\mathbf{k}_5 \sqrt{\omega(k_1)\omega(k_2)\omega(k_3)\omega(k_4)}\omega(k_5)$$
$$e^{-(\mathbf{x}-\mathbf{x}')^2/4a^2}e^{-i(\mathbf{k}_1-\mathbf{k}_2)\cdot\mathbf{x}}e^{-i(\mathbf{k}_3-\mathbf{k}_4)\cdot\mathbf{x}'}[a^\dagger(\mathbf{k}_1)a(\mathbf{k}_2),[a^\dagger(\mathbf{k}_3)a(\mathbf{k}_4),a^\dagger(\mathbf{k}_5)a(\mathbf{k}_5)]]$$
$$= -\frac{\lambda}{2M_N^2}\frac{1}{(2\pi)^3}(4\pi a^2)^{3/2}Tr\rho(t)$$
$$\int d\mathbf{k}_1 d\mathbf{k}_2 d\mathbf{k}_3 d\mathbf{k}_4 d\mathbf{k}_5 \sqrt{\omega(k_1)\omega(k_2)\omega(k_3)\omega(k_4)}\omega(k_5)$$
$$e^{-(\mathbf{k}_1-\mathbf{k}_2)^2 a^2}\delta(-\mathbf{k}_1+\mathbf{k}_2-\mathbf{k}_3+\mathbf{k}_4)$$
$$[a^\dagger(\mathbf{k}_1)a(\mathbf{k}_5)\delta(\mathbf{k}_4-\mathbf{k}_5)\delta(\mathbf{k}_2-\mathbf{k}_3) - a^\dagger(\mathbf{k}_3)a(\mathbf{k}_2)\delta(\mathbf{k}_4-\mathbf{k}_5)\delta(\mathbf{k}_1-\mathbf{k}_5)$$
$$-a^\dagger(\mathbf{k}_1)a(\mathbf{k}_4)\delta(\mathbf{k}_3-\mathbf{k}_5)\delta(\mathbf{k}_2-\mathbf{k}_5) + a^\dagger(\mathbf{k}_3)a(\mathbf{k}_2)\delta(\mathbf{k}_3-\mathbf{k}_5)\delta(\mathbf{k}_1-\mathbf{k}_4)]$$
$$= \frac{\lambda}{2M_N^2}\frac{1}{(2\pi)^3}(4\pi a^2)^{3/2}2Tr\rho(t)$$
$$\int d\mathbf{k}_1 d\mathbf{k}_2 a^\dagger(\mathbf{k}_1)a(\mathbf{k}_1)\omega_1\omega_2(\omega_2 - \omega_1)e^{-(\mathbf{k}_1-\mathbf{k}_2)^2 a^2} \quad (3.4)$$

where, in the second step, the commutation operations have been performed and, in the last step, delta function integrals have been performed, and labels 1 and 2 have been exchanged in a term.

In the non-relativistic limit, $\omega_1\omega_2(\omega_2-\omega_1) \approx M^2[k_2^2-k_1^2]/2M$. The integral over $\mathbf{k}_2$ in (3.4) is then

$$\int d\mathbf{k}_2[k_2^2 - k_1^2]e^{-(\mathbf{k}_2-\mathbf{k}_1)^2 a^2} = \int d\mathbf{k}_2[(\mathbf{k}_2 - \mathbf{k}_1)^2 + 2(\mathbf{k}_2 - \mathbf{k}_1) \cdot \mathbf{k}_1]e^{-(\mathbf{k}_2-\mathbf{k}_1)^2 a^2}$$
$$= \frac{\pi^{3/2}}{a^3}\frac{3}{2a^2}. \qquad (3.5)$$

Inserting (3.5) into Eq. (3.4), we obtain Eq. (3.2) ($N = Tr\rho(t)\int d\mathbf{k} a^\dagger(\mathbf{k})a(\mathbf{k})$).

In the general case, we write (3.4) as

$$\frac{\partial}{\partial t}\bar{H}(t) = \frac{\lambda}{M_N^2}\left(\frac{a^2}{\pi}\right)^{3/2} Tr\rho(t)\int d\mathbf{k}_1 a^\dagger(\mathbf{k}_1)a(\mathbf{k}_1)f(\mathbf{k}_1) \text{ where}$$
$$f(\mathbf{k}_1) \equiv \int d\mathbf{k}_2 \omega_1\omega_2(\omega_2 - \omega_1)e^{-(\mathbf{k}_2-\mathbf{k}_1)^2 a^2} \qquad (3.6)$$

So far, Eq. (3.6) is exact. We shall obtain analytic expressions in two approximate cases.

One is for photons when the density matrix ρ is such that $ka \ll 1$ (here $k \equiv k_1$). Note that, with $a = 100$nm, $2\pi a$ is in the neighborhood of red light's wavelength, so we are considering wavelengths $\gtrsim$infrared.

The other case is for $ka \gg 1$. For photons, we are therefore considering wavelengths $\lesssim$ultraviolet. For electrons, $ka \gg 1$ implies that the energy $(\hbar k)^2/2m_e \gg (\hbar/a)^2/2m_e \approx 3 \times 10^{-6}$eV, which of course means validity in a broad non-relativistic realm as well.

For photons and $ka \ll 1$, in the integral $f(\mathbf{k}_1)$, we write $\omega_2 - \omega_1 \approx k_2$ and $(\mathbf{k}_2 - \mathbf{k}_1)^2 \approx k_2^2$, obtaining $f(\mathbf{k}_1) = k_1(\pi/a^2)^{3/2}(3/2a^2)$.

For $ka \gg 1$ we change the variable of integration to $\Delta \equiv \mathbf{k}_2 - \mathbf{k}_1$. We expand ω_2 to order $\Delta^2 \sim 1/a^2$, thereby omitting terms of order $(k_1 a)^{-2}$ compared to the terms that are retained:

$$f(\mathbf{k}_1) = \omega_1 \int d\Delta\sqrt{M^2 + (\mathbf{k}_1 + \Delta)^2}[\sqrt{M^2 + (\mathbf{k}_1 + \Delta)^2} - \omega_1]e^{-\Delta^2 a^2}$$
$$\approx \omega_1 \int d\Delta\left[\omega_1 + \frac{\mathbf{k}_1 \cdot \Delta}{\omega_1}\right]\left[\frac{\mathbf{k}_1 \cdot \Delta}{\omega_1} + \frac{\Delta^2}{2\omega_1} - \frac{(2\mathbf{k}_1 \cdot \Delta)^2}{8\omega_1^3}\right]e^{-\Delta^2 a^2}$$
$$= \frac{\omega_1}{2}\int d\Delta\left[\Delta^2 + \frac{(\mathbf{k}_1 \cdot \Delta)^2}{\omega_1^2}\right] = \omega_1\frac{\pi^{3/2}}{a^5}\left[\frac{3}{4} + \frac{k_1^2}{4\omega_1^2}\right], \qquad (3.7)$$

so that

$$\frac{\partial}{\partial t}\bar{H}(t) \approx \lambda\left(\frac{\lambda_N}{a}\right)^2 Tr\rho(t) \int d\mathbf{k}_1 a^\dagger(\mathbf{k}_1)a(\mathbf{k}_1)\omega_1\left[\frac{3}{4} + \frac{k_1^2}{4\omega_1^2}\right] \quad (3.8)$$

where $\lambda \equiv \hbar/M_N c \approx 2 \times 10^{-14}$cm is the reduced Compton wavelength of the nucleon.

Once again, we note that the non-relativistic limit Eq. (3.2) is obtained from (3.8), with $\omega_1 \approx Mc^2, k_1/\omega_1 \approx 0$.

Our two approximate expressions are therefore, first, for photons with a density matrix ρ describing photons such that $ka << 1$ and, second, from (3.8), applicable both to massive particles in the relativistic regime ($k_1/\omega_1 \approx 1$) and to photons ($k_1/\omega_1 = 1$), with $ka >> 1$:

$$\frac{\partial}{\partial t}\bar{H}(t) = \lambda\frac{3}{2}\left(\frac{\lambda_N}{a}\right)^2\bar{H}(t) \qquad \text{for } ka << 1. \quad (3.9a)$$

$$\frac{\partial}{\partial t}\bar{H}(t) = \lambda\left(\frac{\lambda_N}{a}\right)^2\bar{H}(t) \qquad \text{for } ka >> 1. \quad (3.9b)$$

The new wrinkle here is that there is exponential growth of the mean energy, not the non-relativistic linear growth (3.2). However, since over the age of the universe T, with $\lambda T \approx 40$, the exponent $\lambda T(\frac{\lambda_N}{a})^2 \approx 10^{-16}$, the exponential growth is effectively linear and there is a negligible fractional contribution of collapse-induced energy to the universe.

In spite of the smallness of this exponent, one should hasten to add that collapse-induced energy effects can have consequences that are not out of the realm of observability, since they can produce anomalous behavior, such as rare but unusual events, which may be experimentally singled out. Non-relativistically, this includes knocking electrons out of atoms [10], breaking up the deuterium nucleus [11], shaking free charged particles so they radiate [12], inducing random walk in small objects [13], contributing to the cosmological constant [14].

It is also worth emphasizing that, for massive particles, the relativistic result (3.8) only applies to free particles. However, the non-relativistic result (3.2) is the energy increase even when there is a potential. The reason is that the potential energy operator for particles in an external potential $V(\mathbf{x})$ and a mutually interacting potential V($\mathbf{x}$-$\mathbf{x}$') is $\int d\mathbf{x}\xi^\dagger(\mathbf{x})\xi(\mathbf{x})V(\mathbf{x}) + \int d\mathbf{x}d\mathbf{x}'\xi^\dagger(\mathbf{x})\xi(\mathbf{x})\xi^\dagger(\mathbf{x}')\xi(\mathbf{x}')V(\mathbf{x} - \mathbf{x}')$. This commutes with the non-relativistic collapse-generating operator $\sim \xi^\dagger(\mathbf{x})\xi(\mathbf{x})$ but does not commute with the relativistic collapse-generating operator $\sim K^{1/2}\xi^\dagger(\mathbf{x})K^{1/2}\xi(\mathbf{x})$. Thus, the energy increase for relativistic particles will be modified from (3.8), which is worth investigating [14].

3.3 Photon Number Decrease

We shall now consider the effect of the collapse dynamics on a laser beam of finite length $\sim \sigma$, for example, a laser pulse.

We shall describe the initial state vector of the laser beam as the coherent state

$$|\psi, 0\rangle \equiv e^{\beta \int d\mathbf{k}\alpha(\mathbf{k})a^\dagger(\mathbf{k})}|0\rangle e^{-\beta^2/2}, \text{ with } \alpha(\mathbf{k}) \equiv \frac{1}{(\pi/\sigma^2)^{3/4}} e^{-(\mathbf{k}-\mathbf{k}_0)^2 \sigma^2/2}. \quad (3.10)$$

β is a positive constant, whose square is the mean photon number, as we shall see in what follows.

For simplicity, (3.10) gives the width of the beam as $\sim \sigma$ also, where of course it is usually quite a bit smaller than the length: this has no consequence, as the only relevant property employed is that the width, like the length, is many times larger than the wavelength.

We define the state of n photons as

$$|n\rangle \equiv \frac{1}{\sqrt{n!}}\left[\int d\mathbf{k}\alpha(\mathbf{k})a^\dagger(\mathbf{k})\right]^n |0\rangle \text{ so } \langle n|m\rangle = \delta_{nm}. \quad (3.11)$$

The initial density matrix is $\rho(0) = |\psi, 0\rangle\langle\psi, 0|$. Thus, the probability that there are n particles in the initial state is the Poisson distribution

$$\langle n|\rho(0)|n\rangle = \frac{\beta^{2n}}{n!} e^{-\beta^2} \text{ from which one finds that the initial mean number of}$$
$$\text{photons is } \bar{n}(0) = \beta^2. \quad (3.12)$$

It follows from Eq. (3.1) for photons ($M = 0$) that, to first order in λ,

$$\langle n|\rho(t)|n\rangle \approx \frac{\beta^{2n}}{n!} e^{-\beta^2} - \frac{\lambda t}{2M_N^2} \frac{1}{(2\pi)^3} (4\pi a^2)^{3/2} \int d\mathbf{k}_1 d\mathbf{k}_2 d\mathbf{k}_3 d\mathbf{k}_4 \sqrt{k_1 k_2 k_3 k_4}$$
$$\cdot e^{-(\mathbf{k}_1-\mathbf{k}_2)^2 a^2} \delta(-\mathbf{k}_1 + \mathbf{k}_2 - \mathbf{k}_3 + \mathbf{k}_4) \cdot \Big[\langle n|a^\dagger(\mathbf{k}_1)a(\mathbf{k}_2)a^\dagger(\mathbf{k}_3)a(\mathbf{k}_4)\rho(0)|n\rangle$$
$$- \langle n|a^\dagger(\mathbf{k}_3)a(\mathbf{k}_4)\rho(0)a^\dagger(\mathbf{k}_1)a(\mathbf{k}_2)|n\rangle + hc\Big]. \quad (3.13)$$

All terms are real, so the bracketed terms in (3.13) are equal to their Hermitian conjugate. Note that the Hamiltonian term makes no contribution to this diagonal matrix element since $|n\rangle$ is very close to being an energy eigenstate, $H|n\rangle \approx nk_0|n\rangle$, so $\langle n|[H, \rho(t)]|n\rangle \approx 0$.

Since $k_0\sigma \gg 1$ is certainly true for a laser pulse, we can readily make the approximations $\alpha(\mathbf{k}_i)\sqrt{k_i} \approx \alpha(\mathbf{k}_i)\sqrt{k_0}$, and $\alpha^2(\mathbf{k}) \approx \delta(\mathbf{k} - \mathbf{k}_0)$.

Using $a(\mathbf{k})|\psi, 0\rangle = \beta\alpha(\mathbf{k})|\psi, 0\rangle$, $a(\mathbf{k})|n\rangle = \sqrt{n}\alpha(\mathbf{k})|n-1\rangle$, $\langle n|\psi, 0\rangle = \frac{\beta^n}{\sqrt{n!}} e^{-\beta^2/2}$, and putting $[a(\mathbf{k}_2), a^\dagger(\mathbf{k}_3)] = \delta(\mathbf{k}_2 - \mathbf{k}_3)$ in the first bracketed term of (3.13) we get:

$$\langle n|\rho(t)|n\rangle \approx \frac{\beta^{2n}}{n!} e^{-\beta^2} \left[1 - \frac{\lambda t}{M_N^2}\left(\frac{a^2}{\pi}\right)^{3/2} \int d\mathbf{k}_1 d\mathbf{k}_2 d\mathbf{k}_3 d\mathbf{k}_4 \sqrt{k_1 k_2 k_3 k_4}\right.$$
$$e^{-(\mathbf{k}_1 - \mathbf{k}_2)^2 a^2} \delta(-\mathbf{k}_1 + \mathbf{k}_2 - \mathbf{k}_3 + \mathbf{k}_4)$$
$$\left.\left[\alpha(\mathbf{k}_1)\alpha(\mathbf{k}_2)\alpha(\mathbf{k}_3)\alpha(\mathbf{k}_4)[n(n-1) - n^2] + \alpha(\mathbf{k}_1)\alpha(\mathbf{k}_4)\delta(\mathbf{k}_2 - \mathbf{k}_3)n\right]\right]. \tag{3.14}$$

For the first bracketed term in (3.14), upon setting $\sqrt{k_1 k_2 k_3 k_4} \approx k_0^2$, the integral may be performed: this is done in Appendix B. The result is $k_0^2 (2\pi/\sigma^2)^{3/2}$, which is $<<$ the second term (see Eqs.(3.15a, 3.15b)) and so may be neglected.

So it is the second bracketed term which is of interest. For the integral involving it, one may obtain a closed expression in the two limits $k_0 a << 1$ and $k_0 a >> 1$:

$$\frac{1}{(\pi/\sigma^2)^{3/2}} \int d\mathbf{k}_1 d\mathbf{k}_2 k_1 k_2 e^{-(\mathbf{k}_1 - \mathbf{k}_2)^2 a^2} e^{-(\mathbf{k}_1 - \mathbf{k}_0)^2 \sigma^2}$$
$$\approx \int d\mathbf{k}_1 d\mathbf{k}_2 k_1 k_2 e^{-(\mathbf{k}_1 - \mathbf{k}_2)^2 a^2} \delta(\mathbf{k}_1 - \mathbf{k}_0) = \int d\mathbf{k}_2 k_2 k_0 e^{-(\mathbf{k}_2 - \mathbf{k}_0)^2 a^2}$$
$$\approx \int d\mathbf{k}_2 k_2 k_0 e^{-k_2^2 a^2} = k_0 \frac{2\pi}{a^4} \quad \text{for } k_0 a << 1, \tag{3.15a}$$
$$\approx \int d\mathbf{k}_2 k_2 k_0 \left(\frac{\pi}{a^2}\right)^{3/2} \delta(\mathbf{k}_2 - \mathbf{k}_0) = k_0^2 \left(\frac{\pi}{a^2}\right)^{3/2} \quad \text{for } k_0 a >> 1. \tag{3.15b}$$

Inserting Eqs.(3.15a, 3.15b) into (3.14), we obtain the results:

$$\langle n|\rho(t)|n\rangle \approx \frac{\beta^{2n}}{n!} e^{-\beta^2} \left[1 - 4\pi^{1/2} n\lambda t \frac{\lambda_N^2}{\lambda_0 a}\right] \quad \text{for } k_0 a << 1, \tag{3.16a}$$

$$\langle n|\rho(t)|n\rangle \approx \frac{\beta^{2n}}{n!} e^{-\beta^2} \left[1 - n\lambda t \left(\frac{\lambda_N}{\lambda_0}\right)^2\right] \quad \text{for } k_0 a >> 1. \tag{3.16b}$$

Setting $\beta^2 = \bar{n}(0)$ and approximating $\beta^2(\beta^2 + 1) \approx \beta^4$, the mean number of photons in such a pulse is calculated to decrease as

$$\bar{n}(t) = \sum_{n=0}^{\infty} n\langle n|\rho(t)|n\rangle$$

$$\approx \bar{n}(0)\left[1 - 4\pi^{1/2}\bar{n}(0)\lambda t \frac{\lambda_N^2}{\lambda_0 a}\right] \quad \text{for } k_0 a << 1 \tag{3.17a}$$

$$\approx \bar{n}(0)\left[1 - \bar{n}(0)\lambda t \left(\frac{\lambda_N}{\lambda_0}\right)^2\right] \quad \text{for } k_0 a >> 1. \tag{3.17b}$$

Eqs.(3.17a, 3.17b) are the result we have been seeking. We see that the rate of photon loss is largest for a large number of photons in a pulse or for a small photon wavelength. Let us consider experimental situations in which these dependencies come to the fore.

Considering the case of a large number of photons, at the Vulcan laser facility [15] there are presently generated high-intensity laser beam pulses containing energy $E_p \approx 500$J, although with a fairly large wavelength, $\lambda_0 = 1053$nm, in the infrared. There are then $\bar{n}(0) = E_p/(hc/\lambda_0) \approx 2.5 \times 10^{21}$ photons in a pulse (pulse length $\sigma \approx .1$mm). In this case $k_0 a \approx .6$ lies between the validity regions of (3.17a) or (3.17b), but (3.17a) gives $\bar{n}(t) \approx 2.5 \times 10^{21}[1 - .75 \times 10^4 \lambda t]$ and (3.17b) gives $\bar{n}(t) \approx 2.5 \times 10^{21}[1 - 10^4 \lambda t]$

Considering the case of energetic photons, the most intense, XFEL (X-ray Free Electron Laser) pulses provide the attendant increase of the $(\lambda_N/\lambda_0)^2$ factor in Eqs.(3.17a, 3.17b), although there is a smaller $\bar{n}(0)$. The LCLSII (Stanford Linear Coherent Light Source) [16] specifies its laser pulses as containing $\approx 10^{12}$ photons, each of 8.3KeV ($\approx$ 1mJ/pulse, pulse length $\sigma \approx .15$mm). Then, $\lambda_N/\lambda_0 \approx 10^{-13}$cm/$10^{-8}$cm $= 10^{-5}$. With these values, (3.17b) becomes

$$\bar{n}(t) = 10^{12}[1 - 100\lambda t]. \qquad (3.18)$$

In both these cases, one might at least imagine an experiment measuring the loss of photons from a pulse with the pulse bouncing back and forth between mirrors many times to be accessible over, say, 1s. This would have to contend with competing loss mechanisms such as attendant loss at each bounce, scattering losses from the gas between the mirrors. Supposing these effects could be compensated for, and the accuracy of the measurement was 1% with no loss observed, this would place a limit $\lambda \lesssim 10^{-4} - 10^{-6}$s^{-1}: the present best upper limit [17] is around $\lambda \leq 10^{-9} - 10^{-10}$s^{-1}.

3.4 Photon Excitation

The operators in the density matrix evolution equation do not change the number of photons. To first order in λ, there is the probability of conversion of a photon of momentum $\mathbf{k}_0$ to one of momentum $\mathbf{k}$. We shall indeed see that this compensates the resulting loss of photons from the beam presented in Sec. 3.3, for which $Tr\rho(t) < 1$. Thus, for the combined processes of photon loss and photon excitation, $Tr\rho(t) = 1$.

We shall also see that the energy increase in Sec. 3.2, Eq. (3.6), is explained, to first order in λ, by replacement of a photon of energy k_0 by one of energy k.

Then we shall consider a consequence of the predicted excited photon distribution.

We need a complete set of orthogonal one-photon states, of which one state is the photon state in the laser beam, $\int d\mathbf{k} \alpha(\mathbf{k}) a^\dagger(\mathbf{k})|0\rangle$. Since we have chosen $\alpha(\mathbf{k})$ to have the form of the ground state of a three-dimensional harmonic oscillator in the variable $\mathbf{k}-\mathbf{k}_0$, the orthogonal set is readily supplied as $\mu_\mathbf{s}^\dagger|0\rangle \equiv \int d\mathbf{k}\chi_\mathbf{s}(\mathbf{k})a^\dagger(\mathbf{k})|0\rangle$. Here $\chi_\mathbf{s}(\mathbf{k}) \equiv N_s H_{s_1} H_{s_2} H_{s_3} \alpha(\mathbf{k})$, where H_{s_i} is a Hermite polynomials in $k_i - k_{0i}$, and the three indices $\mathbf{s} \equiv (s_1, s_2, s_3)$ take on all integer values ≥ 0 (so $\chi_{0,0,0}(\mathbf{k}) = \alpha(\mathbf{k})$).

We wish to consider the expectation value of the density matrix for an $n + 1$ particle state (i.e., the probability that this state is occupied), where n particles comprise the state $|n\rangle$ and one more particle is "almost" in the momentum eigenstate $|\mathbf{k}\rangle = a^\dagger(\mathbf{k})|0\rangle$. By "almost" is meant that the state is orthogonal to $\mu_{0,0,0}^\dagger|0\rangle$. To this end, we define the projection operator $P \equiv 1 - \mu_{0,0,0}^\dagger|0\rangle\langle 0|\mu_{0,0,0}$. We shall also find it useful to define $\gamma_n^\dagger \equiv \frac{1}{\sqrt{n!}}[\int d\mathbf{k}\alpha(\mathbf{k})a^\dagger(\mathbf{k})]^n$ so $\gamma_n^\dagger|0\rangle = |n\rangle$. Thus, the $n + 1$ particle state is $\gamma_n^\dagger P|\mathbf{k}\rangle$.

Since $1 = \sum_\mathbf{s} \mu_\mathbf{s}^\dagger|0\rangle\langle 0|\mu_\mathbf{s}$, it follows that $|\mathbf{k}\rangle = \sum_\mathbf{s} \chi_\mathbf{s}(\mathbf{k})\mu_\mathbf{s}^\dagger|0\rangle$ and $P|\mathbf{k}\rangle = \sum_{\mathbf{s}\neq(0,0,0)} \chi_\mathbf{s}(\mathbf{k})\mu_\mathbf{s}^\dagger|0\rangle$ so $\langle 0|\gamma_1 P|\mathbf{k}\rangle = 0$. Therefore, $\langle m|\gamma_n^\dagger P|\mathbf{k}\rangle = \delta_{m,n+1}\frac{1}{\sqrt{n+1}}\langle 0|\gamma_1 P|\mathbf{k}\rangle = 0$, and so $\rho(0)\gamma_n^\dagger P|\mathbf{k}\rangle = 0$.

Then, the density matrix diagonal element is, to first order in λ,

$$\langle \mathbf{k}|P\gamma_n \rho(t)\gamma_n^\dagger P|\mathbf{k}\rangle$$
$$= \frac{\lambda t}{2M_N^2}\left(\frac{a^2}{\pi}\right)^{3/2} \int d\mathbf{k}_1 d\mathbf{k}_2 d\mathbf{k}_3 d\mathbf{k}_4 \sqrt{k_1 k_2 k_3 k_4} e^{-(\mathbf{k}_1-\mathbf{k}_2)^2 a^2} \delta(-\mathbf{k}_1+\mathbf{k}_2-\mathbf{k}_3+\mathbf{k}_4)$$
$$\cdot \sum_{\mathbf{s},\mathbf{s}'\neq(0,0,0)} \chi_\mathbf{s}(\mathbf{k})\langle 0|\mu_\mathbf{s}\gamma_n a^\dagger(\mathbf{k}_3)a(\mathbf{k}_4)\rho(0)a^\dagger(\mathbf{k}_1)a(\mathbf{k}_2)\gamma_n^\dagger \mu_{\mathbf{s}'}^\dagger|0\rangle \chi_{\mathbf{s}'}(\mathbf{k}) + hc\Big]$$
$$= \frac{\lambda t}{M_N^2}\left(\frac{a^2}{\pi}\right)^{3/2} \int d\mathbf{k}_1 d\mathbf{k}_2 d\mathbf{k}_3 d\mathbf{k}_4 \sqrt{k_1 k_2 k_3 k_4} e^{-(\mathbf{k}_1-\mathbf{k}_2)^2 a^2} \delta(-\mathbf{k}_1+\mathbf{k}_2-\mathbf{k}_3+\mathbf{k}_4)$$
$$\cdot \beta^2 \frac{\beta^{2n}}{n!} e^{-\beta^2} \sum_{\mathbf{s},\mathbf{s}'\neq(0,0,0)} \chi_\mathbf{s}(\mathbf{k}_3)\chi_\mathbf{s}(\mathbf{k})\chi_{\mathbf{s}'}(\mathbf{k})\chi_{\mathbf{s}'}(\mathbf{k}_2)\alpha(\mathbf{k}_1)\alpha(\mathbf{k}_4)$$
$$= \frac{\lambda t}{M_N^2}\left(\frac{a^2}{\pi}\right)^{3/2} \int d\mathbf{k}_1 d\mathbf{k}_2 d\mathbf{k}_3 d\mathbf{k}_4 \sqrt{k_1 k_2 k_3 k_4} e^{-(\mathbf{k}_1-\mathbf{k}_2)^2 a^2} \delta(-\mathbf{k}_1+\mathbf{k}_2-\mathbf{k}_3+\mathbf{k}_4)$$
$$\cdot \beta^2 \frac{\beta^{2n}}{n!} e^{-\beta^2}[\delta(\mathbf{k}-\mathbf{k}_3) - \alpha(\mathbf{k})\alpha(\mathbf{k}_3)][\delta(\mathbf{k}-\mathbf{k}_2) - \alpha(\mathbf{k})\alpha(\mathbf{k}_2)]\alpha(\mathbf{k}_1)\alpha(\mathbf{k}_4). \quad (3.19)$$

(First, the matrix elements were evaluated. Second, the completeness relation $\sum_{\mathbf{s}\neq(0,0,0)} \chi_\mathbf{s}(\mathbf{k})\chi_\mathbf{s}(\mathbf{k}') + \alpha(\mathbf{k})\alpha(\mathbf{k}') = \delta(\mathbf{k}-\mathbf{k}')$ was employed.)

It is the term $\sim \delta(\mathbf{k}-\mathbf{k}_3)\delta(\mathbf{k}-\mathbf{k}_2)$ that provides the important contribution. Summing (3.19) over all n (setting $\beta^2 = \bar{n}(0)$) gives the probability density for the presence of a photon of momentum $\mathbf{k}$:

$$\mathcal{P}(\mathbf{k}) = \bar{n}(0)\frac{\lambda t}{M_N^2}\left(\frac{a^2}{\pi}\right)^{3/2} \int d\mathbf{k}_1 k_1 k e^{-(\mathbf{k}_1-\mathbf{k})^2 a^2} \left(\frac{\sigma^2}{\pi}\right)^{3/2} e^{-(\mathbf{k}_1-\mathbf{k}_0)^2\sigma^2} + R(\mathbf{k}) \quad (3.20)$$

with the contribution of the other terms being

$$R(\mathbf{k}) \equiv \bar{n}(0) \frac{\lambda t}{M_N^2} \left(\frac{a^2}{\pi}\right)^{3/2} \int d\mathbf{k}_1 d\mathbf{k}_2 d\mathbf{k}_3 d\mathbf{k}_4 \sqrt{k_1 k_2 k_3 k_4}$$

$$\delta(-\mathbf{k}_1 + \mathbf{k}_2 - \mathbf{k}_3 + \mathbf{k}_4) e^{-(\mathbf{k}_1-\mathbf{k}_2)^2 a^2}$$

$$\alpha(\mathbf{k}_1)\alpha(\mathbf{k}_2)\alpha(\mathbf{k}_3)\alpha(\mathbf{k}_4)\left[\alpha^2(\mathbf{k}) - 2\delta(\mathbf{k}_2 - \mathbf{k}))\right]$$

$$\approx \bar{n}(0) \frac{\lambda t}{M_N^2} \left(\frac{a^2}{\pi}\right)^{3/2} \left[k_0^2 2^{3/2} e^{-(\mathbf{k}-\mathbf{k}_0)^2 \sigma^2} - 2k_0^2 \left(\frac{4}{3}\right)^{3/2} e^{-(\mathbf{k}-\mathbf{k}_0)^2 2\sigma^2/3}\right], \quad (3.21)$$

where the result in the last line of (3.21) is obtained in Appendix B. We can neglect $R(\mathbf{k})$ with respect to the first term. For (3.21) only makes a contribution for $\mathbf{k} = \mathbf{k}_0 + o(1/\sigma)$, which cannot be distinguished from a photon in the laser pulse. Moreover, we note that the integrated probability contribution of the term in the bracket in (3.21) is $-k_0^2 (2\pi/\sigma^2)^{3/2}$, which precisely cancels (B.48), the term dropped from the photon loss expression (3.14) because it is smaller than the term kept.

Employing the approximation $\left(\frac{\sigma^2}{\pi}\right)^{3/2} e^{-(\mathbf{k}_1-\mathbf{k}_0)^2 \sigma^2} \approx \delta(\mathbf{k}_1 - \mathbf{k}_0)$ in (3.20), the result we have been seeking, the probability density of the existence of collapse-excited photons, is:

$$\mathcal{P}(\mathbf{k}) = \bar{n}(0) \frac{\lambda t}{M_N^2} \left(\frac{a^2}{\pi}\right)^{3/2} k_0 k e^{-(\mathbf{k}_0-\mathbf{k})^2 a^2}. \quad (3.22)$$

First, to connect with the result of Section 3.3, to verify that the loss of photons there and the gain (3.22) here account for all photons. The trace of the density matrix over the laser beam states is found from Eq. (3.14)'s second bracketed term (with the integral involved replaced by the last term in the second line of (3.15a)):

$$\sum_n \langle n|\rho(t)|n\rangle \approx \left[1 - \bar{n}(0) \frac{\lambda t}{M_N^2} \left(\frac{a^2}{\pi}\right)^{3/2} \int d\mathbf{k}_1 k_1 k_0 e^{-(\mathbf{k}_1-\mathbf{k}_0)^2 a^2}\right]. \quad (3.23)$$

When we add this to the trace over the excited photon states, $\int d\mathbf{k} \mathcal{P}(\mathbf{k})$, we obtain the correct result that the trace over all states is 1 to order λt.

Second, to connect with the result of Section 3.2, to verify that result gives the energy increase in this instance. $\mathcal{P}(\mathbf{k})/\bar{n}(0)$ is the probability that a single photon has been converted to momentum $\mathbf{k}$ from momentum $\mathbf{k}_0$. The energy change for such a photon is therefore $k - k_0$ and for $\bar{n}(0)$ photons is therefore, using (3.22),

$$\bar{E}(t) - \bar{E}(0) = \bar{n}(0) \frac{\lambda t}{M_N^2} \left(\frac{a^2}{\pi}\right)^{3/2} \int d\mathbf{k}(k - k_0) k_0 k e^{-(\mathbf{k}_0-\mathbf{k})^2 a^2}. \quad (3.24)$$

On the other hand, Eq. (3.6) specialized to photons is

$$\frac{\partial}{\partial t}\bar{H}(t) = \frac{\lambda}{M_N^2}\left(\frac{a^2}{\pi}\right)^{3/2} Tr\rho(t) \int d\mathbf{k}_1 a^\dagger(\mathbf{k}_1)a(\mathbf{k}_1) \int d\mathbf{k}_2 k_1 k_2 (k_2 - k_1) e^{-(\mathbf{k}_2-\mathbf{k}_1)^2 a^2}. \quad (3.25)$$

To first order in λ, $\rho(t) \to \rho(0)$ and, since $\int d\mathbf{k}_1 a^\dagger(\mathbf{k}_1)a(\mathbf{k}_1)$ is the beam photon number operator, the trace is $\bar{n}(0)$ and we see that (3.25) is then identical to (3.24).

Now let's turn to an application of (3.22). Among the most energetic of CW lasers is the carbon dioxide laser, with a wavelength $\lambda_0 \approx 1000$nm, with a few-hundred-kw beam achieved [18]. We shall consider a 1 megawatt beam, which has been suggested achieved for military purposes.

Since $\lambda_0 >> a = 100$nm, the probability density distribution (3.22) is well approximated by

$$\mathcal{P}(\mathbf{k}) \approx \bar{n}(0) \frac{\lambda t}{M_N^2} \left(\frac{a^2}{\pi}\right)^{3/2} k_0 k e^{-k^2 a^2}. \quad (3.26)$$

Thus, each photon in the beam has a probability of being converted to an excited photon of some wavelength characterized by the scale $\lesssim 2\pi a$, and the photons sprayed out have a spherically symmetric distribution.

So, suppose we consider an experiment continuously monitoring a 3m length of such a beam (traveling from the laser to some kind of absorber), looking for sprayed photons emerging from the beam in the suggested energy range.

The probability of any such photon appearing is

$$\Gamma \bar{n}(0)t \equiv \int d\mathbf{k}\mathcal{P}(\mathbf{k}) \approx \bar{n}(0) \frac{\lambda t \hbar^2}{M_N^2 c^2}\left(\frac{a^2}{\pi}\right)^{3/2} k_0 \frac{2\pi}{a^4} = 4\pi^{1/2}\bar{n}(0)\lambda t \frac{\lambda_N^2}{\lambda_0 a}. \quad (3.27)$$

Although our calculation has been to only first order in λ, and therefore only holds for small values of the probability, if we regard Γ defined earlier as a time-translation-invariant rate of production of anomalous photons per laser photon, we may allow $\Gamma \bar{n}(0)t$, the number of photons produced in time t, to exceed 1.

The energy in a 3m beam length of a 1 megawatt beam is $3 \times 10^6/c = 10^{-2}J$. The energy in a single photon is $hc/\lambda_0 \approx 2 \times 10^{-19}J$. Thus there are $\bar{n}(0) \approx 5 \times 10^{16}$ photons in that length and so, from (3.27), $\Gamma \bar{n}(0)t \approx .14\lambda t$. In one year$\approx 3 \times 10^7$s, according to Eq. (3.27), one expects $\approx 4 \times 10^6 \lambda$ anomalous "sprayed" photons. If no photons are seen, and there is 5% probability of experimental error, that would place a limit $\lambda \lesssim 10^{-8}$s^{-1}.

3.5 Effect on Cosmic Blackbody Radiation

We wish to examine how the distribution of blackbody photons is altered by the collapse excitation mechanism and apply the result to the cosmic blackbody spectrum.

3.5.1 Effect of Collapse on Blackbody Radiation

Blackbody radiation is described by the thermal density matrix $\rho(0) = e^{-\beta H}/Tre^{-\beta H}$, where $\beta \equiv 1/k_B T$ and $H = \int dk k a^\dagger(\mathbf{k})a^\dagger(\mathbf{k})$. It is traditional to work with box-normalized momenta rather than with continuous momenta, so we shall do that, writing $\mathbf{k}_i = (2\pi/L)\mathbf{n}_i = (2\pi/L)(n_{ix}, n_{iy}, n_{iz})$ with the n_{ij} as integers $-\infty < n_{ij} < \infty$. Likewise, we write $\int d\mathbf{k}_i = (2\pi/L)^3 \sum_{\mathbf{n}_i}$, and $a(\mathbf{k}_i) = (L/2\pi)^{3/2} a_\mathbf{i}$, $\delta(\mathbf{k}_\mathbf{i} - \mathbf{k}_{\mathbf{i}'}) = (L/2\pi)^3 \delta_{\mathbf{n}_\mathbf{i},\mathbf{n}'_\mathbf{i}}$, so $[a_i, a^\dagger_{i'}] = \delta_{\mathbf{n}_\mathbf{i},\mathbf{n}'_\mathbf{i}}$ and $H = \sum_m k_m a^\dagger_m a_m$.

We may now write the expression for the time rate of change of the mean energy $\bar{\epsilon}_s$ in a single mode (where the energy operator for the mode is $\epsilon_s \equiv k_s a^\dagger_s a_s$) to first order in λ. This is essentially Eq. (3.4) written in terms of box-normalized momenta but without summing over all modes and with an extra factor of 2 because there are two polarizations:

$$\frac{\partial}{\partial t}\bar{\epsilon}_s = -\frac{\lambda}{M_N^2}\left(\frac{a^2}{\pi}\right)^{3/2} Tr\rho(0)\left(\frac{2\pi}{L}\right)^3 \sum_{\mathbf{n}_1,\mathbf{n}_2,\mathbf{n}_3,\mathbf{n}_4} k_s \sqrt{k_1 k_2 k_3 k_4} k_s e^{-(\mathbf{k}_1-\mathbf{k}_2)^2 a^2}$$

$$\delta_{-\mathbf{n}_1+\mathbf{n}_2-\mathbf{n}_3+\mathbf{n}_4}[a^\dagger_1 a_2, [a^\dagger_3 a_4, a^\dagger_s a_s]]$$

$$= -\frac{2\lambda}{M_N^2}\left(\frac{a^2}{\pi}\right)^{3/2} Tr\rho(0)\left(\frac{2\pi}{L}\right)^3 k_s^2 \sum_{\mathbf{n}_1} k_1 [a^\dagger_s a_s - a^\dagger_1 a_1] e^{-(\mathbf{k}_s-\mathbf{k}_1)^2 a^2}. \quad (3.28)$$

As a check, we confirm that the mean number of photons is not changed by the collapse process, as follows. We note that $n_s = \epsilon_s/k_s$, so if we divide (3.28) by k_s and all modes are summed over, the left side of (3.28) is $\frac{\partial}{\partial t}\bar{n}$, and with one less factor of k_s, the sum over $\mathbf{n}_s$ of the right side of (3.28) vanishes.

We now may take the trace in Eq. (3.28). Since

$$Tre^{-\beta k_m a^\dagger_m a_m} = \sum_{j=0}^\infty \langle 0|\frac{a_m^j}{\sqrt{j!}}e^{-\beta k_m a^\dagger_m a_m}\frac{a_m^{\dagger j}}{\sqrt{j!}}|0\rangle = \sum_{j=0}^\infty e^{-\beta k_m j} = \frac{1}{1-e^{-\beta k_m}}$$

and so

$$Tr\rho(0) a^\dagger_m a_m = [1 - e^{-\beta k_m}]\sum_{j=0}^\infty e^{-\beta k_m j} j = \frac{1}{e^{\beta k_m} - 1},$$

we obtain

$$\bar{\epsilon}_s(t) = \frac{2k_s}{e^{\beta k_s} - 1} - \frac{2\lambda t}{M_N^2}\left(\frac{a^2}{\pi}\right)^{3/2}\left(\frac{2\pi}{L}\right)^3 k_s^2 \sum_{\mathbf{n}_1} k_1 \left[\frac{1}{e^{\beta k_s} - 1} - \frac{1}{e^{\beta k_1} - 1}\right] e^{-(\mathbf{k}_s-\mathbf{k}_1)^2 a^2}.$$

$$= \frac{2k_s}{e^{\beta k_s} - 1}\left[1 - \frac{\lambda t}{M_N^2}\left(\frac{a^2}{\pi}\right)^{3/2} k_s \int d\mathbf{k}_1 k_1 e^{-(\mathbf{k}_s-\mathbf{k}_1)^2 a^2}\right]$$

$$+ \frac{2\lambda t}{M_N^2}\left(\frac{a^2}{\pi}\right)^{3/2} k_s^2 \int d\mathbf{k}_1 k_1 e^{-(\mathbf{k}_s-\mathbf{k}_1)^2 a^2} \frac{1}{e^{\beta k_1} - 1} \quad (3.29)$$

Dynamical Collapse for Photons 37

where we have replaced the sum by an integral, returning to the continuum momentum for that variable, but so far keeping the discrete momentum for k_s.

According to the first line of Eq. (3.29), as time progresses, photons are kicked out of the mode with momentum k_s. According to the second line, photons are also kicked into this mode from all the other modes. As we shall see, loss from high-probability modes is the rule, since the photons lost are most likely kicked to higher-energy, low-probability modes, characterized by $k \sim 1/a$.

(As a final check, we note that the total mean energy increase calculated in Section 3.2 agrees with the result here. If we apply Eq. (3.4) to first order in λ, where $a^\dagger(\mathbf{k}_1)a(\mathbf{k}_1)$ is replaced by $(L/2\pi)^3 a_1^\dagger a_1$, so that the trace equation delineated earlier applies, with the extra factor of 2 for the two polarizations, the result is

$$\bar{E}(t) = \left(\frac{L}{2\pi}\right)^3 \left[\int d\mathbf{k} \frac{2k}{e^{\beta k} - 1}\right.$$
$$\left. + \frac{2\lambda}{M_N^2}\left(\frac{a^2}{\pi}\right)^{3/2} \int d\mathbf{k}_1 d\mathbf{k}_2 k_1 k_2 (k_2 - k_1) \frac{1}{e^{\beta k_1} - 1} e^{-(\mathbf{k}_1 - \mathbf{k}_2)^2 a^2}\right]. \quad (3.30)$$

This is the same as the integral of Eq. (3.29), with appropriate renaming of integration variables.)

3.5.2 Effect of Collapse on Cosmic Microwave Radiation

We wish to use (3.29) to calculate the change, due to the collapse process, in photon number in the modes of wavelength $\approx .05$cm to 50cm in the cosmological microwave radiation we receive, as the radiation travels toward us.

The distance of the radiation at any time t after recombination is characterized by the redshift parameter $Z(t)$. Taking $t = 0$ as the time of recombination, which occurred $\approx 400,000$yr after the big bang, it is found that $Z(0) \approx 1000$. Denoting by a subscript 0 the present value of a quantity, the radiation temperature T(t) is related to the present radiation temperature $T_0 \approx 2.7°K$ by the relation $T(t) = (1+Z(t))T_0$. Thus, at recombination, the temperature was $T(0) \approx 1000 T_0 \approx 3000°K$.

If we neglect the time differences between the Hubble time $\approx 14 \times 10^9$yr, the age of the universe, and the time since recombination t_0, the time evolution of Z, according to Hubble's law, is $Z(t) \approx 1000[1 - \frac{t}{t_0}]$. Any length ℓ, such as a wavelength λ or the length L of the side of the normalization box behaves as $\ell(t) = \ell_0/(1 + Z(t))$.

We shall simplify (3.29) by the approximation in the integrals $e^{-(\mathbf{k}_s - \mathbf{k}_1)^2 a^2} \approx e^{-k_1^2 a^2}$. This is certainly valid at present since the high-probability modes' wavelengths are much larger than a. However, we must consider this approximation at all times. We note that $(k_s - k_1)a = (k_{0s} - k_{01})(1 + Z)a$, and $(1 + Z)a$ ranges from its its value at recombination, 10^{-2}cm, to its present value, 10^{-5}cm. For this approximation to be valid, then $k_{0s} << [(1+Z)a]^{-1}$ for the full range of Z. This certainly is not true for the shorter wavelengths at recombination time, e.g., for $\lambda_{0s} \approx .05$cm, then $k_{0s} \approx 100$ is equal to $[(1+Z)a]^{-1}$. So we may just take it as surely valid for longer wavelengths, say $\lambda_0 \gtrsim .5$ cm, or consider that, since it is valid for at least part of the photon journey, the result may be at least approximately applied to shorter wavelengths.

With this approximation we may immediately evaluate the integral in the bracket in Eq. (3.29), $\int d\mathbf{k}_1 k_1 e^{-k_1^2 a^2} = 2\pi/a^4$.

The other integral in Eq. (3.29), is approximated as

$$\int d\mathbf{k}_1 k_1 e^{-k_1^2 a^2} \frac{1}{e^{\beta k_1}-1} \approx \int d\mathbf{k}_1 k_1 \frac{1}{e^{\beta k_1}-1} = 4\pi \Gamma(4)\zeta(4)/\beta^4 \approx 24\pi/\beta^4$$
$$= 12(2\pi)^5/\lambda_{Th}^4$$

where we write $2\pi\beta = hc/k_B T \equiv \lambda_{Th}$ is the thermal wavelength, with $\lambda_{Th0} \approx .5$cm. Here we have made the approximation $[e^{\beta k_1}-1]^{-1} e^{-k_1^2 a^2} \approx [e^{\beta k_1}-1]$, i.e., we may set $e^{-k_1^2 a^2} \approx 1$. To see this, first note that $\beta k = \beta_0 k_0 = \lambda_{Th}/\lambda_0$. Thus, $e^{-\beta k_1} e^{-k_1^2 a^2} \approx e^{-.1 k_{10}} e^{-(k_{10} Z 10^{-5})^2}$, and so $e^{-.1 k_{10}}$ dominates the integral for the full range of Z. We have also approximated the Riemann zeta function $\zeta(4) = 1.08... \approx 1$.

Therefore, (3.29) becomes

$$\bar{\epsilon}_s(t) = \frac{2k_s}{e^{\beta k_s}-1}\left[1 - 4\pi^{1/2}\lambda t \frac{\lambda_N^2}{a\lambda_s}\right] + k_s \frac{24(2\pi)^6}{\pi^{3/2}}\lambda t \frac{\lambda_N^2 a^3}{\lambda_{Th}^4 \lambda_s}. \quad (3.31)$$

We discard the gain (last) term of (3.31) as loss of photons massively predominates: the ratio of the loss term (second term in the bracket $\times 2k_s$) to the gain term is $\approx 3 \times 10^{14}$ (it is dominated by a^{-4}).

We want the number of photons in each mode, not the energy, so we divide by k_s. Also, we are interested in the number/volume in any mode with the same frequency, so we multiply the right-hand side of (3.31) by

$$1 = \left(\frac{L}{2\pi}\right)^3 \int_\Omega d\mathbf{k}_s = V\frac{1}{(2\pi)^3}4\pi k^2 dk = V\frac{4\pi}{c^3}\nu^2 d\nu$$

where, since we are returning to continuum variables, we replace k_s by k, we denote by V the (normalization) volume containing the radiation, and then have converted from variable k to frequency ν. Denoting by $\bar{n}(\nu,t)d\nu$ the number of photons with energy between $h\nu$ and $h(\nu+d\nu)$ in the volume V, we therefore have

$$\bar{n}(\nu,t)d\nu = V\frac{8\pi \nu^2 d\nu}{c^3[e^{h\nu/k_B T}-1]}\left[1 - 4\pi^{1/2}\lambda t\nu\frac{\lambda_N^2}{ac}\right], \quad (3.32a)$$

$$\frac{d}{dt}\bar{n}(\nu,t)d\nu = -V\frac{32\pi^{3/2}\nu^3 d\nu}{c^3[e^{h\nu/k_B T}-1]}\lambda\frac{\lambda_N^2}{ac}. \quad (3.32b)$$

We have taken the time derivative in (3.32b) to obtain the expression for the rate of photon loss over a short time interval. To get the total photon loss over t_0, we first express (3.32b) in terms of present variables, and then integrate over t. We note that $V(t) = V_0/[1 + Z(t)]^3$, $\nu(t) = \nu_0[1 + Z(t)]^3$ and $h\nu(t)/k_B T(t) = h\nu_0/k_B T_0$. Famously, the no-loss blackbody spectrum (the factor multiplying the bracket in (3.32a)) is time independent, but the loss term has an extra ν factor and so acquires an extra $1 + Z(t)$ factor:

$$\bar{n}(\nu_0, t_0)/V_0 = \frac{8\pi \nu_0^2}{c^3[e^{h\nu_0/k_B T_0} - 1]}\left[1 - 4\pi^{1/2}\nu_0\frac{\lambda_N^2}{ac}\lambda \int_0^{t_0} dt[1 + Z(t)]\right]$$

$$= \frac{8\pi \nu_0^2}{c^3[e^{h\nu_0/k_B T_0} - 1]}\left[1 - 4\pi^{1/2}\nu_0\frac{\lambda_N^2}{ac}500\lambda t_0\right]$$

$$\approx \frac{8\pi \nu_0^2}{c^3[e^{h\nu_0/k_B T_0} - 1]}\left[1 - .6\frac{\lambda}{(\lambda_0/.1)}\right]. \quad (3.33)$$

In the last line of (3.33), the unit of λ is sec^{-1} and λ_0 is cm: we note that the peak of the energy per unit wavelength spectrum is $\approx .1$cm. In going from the second to the third line, we have used $t_0 \approx 40 \times 10^{16}$s and $\lambda_N \approx .5$cm.

Eq. (3.33) is the result we have sought. We see that the spectrum of the radiation received is altered, in that over its travel to us, collapse reduces the mean number of photons/volume in each mode by an amount inversely proportional to the wavelength.

The cosmic microwave radiation is experimentally found to have the blackbody form (excepting the famous anisotropies). The temperature of the radiation is quoted as [19] $2.72548 \pm .00057°K$, so error/temperature $\Delta \approx 2 \times 10^{-4}$. If the temperature of the pure blackbody spectrum is $T_0[1 - \Delta]$, it may be written as

$$\bar{n}(\nu_0, t_0)/V_0 = \frac{8\pi \nu_0^2}{c^3[e^{h\nu_0/k_B T_0[1-\Delta]} - 1]} \approx \frac{8\pi \nu_0^2}{c^3[e^{h\nu_0/k_B T_0} - 1]}\left[1 - \frac{e^{\lambda_{Tho}/\lambda_0}(\lambda_{Tho}/\lambda_0)\Delta}{[e^{\lambda_{Tho}/\lambda_0} - 1]}\right]$$

(3.34)

$(h\nu_0/k_B T_0 = \lambda_{Tho}/\lambda_0)$.

For wavelengths short enough that $e^{\lambda_{Tho}/\lambda_0} - 1 \approx e^{\lambda_{Tho}/\lambda_0}$ (say for $\lambda_0 \lesssim .25$cm, since $\lambda_{Tho} \approx .5$cm), for which the validity of (3.33) is perhaps problematic (since, as mentioned, the approximation made in its evaluation is best for longer wavelengths), it is nonetheless interesting to note that the bracket in (3.34) becomes $[1-(\lambda_{Tho}/\lambda_0)\Delta]$. This is identical in form to the bracket in (3.33), a constant divided by λ_0, so the spectrum with the collapse diminution of photons is indistinguishable from a blackbody spectrum with a lower temperature such that $\Delta = .06\lambda/\lambda_{Tho} \approx .1\lambda$. This provides no limit on λ.

For wavelengths long enough that $e^{\lambda_{Tho}/\lambda_0} - 1 \approx \lambda_{Tho}/\lambda_0$ (say $\lambda_0 \gtrsim 1$cm), the bracket in (3.34) becomes $\approx [1 - \Delta]$. Since there is not the collapse dependence $\sim 1/\lambda_0$, for large enough λ one would see an alteration of the spectrum shape that could be distinguished from a blackbody spectrum with a different temperature. Since one does not see such an alteration, if one supposes that it is there, then the error masks it. In that case, from (3.34), one has $.06\lambda/\lambda_0 \lesssim \Delta$ which, for $\lambda_0 = 1$cm, gives the limitation $\lambda \lesssim 3 \times 10^{-3} \text{s}^{-1}$.

One might do another calculation, considering that the photons lost from the blackbody radiation were converted to higher-energy photons, on the wavelength scale $2\pi a$, and calculate the distribution of this anomalous radiation arriving along with the unaltered photons comprising the bulk of the blackbody radiation. However, there are many sources of infrared radiation [20], among which one could not detect this contribution.

A concluding remark.

There are two reasons why the excitation effect of collapse is small on the cosmic radiation and, indeed, upon all the photon collections considered here, compared say to the excitation of electrons in atoms. One is that the effect increases with mass-energy, and the photons considered here have so much less energy than the mass-energy of electrons. The other is that one can experimentally compensate for the small rate of excitation of particles by observing many particles over a long period of time, but, while it is possible to conveniently amass and observe large numbers of atoms, for a long time, photons are not readily accumulated in large amounts, as they insist upon scurrying away.

References

[1] P. Pearle, *Reduction of the state vector by a nonlinear Schrödinger equation*, Phys. Rev. D **13**, 857 (1976).

[2] P. Pearle, *Combining stochastic dynamical state-vector reduction with spontaneous localization*, Physical Review A **39**, 2277 (1989).

[3] G. C. Ghirardi, P. Pearle, and A. Rimini, *Markov processes in Hilbert space and continuous spontaneous localization of systems of identical particles*, Physical Review A **42**, 78 (1990).

[4] For a recent review, see A. Bassi, K. Lochan, S. Satin, T. P. Singh, and H. Ulbricht, *Models of wave-function collapse, underlying theories, and experimental tests*, Rev. Mod. Phys. **85**, 471 (2013). Earlier reviews are P. Pearle, *Collapse models*, in *Open Systems and Measurement in Relativistic Quantum Theory*, edited by F. Petruccione and H. P. Breuer (Springer, Berlin, 1999), p. 195; A. Bassi and G. C. Ghirardi, *Dynamical reduction models*, Phys. Rep. **379**, 257 (2003); P. Pearle, *How stands collapse I*, J. Phys. A **40**, 3189 (2007) and continued as *How stands collapse II*, in *Quantum Reality, Relativistic Causality and Closing the Epistemic Circle*, edited by W. Myrvold and J. Christian (Springer, New York, 2009), p. 257.

[5] P. Pearle and E. Squires, *Bound state excitation, nucleon decay experiments, and models of wave function collapse*, Phys. Rev. Lett. **73**, 1 (1994).

[6] G. C. Ghirardi, A. Rimini, and T. Weber, *Unified dynamics for microscopic and macroscopic systems*, Phys. Rev. D **34**, 470 (1986); *Disentanglement of quantum wave functions: Answer to "Comment on 'Unified dynamics for microscopic and macroscopic systems,'"* Phys. Rev. D **36**, 3287 (1987); *The puzzling entanglement of Schrödinger's wave function*, Found. Physics **18**, 1, (1988).

[7] C. Dove and E. J. Squires, *Symmetric versions of explicit wave function collapse models*, Found. Phys. **25** 1267 (1995); R. Tumulka, *A relativistic version of the Ghirardi–Rimini–Weber Model*, Journ. Stat. Phys. **125**, 825 (2006).

[8] Work on relativistic CSL appears in P. Pearle, *Toward a relativistic theory of statevector reduction* in *Sixty-Two Years of Uncertainty*, edited by A. Miller (Plenum, New York, 1990), p. 193; G. C. Ghirardi, R. Grassi, and P. Pearle, *Relativistic dynamical reduction models: general framework and examples*, Found. Phys. **20**, 1271 (1990); P. Pearle, *Relativistic collapse model with tachyonic features*, Phys. Rev. A **59**, 80 (1999); O. Nicrosini and A. Rimini, *Relativistic spontaneous localization: a proposal*, Found. Phys. **33**, 1061 (2003); D. J. Bedingham, *Stochastic particle annihilation: a model of state reduction in relativistic quantum field theory*, J. Phys. A: Math. Theor. **40**, 647 (2007), *Relativistic state reduction dynamics*, Found. Phys. **41**, 686 (2010); *Relativistic state reduction model*, J. Phys. Conference Series **306**, 012034 (2011); D. Bedingham, D. Dürr, G. C. Ghirardi, S. Goldstein, R. Tumulka, and N. Zanghi, *Matter density and relativistic models of wave function collapse*, J. Stat. Phys. **154**, 623 (2014); P. Pearle, *Relativistic dynamical collapse model*, Phys. Rev. D **91**, 105012 (2015).

[9] P. Pearle, *Collapse miscellany*, in *Quantum Theory: A Two Time Success Story: Yakir Aharonov Festschrift*, D. Struppa and J. Tollakson (eds.), (Springer, Milan 2013), p. 131.

[10] B. Collett, P. Pearle, F. Avignone, and S. Nussinov, *Constraint on collapse models by limit on spontaneous x-ray emission in Ge*, Found. Phys. **25**, 1399 (1995); P. Pearle, J. Ring, J. I. Collar, and F. T. Avignone, *The CSL collapse model and spontaneous radiation: an update*, Found. Phys. **29**, 465 (1999).

[11] G. Jones, P. Pearle, and J. Ring, *Consequence for wavefunction collapse model of the Sudbury Neutrino Observatory experiment*, Found. Phys. **34**, 1467 (2004).

[12] Q. Fu, *Spontaneous radiation of free electrons in a nonrelativistic collapse model*, Phys. Rev. A **56**, 1806 (1997); S. L. Adler, *Lower and upper bounds on CSL parameters from latent image formation and IGM heating*, Journal of Physics A **40**, 2935 (2007) and corrigendum, Journal of Physics A **40** 1 (2007); C. Curceanu, S. Bartolucci, A. Bassi, et al., *Spontaneously emitted X-rays: an experimental signature of the dynamical reduction models*, Found. Phys. **46**, 263 (2016).

[13] B. Collett and P. Pearle, *Wavefunction collapse and random walk*, Found. Phys. **33**, 1495 (2003); D. Goldwater, M. Paternostro, and P. F. Barker, *Testing wavefunction collapse models using parametric heating of a trapped nanosphere*, Phys. Rev. A **94**, 010104 (2016); J. Li, S. Zippilli, J.Zhang, and D. Vitali, *Discriminating the effects of collapse models from environmental diffusion with levitated nanospheres*, Phys. Rev. A **93**, 050102 (2016); Y. Li, A. M. Steane, D. Bedingham, and G.A.D. Briggs, *Detecting continuous spontaneous localisation with charged bodies in a Paul trap*, Phys. Rev. A **95**, 032112 (2017).

[14] T. Josset, A. Perez, and D. Sudarsky, *Dark energy as the weight of violating energy conservation*, Phys. Rev. Lett. **118**, 021102 (2017).

[15] Vulcan laser facility, http://www.clf.stfc.ac.uk/CLF/Facilities/Vulcan/Vulcan+laser/12250.aspx.

[16] Stanford LCLS facility, https://portal.slac.stanford.edu/sites/lcls_public/instruments/cxi/Documents/CXItechspecs.pdf
[17] W. Feldmann and R. Tumulka, *Parameter diagrams of the GRW and CSL theories of wave function collapse*, Journ. Phys. A, **45**, 065304 (2012); see A. Bassi et. al.[4]; Also see F. Laloe, W. Mullin, and P. Pearle, *Heating of trapped ultracold atoms by collapse dynamics*, Phys. Rev. A **90**, 052119 (2014), reference 7.
[18] See e.g., Wikipedia's article, https://en.wikipedia.org/wiki/Carbon_dioxide_laser.
[19] D. J. Fixsen, *The temperature of the cosmic microwave background*, APJ **707**, 916 (2009).
[20] A. Franceschini, G. Rodighiero, M. Vaccari, *Extragalactic optical-infrared background radiation, its time evolution and the cosmic photon-photon opacity*, A&A**487**, 837 (2008).

Appendix A Collapse

Here we apply the density matrix evolution Eq. (3.1) to examine the collapse of the initial superposition state $|\psi, 0\rangle = \frac{1}{\sqrt{2}}[|L\rangle + |R\rangle]$, of two spatially displaced states, each consisting of N superposed particles in a volume $\sim \sigma^3$, all moving with the very well-defined momentum $\mathbf{k}_0$ (we assume $k_0 \sigma \gg k_0 a \gg 1$) in a direction orthogonal to the vector $\mathbf{x}_L - \mathbf{x}_R$ (we assume $|\mathbf{x}_L - \mathbf{x}_R| > \sigma$) connecting them:

$$|L\rangle \equiv \frac{1}{\sqrt{N!}} \left[\int d\mathbf{x} \xi^\dagger(\mathbf{x}) \frac{1}{(2\pi\sigma^2)^{3/4}} e^{-(\mathbf{x}-\mathbf{x}_L)^2/4\sigma^2} e^{i\mathbf{k}_0 \cdot \mathbf{x}} \right]^N |0\rangle$$

$$= \frac{1}{\sqrt{N!}} \left[\int d\mathbf{k} a^\dagger(\mathbf{k}) \left(\frac{2\sigma^2}{\pi}\right)^{3/4} e^{-(\mathbf{k}-\mathbf{k}_0)^2 \sigma^2} e^{-i\mathbf{k} \cdot \mathbf{x}_L} \right]^N |0\rangle, \qquad (A.35)$$

and similarly for $|R\rangle$, with $\mathbf{x}_L \to \mathbf{x}_R$. We want $\langle L|R\rangle = e^{-N|\mathbf{x}_L - \mathbf{x}_R|^2/8\sigma^2} \approx 0$, so we assume $N|\mathbf{x}_L - \mathbf{x}_R|^2 \gg 8\sigma^2$. We shall denote $\alpha_L^*(\mathbf{k}) = \left(\frac{2\sigma^2}{\pi}\right)^{3/4} e^{-(\mathbf{k}-\mathbf{k}_0)^2 \sigma^2} e^{-i\mathbf{k} \cdot \mathbf{x}_L}$ and $\omega(k) \equiv \sqrt{k^2 + M^2}$.

We note that each of these states is an approximate energy eigenstate to high accuracy:

$$H|L\rangle = \int d\mathbf{k} \omega(k) a^\dagger(\mathbf{k}) a(\mathbf{k}) |L\rangle$$

$$= N \frac{1}{\sqrt{N!}} \left[\int d\mathbf{k} a^\dagger(\mathbf{k}) \alpha_L^*(\mathbf{k}) \right]^{N-1} \int d\mathbf{k} a^\dagger(\mathbf{k}) \omega(k) \alpha_L^*(\mathbf{k}) |0\rangle \approx N\omega(k_0)|L\rangle, \qquad (A.36)$$

since the gaussian in $\alpha(\mathbf{k})$ implies $k = k_0 + o(1/\sigma)$.

We wish to show that the off-diagonal elements of the density matrix decay, signaling collapse to one or the other state, to first order in λ. When we calculate $\langle L|\rho(t)|R\rangle$ using (3.1), the Hamiltonian term makes no contribution since

Dynamical Collapse for Photons 43

$\langle L|[H, \rho(t)]|R\rangle \approx N\omega(\mathbf{k}_0)[\langle L|\rho(t)|R\rangle - \langle L|\rho(t)|R\rangle] = 0$, so we get, with $\rho(0) = \frac{1}{2}[|L\rangle + |R\rangle][\langle L| + \langle R|]$ (and noting that the matrix elements of the energy density operators between $\langle L|$ and $|R\rangle$ essentially vanish because the overlap integral between left and right states when there is one or two less particles is still negligibly small):

$$\langle L|\rho(t)|R\rangle = \frac{1}{2} - \frac{\lambda t}{2M_N^2} \int d\mathbf{x} \int d\mathbf{x}' e^{-(\mathbf{x}-\mathbf{x}')^2/4a^2}$$

$$\left[\langle L|K^{1/2}\xi^\dagger(\mathbf{x})K^{1/2}\xi(\mathbf{x})K^{1/2}\xi^\dagger(\mathbf{x}')K^{1/2}\xi(\mathbf{x}')|L\rangle \frac{1}{2} \right.$$

$$+ \frac{1}{2}\langle R|K^{1/2}\xi^\dagger(\mathbf{x})K^{1/2}\xi(\mathbf{x})K^{1/2}\xi^\dagger(\mathbf{x}')K^{1/2}\xi(\mathbf{x}')|R\rangle$$

$$\left. -2\frac{1}{2}\langle L|K^{1/2}\xi^\dagger(\mathbf{x})K^{1/2}\xi(\mathbf{x})|L\rangle\langle R|K^{1/2}\xi^\dagger(\mathbf{x}')K^{1/2}\xi(\mathbf{x}')|R\rangle \right]$$

$$= \frac{1}{2} - \frac{\lambda t}{4M_N^2}\left(\frac{a^2}{\pi}\right)^{3/2} \int d\mathbf{k}_1 d\mathbf{k}_2 d\mathbf{k}_3 d\mathbf{k}_4 \sqrt{\omega(k_1)\omega(k_2)\omega(k_3)\omega(k_4)}$$

$$e^{-(\mathbf{k}_1-\mathbf{k}_2)^2 a^2}\delta(-\mathbf{k}_1 + \mathbf{k}_2 - \mathbf{k}_3 + \mathbf{k}_4)$$

$$\left[\langle L|a^\dagger(\mathbf{k}_1)[a^\dagger(\mathbf{k}_3)a(\mathbf{k}_2) + \delta(\mathbf{k}_2 - \mathbf{k}_3)]a(\mathbf{k}_4)|L\rangle \right.$$

$$+ \langle R|a^\dagger(\mathbf{k}_1)[a^\dagger(\mathbf{k}_3)a(\mathbf{k}_2) + \delta(\mathbf{k}_2 - \mathbf{k}_3)]a(\mathbf{k}_4)|R\rangle$$

$$\left. -2\langle L|a^\dagger(\mathbf{k}_1)a(\mathbf{k}_2)|L\rangle\langle R|a^\dagger(\mathbf{k}_3)a(\mathbf{k}_4)|R\rangle \right]$$

$$= \frac{1}{2} - \frac{\lambda t}{4M_N^2}\left(\frac{a^2}{\pi}\right)^{3/2} \int d\mathbf{k}_1 d\mathbf{k}_2 d\mathbf{k}_3 d\mathbf{k}_4 \sqrt{\omega(k_1)\omega(k_2)\omega(k_3)\omega(k_4)}$$

$$e^{-(\mathbf{k}_1-\mathbf{k}_2)^2 a^2}\delta(-\mathbf{k}_1 + \mathbf{k}_2 - \mathbf{k}_3 + \mathbf{k}_4)$$

$$2\Big[N(N-1)\alpha(\mathbf{k}_1)\alpha(\mathbf{k}_2)\alpha(\mathbf{k}_3)\alpha(\mathbf{k}_4) + N\alpha(\mathbf{k}_1)\alpha(\mathbf{k}_4)\delta(\mathbf{k}_2 - \mathbf{k}_3)$$

$$-N^2 a_L^*(\mathbf{k}_1)\alpha_L(\mathbf{k}_2)\alpha_R^*(\mathbf{k}_3)\alpha_R(\mathbf{k}_4)\Big], \tag{A.37}$$

where we have set $\alpha(\mathbf{k}) \equiv |\alpha_{L,R}(\mathbf{k})|$.

We shall now evaluate the three integrals involving the bracketed terms in Eq. (A.37).

For the first term, again, we utilize the excellent approximation $f(\mathbf{k})\alpha(\mathbf{k}) \approx f(\mathbf{k}_0)\alpha(\mathbf{k})$ (since the factor $\alpha(\mathbf{k})$ implies $\mathbf{k} = \mathbf{k}_0$ to order $1/\sigma$):

$$I_1 \equiv \left(\frac{a^2}{\pi}\right)^{3/2} \int d\mathbf{k}_1 d\mathbf{k}_2 d\mathbf{k}_3 d\mathbf{k}_4 \sqrt{\omega(k_1)\omega(k_2)\omega(k_3)\omega(k_4)}$$

$$e^{-(\mathbf{k}_1-\mathbf{k}_2)^2 a^2}\delta(-\mathbf{k}_1 + \mathbf{k}_2 - \mathbf{k}_3 + \mathbf{k}_4)\alpha(\mathbf{k}_1)\alpha(\mathbf{k}_2)\alpha(\mathbf{k}_3)\alpha(\mathbf{k}_4)$$

$$\approx \omega^2(k_0)\left(\frac{a^2}{\pi}\right)^{3/2} \int d\mathbf{k}_1 d\mathbf{k}_2 d\mathbf{k}_3 d\mathbf{k}_4 \delta(-\mathbf{k}_1 + \mathbf{k}_2 - \mathbf{k}_3 + \mathbf{k}_4)\alpha(\mathbf{k}_1)\alpha(\mathbf{k}_2)\alpha(\mathbf{k}_3)\alpha(\mathbf{k}_4)$$

$$= \omega^2(k_0)\left(\frac{a^2}{\pi}\right)^{3/2} \int d\mathbf{k}_1 d\mathbf{k}_2 d\mathbf{k}_3 d\mathbf{k}_4 \frac{1}{(2\pi)^3} \int d\mathbf{x} e^{i\mathbf{x}\cdot(-\mathbf{k}_1+\mathbf{k}_2-\mathbf{k}_3+\mathbf{k}_4)}$$
$$\alpha(\mathbf{k}_1)\alpha(\mathbf{k}_2)\alpha(\mathbf{k}_3)\alpha(\mathbf{k}_4)$$
$$= \omega^2(k_0)\left(\frac{a^2}{\pi}\right)^{3/2} \frac{1}{(2\pi)^3} \int d\mathbf{x} \left(\frac{2\sigma^2}{\pi}\right)^3 \left[\left(\frac{\pi}{\sigma^2}\right)^{3/2} e^{-\mathbf{x}^2/4\sigma^2}\right]^4$$
$$= \omega^2(k_0)\left(\frac{a}{\sigma}\right)^3. \tag{A.38}$$

The second integral is, using the same approximation,

$$I_2 \equiv \left(\frac{a^2}{\pi}\right)^{3/2} \int d\mathbf{k}_1 d\mathbf{k}_2 d\mathbf{k}_3 d\mathbf{k}_4 \sqrt{\omega(k_1)\omega(k_2)\omega(k_3)\omega(k_4)}$$
$$e^{-(\mathbf{k}_1-\mathbf{k}_2)^2 a^2} \delta(-\mathbf{k}_1+\mathbf{k}_2-\mathbf{k}_3+\mathbf{k}_4)\alpha(\mathbf{k}_1)\alpha(\mathbf{k}_4)\delta(\mathbf{k}_2-\mathbf{k}_3)$$
$$= \left(\frac{a^2}{\pi}\right)^{3/2} \int d\mathbf{k}_1 d\mathbf{k}_2 \omega(k_1)\omega(k_2) e^{-(\mathbf{k}_1-\mathbf{k}_2)^2 a^2} \alpha^2(\mathbf{k}_1)$$
$$\approx \omega(k_0)\left(\frac{a^2}{\pi}\right)^{3/2} \int d\mathbf{k}_1 d\mathbf{k}_2 \omega(k_2) e^{-(\mathbf{k}_0-\mathbf{k}_2)^2 a^2} \alpha^2(\mathbf{k}_1). \tag{A.39}$$

We may likewise use the approximation $\omega(k_2)e^{-(\mathbf{k}_0-\mathbf{k}_2)^2 a^2} \approx \omega(k_0)e^{-(\mathbf{k}_0-\mathbf{k}_2)^2 a^2}$, since the gaussian implies $\mathbf{k}_2 = \mathbf{k}_0$ to order $1/a$. Then,

$$I_2 \approx \omega^2(k_0)\left(\frac{a^2}{\pi}\right)^{3/2} \int d\mathbf{k}_1 d\mathbf{k}_2 e^{-(\mathbf{k}_0-\mathbf{k}_2)^2 a^2} \left(\frac{2\sigma^2}{\pi}\right)^{3/2} e^{-(\mathbf{k}_1-\mathbf{k}_0)^2 2\sigma^2}$$
$$= \omega^2(k_0). \tag{A.40}$$

The last integral, with the already employed approximations, is

$$I_3 \approx \omega^2(k_0)\left(\frac{a^2}{\pi}\right)^{3/2} \int d\mathbf{k}_1 d\mathbf{k}_2 d\mathbf{k}_3 d\mathbf{k}_4 \delta(-\mathbf{k}_1+\mathbf{k}_2-\mathbf{k}_3+\mathbf{k}_4)$$
$$\alpha(\mathbf{k}_1)\alpha(\mathbf{k}_2)\alpha(\mathbf{k}_3)\alpha(\mathbf{k}_4) e^{i\mathbf{x}_L\cdot(\mathbf{k}_2-\mathbf{k}_1)} e^{i\mathbf{x}_R\cdot(\mathbf{k}_4-\mathbf{k}_3)}$$
$$= \omega^2(k_0)\left(\frac{a^2}{\pi}\right)^{3/2} \frac{1}{(2\pi)^3} \int d\mathbf{x} \int d\mathbf{k}_1 d\mathbf{k}_2 d\mathbf{k}_3 d\mathbf{k}_4 e^{i\mathbf{x}\cdot(-\mathbf{k}_1+\mathbf{k}_2-\mathbf{k}_3+\mathbf{k}_4)}$$
$$\alpha(\mathbf{k}_1)\alpha(\mathbf{k}_2)\alpha(\mathbf{k}_3)\alpha(\mathbf{k}_4) e^{i(\mathbf{x}_L-\mathbf{x}_R)\cdot(\mathbf{k}_2-\mathbf{k}_1)}$$
$$= \omega^2(k_0)\left(\frac{a^2}{\pi}\right)^{3/2} \frac{1}{(2\pi)^3} \int d\mathbf{x} \int d\mathbf{k}_1 d\mathbf{k}_2 e^{i\mathbf{x}\cdot(-\mathbf{k}_1+\mathbf{k}_2)}$$
$$\alpha(\mathbf{k}_1)\alpha(\mathbf{k}_2) e^{i(\mathbf{x}_L-\mathbf{x}_R)\cdot(\mathbf{k}_2-\mathbf{k}_1)} \left(\frac{2\pi}{\sigma^2}\right)^{3/2} e^{-\frac{\mathbf{x}^2}{2\sigma^2}}$$
$$= \omega^2(k_0)\left(\frac{a^2}{\pi}\right)^{3/2} \int d\mathbf{k}_1 d\mathbf{k}_2 \alpha(\mathbf{k}_1)\alpha(\mathbf{k}_2) e^{i(\mathbf{x}_L-\mathbf{x}_R)\cdot(\mathbf{k}_2-\mathbf{k}_1)} e^{-(\mathbf{k}_1-\mathbf{k}_2)^2 \sigma^2/2}. \tag{A.41}$$

Upon making the replacement $\mathbf{k}'_i = (\mathbf{k}_i - \mathbf{k}_0)$, and then removing the primes, we continue:

$$I_3 = \omega^2(k_0)\left(\frac{a^2}{\pi}\right)^{3/2}\left(\frac{2\sigma^2}{\pi}\right)^{3/2}\int d\mathbf{k}_1 d\mathbf{k}_2 e^{-k_1^2\sigma^2}e^{-k_2^2\sigma^2}e^{i(\mathbf{x}_L-\mathbf{x}_R)\cdot(\mathbf{k}_2-\mathbf{k}_1)}e^{-(\mathbf{k}_1-\mathbf{k}_2)^2\sigma^2/2}$$
$$= \omega^2(k_0)\left(\frac{a^2}{\pi}\right)^{3/2}\left(\frac{2\sigma^2}{\pi}\right)^{3/2}\int d\mathbf{k}_1 d\mathbf{k}_2 e^{-(\mathbf{k}_1+\mathbf{k}_2)^2\sigma^2/2}e^{i(\mathbf{x}_L-\mathbf{x}_R)\cdot(\mathbf{k}_2-\mathbf{k}_1)}e^{-(\mathbf{k}_1-\mathbf{k}_2)^2\sigma^2}.$$
(A.42)

With change of variables $\mathbf{k}' \equiv \mathbf{k}_1 + \mathbf{k}_2, \mathbf{k} \equiv \mathbf{k}_1 - \mathbf{k}_2$ and so $d\mathbf{k}_1 d\mathbf{k}_2 = \frac{1}{8}d\mathbf{k}d\mathbf{k}'$:

$$I_3 = \omega^2(k_0)\left(\frac{a^2}{\pi}\right)^{3/2}\left(\frac{\sigma^2}{2\pi}\right)^{3/2}\int d\mathbf{k}d\mathbf{k}'e^{-k'^2\sigma^2/2}e^{-i(\mathbf{x}_L-\mathbf{x}_R)\cdot\mathbf{k}}e^{-k^2\sigma^2}$$
$$= \omega^2(k_0)\left(\frac{a}{\sigma}\right)^3 e^{-(\mathbf{x}_L-\mathbf{x}_R)^2/4\sigma^2}.$$
(A.43)

Putting (A.38), (A.40) and (A.43) into (A.37), we obtain the expression for the off-diagonal matrix element,

$$\langle L|\rho(t)|R\rangle \approx \frac{1}{2} - \frac{N\lambda t\omega^2(k_0)}{2M_N^2}\left[N\left(\frac{a}{\sigma}\right)^3\left(1 - e^{-(\mathbf{x}_L-\mathbf{x}_R)^2/4\sigma^2}\right) + \left(1 - \left(\frac{a}{\sigma}\right)^3\right)\right]$$
$$\approx \frac{1}{2} - \frac{N\lambda t\omega^2(k_0)}{2M_N^2}\left[N\left(\frac{a}{\sigma}\right)^3\left(1 - e^{-(\mathbf{x}_L-\mathbf{x}_R)^2/4\sigma^2}\right) + 1\right],$$
(A.44)

cited in Section 3.1.

Appendix B Integrals

B.1 Integral Involved in the First Bracketed Term in Eq. (3.14)

We wish to evaluate the integral

$$I \equiv \int d\mathbf{k}_1 d\mathbf{k}_2 d\mathbf{k}_3 d\mathbf{k}_4 \sqrt{k_1 k_2 k_3 k_4} e^{-(\mathbf{k}_1-\mathbf{k}_2)^2 a^2}$$
$$\delta(-\mathbf{k}_1 + \mathbf{k}_2 - \mathbf{k}_3 + \mathbf{k}_4)\alpha(\mathbf{k}_1)\alpha(\mathbf{k}_2)\alpha(\mathbf{k}_3)\alpha(\mathbf{k}_4)$$
$$\approx k_0^2 \int d\mathbf{k}_1 d\mathbf{k}_2 d\mathbf{k}_3 d\mathbf{k}_4 e^{-(\mathbf{k}_1-\mathbf{k}_2)^2 a^2}\frac{1}{(2\pi)^3}\int d\mathbf{x}e^{i\mathbf{x}\cdot[-\mathbf{k}_1+\mathbf{k}_2-\mathbf{k}_3+\mathbf{k}_4]}$$
$$\frac{1}{(\pi/\sigma^2)^3}e^{-(\mathbf{k}_1-\mathbf{k}_0)^2\sigma^2/2}e^{-(\mathbf{k}_2-\mathbf{k}_0)^2\sigma^2/2}e^{-(\mathbf{k}_3-\mathbf{k}_0)^2\sigma^2/2}e^{-(\mathbf{k}_4-\mathbf{k}_0)^2\sigma^2/2}$$
(B.45)

We make the change of variables $\mathbf{k}'_i \equiv \mathbf{k}_i - \mathbf{k}_0$, thereafter unpriming the $\mathbf{k}$s, and perform the integrals over $\mathbf{k}_3, \mathbf{k}_4$:

$$I \equiv k_0^2 \int d\mathbf{k}_1 d\mathbf{k}_2 e^{-(\mathbf{k}_1-\mathbf{k}_2)^2 a^2}\frac{1}{(2\pi)^3}\int d\mathbf{x}e^{i\mathbf{x}\cdot[-\mathbf{k}_1+\mathbf{k}_2]}\left(\frac{2\pi}{\sigma^2}\right)^3$$
$$e^{-\mathbf{x}^2/\sigma^2}\frac{1}{(\pi/\sigma^2)^3}e^{-k_1^2\sigma^2/2}e^{-k_2^2\sigma^2/2}.$$
(B.46)

Changing variables to $\mathbf{k} \equiv \mathbf{k}_1 - \mathbf{k}_2, \mathbf{k}' \equiv \mathbf{k}_1 + \mathbf{k}_2$, using $d\mathbf{k}d\mathbf{k}' = 8d\mathbf{k}_1 d\mathbf{k}_2$:

$$I \equiv \frac{1}{(2\pi)^3} k_0^2 \int d\mathbf{x} e^{-\mathbf{x}^2/\sigma^2} \int d\mathbf{k} e^{-\mathbf{k}^2(a^2+\sigma^2/4)} e^{-i\mathbf{x}\cdot\mathbf{k}} \int d\mathbf{k}' e^{-\mathbf{k}'^2 \sigma^2/4}. \quad (B.47)$$

Since $\sigma \gg a$, we can make the approximation $a^2 + \sigma^2/4 \approx \sigma^2/4$, and so obtain:

$$I \approx \frac{1}{(2\pi)^3} k_0^2 \int d\mathbf{x} e^{-\mathbf{x}^2/\sigma^2} \left(\frac{4\pi}{\sigma^2}\right)^{3/2} e^{-\mathbf{x}^2/\sigma^2} \left(\frac{4\pi}{\sigma^2}\right)^{3/2} = k_0^2 \frac{(2\pi)^{3/2}}{\sigma^3}. \quad (B.48)$$

As noted, (B.48) $\sim \frac{k_0^2}{\sigma^3}$ is small compared to the second bracketed term in Eq. (3.14) which is, according to (3.15a) or (3.15b) $\sim \frac{k_0}{a^4}$ or $\sim \frac{k_0^2}{a^3}$, and so (B.48) may be neglected.

B.2 Integrals Involved in Eq. (3.21)

We wish to evaluate the integral

$$I' \equiv \int d\mathbf{k}_1 d\mathbf{k}_2 d\mathbf{k}_3 d\mathbf{k}_4 \sqrt{k_1 k_2 k_3 k_4} e^{-(\mathbf{k}_1 - \mathbf{k}_2)^2 a^2}$$
$$\delta(-\mathbf{k}_1 + \mathbf{k}_2 - \mathbf{k}_3 + \mathbf{k}_4) \alpha(\mathbf{k}_1) \alpha(\mathbf{k}_2) \alpha(\mathbf{k}_3) \alpha(\mathbf{k}_4)$$
$$\cdot \left[\alpha^2(\mathbf{k}) - 2\delta(\mathbf{k}_2 - \mathbf{k}))\right] \quad (B.49)$$

The integral multiplying $\alpha^2(\mathbf{k})$ is the result (B.48), so the first bracketed term in (B.48) leads to

$$I_1 = k_0^2 2^{3/2} e^{-(\mathbf{k}-\mathbf{k}_0)^2 \sigma^2}. \quad (B.50)$$

As for the second bracketed term in (B.49), this is the integral (B.46) with $-2\delta(\mathbf{k}_2 - \mathbf{k})$ inserted:

$$I_2 \equiv -2 \frac{1}{\pi^3} k_0^2 \int d\mathbf{k}_1 e^{-(\mathbf{k}_1 - \mathbf{k})^2 a^2} \int d\mathbf{x} e^{i\mathbf{x}\cdot[-\mathbf{k}_1+\mathbf{k}]} e^{-\mathbf{x}^2/\sigma^2} e^{-\mathbf{k}_1^2 \sigma^2/2} e^{-\mathbf{k}^2 \sigma^2/2}$$
$$= -2\left(\frac{\sigma^2}{\pi}\right)^{3/2} k_0^2 \int d\mathbf{k}_1 e^{-(\mathbf{k}_1 - \mathbf{k})^2 (a^2 + \sigma^2/4)} e^{-\mathbf{k}_1^2 \sigma^2/2} e^{-\mathbf{k}^2 \sigma^2/2}$$
$$\approx -2\left(\frac{\sigma^2}{\pi}\right)^{3/2} k_0^2 \int d\mathbf{k}_1 e^{-(\mathbf{k}_1 - \mathbf{k})^2 \sigma^2/4} e^{-\mathbf{k}_1^2 \sigma^2/2} e^{-\mathbf{k}^2 \sigma^2/2}$$
$$= -2 k_0^2 \left(\frac{4}{3}\right)^{3/2} e^{-(\mathbf{k}-\mathbf{k}_0)^2 2\sigma^2/3}. \quad (B.51)$$

4
Quantum State Reduction

DORJE C. BRODY AND LANE P. HUGHSTON

We propose an energy-driven stochastic master equation for the density matrix as a dynamical model for quantum state reduction. In contrast, most previous studies of state reduction have considered stochastic extensions of the Schrödinger equation, and have introduced the density matrix as the expectation of the random pure projection operator associated with the evolving state vector. After working out properties of the reduction process, we construct a general solution to the energy-driven stochastic master equation. The solution is obtained by the use of nonlinear filtering theory and takes the form of a completely positive stochastic map.

4.1 Introduction

Many physicists have expressed the view that quantum mechanics needs to be modified to provide a mechanism for "collapse of the wave function" (Pearle 1976, Penrose 1986, Bell 1987, Diósi 1989, Ghirardi 2000, Adler 2003a, Weinberg 2012). Among the ways forward that have been proposed, perhaps the most fully developed, at least from a mathematical point of view, are the so-called stochastic models for state reduction, in connection with which there is now a substantial body of literature. In such models, the quantum system is usually taken to be in a pure state, represented by a vector in Hilbert space, evolving as a stochastic process. The state of the system evolves randomly in such a way that it eventually approaches an eigenstate of a preferred observable, such as position or energy. In the situation where the reduction is to a state of definite energy, which is the case that will concern us here, the setup is as follows. The Hilbert space $\mathcal{H}$ is taken to be of finite dimension N, and the state-vector process $\{|\psi_t\rangle\}_{t \geq 0}$ is assumed to satisfy an Ito-type stochastic differential equation of the form

$$d|\psi_t\rangle = -i\hbar^{-1}\hat{H}|\psi_t\rangle\, dt - \tfrac{1}{8}\sigma^2(\hat{H} - H_t)^2|\psi_t\rangle\, dt + \tfrac{1}{2}\sigma(\hat{H} - H_t)|\psi_t\rangle\, dW_t. \quad (4.1)$$

Here $\{W_t\}_{t \geq 0}$ is a standard Brownian motion, and $|\psi_t\rangle \in \mathcal{H}$ is the state vector at time t. The initial state vector $|\psi_0\rangle$ is an input of the model. We write

$$H_t = \frac{\langle \psi_t | \hat{H} | \psi_t \rangle}{\langle \psi_t | \psi_t \rangle} \quad (4.2)$$

for the expectation value of the Hamiltonian operator $\hat{H}$ in the state $|\psi_t\rangle$. The reduction parameter σ, which has dimensions such that

$$\sigma^2 \approx [\text{energy}]^{-2}[\text{time}]^{-1}, \quad (4.3)$$

determines the characteristic timescale τ_R associated with the reduction of the state, which is of the order $\tau_R \approx 1/(\sigma \Delta H)^2$, where ΔH is the initial uncertainty of the energy. Thus a state with high initial energy uncertainty has a shorter characteristic reduction timescale than a state with low energy uncertainty. After a few multiples of τ_R, the system will be nearly in an eigenstate of energy. The determination of σ is an empirical matter. One intriguing possibility suggested by a number of authors (Karolyhazy 1966, Karolyhazy et al. 1986, Penrose 1986, 1996, Diósi 1989, Percival 1994, Hughston 1996) is that state reduction is determined in some way by gravitational phenomena. In that case we might suppose that σ is given by a relation of the form $\sigma^2 \approx E_P^{-2} T_P^{-1}$, where E_P is the Planck energy and T_P is the Planck time, and hence of the order

$$\sigma^2 \approx \sqrt{G\hbar^{-3}c^{-5}}. \quad (4.4)$$

A surprising feature of this expression is that the large numbers associated with the various physical constants cancel out, and we are left with a reduction timescale that is in principle observable in the laboratory, given by

$$\tau_R \approx \left(\frac{2.8\,\text{MeV}}{\Delta H}\right)^2 \text{s}. \quad (4.5)$$

Going forward, we shall not make any specific assumptions regarding the magnitude of the reduction parameter. Nevertheless, to get a feeling for the numbers involved, we note that the binding energies per nucleon of low-mass nuclei are of the order of 1.1 MeV for the deuteron, 2.6 MeV for He^3, and 7.1 MeV for He^4. Since the fusion reactions leading to the production of such nuclei are essential in normal stellar evolution, it is not unreasonable to suppose that some form of observer-free "objective" state reduction is involved in the process and that gravitational effects play a role as well.

No attempt will be made to review the extensive literature of dynamical collapse models, of which the energy-driven model described above is an example, or to discuss in any detail the relative merits of the various models that have been proposed. See Bassi and Ghirardi (2003), Bassi (2007), Pearle (2007, 2009), Bassi et al. (2013), Ghirardi (2016) for surveys. For aspects of the energy-driven models, we refer the reader to Gisin (1989), Ghirardi et al. (1990), Percival (1994, 1998), Hughston (1996), Pearle (1999, 2004), Adler and Horwitz (2000), Adler and Mitra (2000), Adler et al. (2001), Brody and Hughston (2002a,b, 2005, 2006), Adler (2003a,b, 2004), Brody et al. (2003, 2006), Gao (2013), Mengütürk (2016). Adler (2002), in an empirical study of energy-driven models, concludes thus:

Our analysis supports the suggestion that a measurement takes place when the different outcomes are characterized by sufficiently distinct environmental interactions for the reduction process to be rapidly driven to completion.

Although other collapse models have been considered at length in the literature, including, for example, the GRW model (Ghirardi, Rimini and Weber 1986) and so-called *Continuous Spontaneous Localization* (CSL) models (Diósi 1989, Pearle 1989, Ghirardi, Pearle and Rimini 1990), the energy-driven reduction models stand out, in our view, on account of (a) their parsimonious mathematical structure, and (b) the fact that they are universal. By "universal," we mean applicable to any quantum system. We point out that energy-driven models maintain the conservation of energy in a well-defined probabilistic sense, as an extension of the Ehrenfest theorem, whereas models driven by observables that do not commute with the energy, such as position, do not conserve energy (Pearle 2000, Bassi, Ippoliti and Vacchini 2005). Furthermore, energy-driven models give the Born rule and the Lüders projection postulate as exact results (Adler and Horwitz 2000, Adler et al. 2001, Adler 2003b), whereas other models do not. For these reasons, we emphasize here the role of energy-driven models. This is not to say that energy-driven models are the only ones to be taken seriously. But if one wishes to propose a stochastic reduction model that is applicable to *any* nonrelativistic system, without qualification, *including finite dimensional systems*, then must be an energy-driven model.

In that case, is the dynamics necessarily of the form (4.1) given above? Clearly not, since, for a start, one could consider the possibility that other forms of noise than Brownian motion act as a basis for the stochastic dynamics of the state, and indeed there is a sizable literature dealing with dynamical reduction models based on other types of noise. To keep the discussion focussed, we stick here with models based on Brownian motion, though in the final section of the chapter we comment briefly on a generalization to models based on Lévy noise. One might also introduce time-dependent coupling (Brody and Hughston 2005, 2006, Brody et al. 2006, Mengütürk 2016), which offers an approach to the "tail problem" (Shimony 1990, Pearle 2009). Again, we pass over such considerations for the present.

There is, however, an important aspect of the dynamical equation (4.1) that seems to build in what might be viewed as an unnecessary assumption, even if one accepts the principle that reduction must be energy driven, and even if one narrows the scope to models based on a Brownian filtration. This concerns the issue of what constitutes a "state" in quantum mechanics. The physics community seems to be divided on the matter. It is worth recalling that in von Neumann's highly influential 1932 book, the term "state" is reserved for *pure* states, and the statistical operator is introduced to describe *mixtures*. He introduces the notion of a statistical ensemble, corresponding to a countable collection of quantum systems, each of which is in a pure state, and he distinguishes two cases. In the first case, the individual systems of the ensemble can be in different states, and the statistical operator is determined by their relative frequencies. In the second case, which he calls a homogeneous ensemble, the various individual systems are in the same state. The statistical operator for a homogeneous ensemble is identical to the state of any one of its elements and takes the form of a pure projection operator.

In his consideration of statistical ensembles, von Neumann (1932) was motivated in part by the frequentist theories of von Mises (1919, 1928). In particular, von Neumann identifies his concept of ensemble with von Mises's idea of a "Kollektiv" (random sequence):

Such ensembles, called collectives, are in general necessary for establishing probability theory as the theory of frequencies. They were introduced by Richard von Mises, who discovered their meaning for probability theory, and who built up a complete theory on this foundation.

According to von Mises, "Erst das Kollektiv, dann die Wahrscheinlichkeit." At about the same time that these developments were under way, Kolmogorov (1933) revolutionized classical probability theory by giving it a set-theoretic foundation and providing it with a subtle measure-theoretic definition of conditional expectation that allows one to handle in a satisfactory way the logical issues associated with conditioning on events of probability zero. The mathematics community took on board Kolmogorov's innovations, and success followed success, with the introduction of many further new ideas, including, among others, martingales, stochastic calculus, and nonlinear filtering. Von Mises's theory, despite its attractive features, was eventually dropped by mathematicians, even though the ensemble concept (and elements of the frequentist thinking underpinning it) has been kept alive by physicists and is still taught to students (Isham 1995, for instance, gives a good treatment of the relevant material). See van Lambalgen (1999) for a rather detailed discussion of where von Mises's ideas stand today. It appears that the more general use of the term "state" (to include mixed as well as pure states) was introduced by Segal (1947), in his postulates for general quantum mechanics. Segal's point of view was adopted by Haag and Kastler (1964) and also by Davies (1976), who says:

The states are defined as the non-negative trace class operators of trace one, elsewhere called mixed states or density matrices.

If the matter were purely one of terminology, there would be no point in worrying about it very much. The problem is that in the language physicists use there can be assumptions that are implicit in the choice of words, and these in turn can guide the direction of the subject as it moves forward. The issue of what exactly constitutes a "state" is such a case.

The point that concerns us here is that most of the models that have been developed in detail in the collapse literature treat the quantum system as a randomly evolving *pure* state. This point of view is represented, for example, in Ghirardi, Pearle and Rimini (1990) in the context of their development of the CSL model, where we find the following succinct account of their stance on the matter:

The theory discussed here allows one to describe naturally quantum measurement processes by dynamical equations valid for all physical systems. It is worthwhile repeating, that, in this theoretical scheme, any member of the statistical ensemble has at all times a definite wave function. As a consequence, the wave function itself can be interpreted as a real property of a single closed physical system.

The emphasis placed on the role of pure states reflects a view held by many physicists that pure states should be treated as being fundamental. See, for example, Penrose (2016), who argues persuasively concerning the preferred status of pure states. According to this view, which, as we have indicated, is generally in line with that of von Neumann (1932), individual systems are represented by pure states. Physicists are likewise divided on the issue of the status of statistical ensembles. Are they essential to the theory? Mielnik (1974) offers the following:

It is an old question whether the formalism of quantum theory is adequate to describe the properties of single systems. What is verified directly in the most general quantum experiment are rather the properties of statistical ensembles.

Although our brief remarks cannot do justice to the deep insights of the authors mentioned, one will be impressed by the diversity of opinion held by physicists on the nature of quantum states and the role of statistical ensembles. It should be emphasized, nevertheless, that, as far as we can see, there is no empirical basis for assuming that individual quantum systems are necessarily in pure states. Nor is there any evidence showing that density matrices necessarily have to be interpreted as representing ensembles. In fact, it seems to be accepted in the quantum information community that the state of an individual system should be represented, in certain circumstances, by a higher-rank density matrix. This can happen, for example, if the system is entangled with another system and the state of the composite system is pure, in which case the state of the first system is obtained by taking the reduced density matrix of the system as a whole, where we trace out the degrees of freedom associated with the second system. It thus seems reasonable to take matters a step further and drop altogether the assumption that individual systems are necessarily in pure states. It also seems reasonable to drop the assumption that statistical ensembles play a fundamental role in the theory. In our approach, therefore, we make no use of frequentist thinking, and we avoid reference to observers, measurements, and ensembles. We regard state reduction as an entirely *objective* phenomenon, and even in the case of an individual system we model the state as a randomly evolving density matrix. We denote the density matrix process by $\{\hat{\rho}_t\}_{t\geq 0}$, and we require that $\hat{\rho}_t$ should be nonnegative definite for all t and such that tr $\hat{\rho}_t = 1$. The dynamical equation generalizing (4.1) then takes the following form:

Definition 1 *We say that the state $\{\hat{\rho}_t\}_{t\geq 0}$ of an isolated quantum system with Hamiltonian $\hat{H}$ satisfies an energy-driven stochastic master equation with parameter σ if*

$$\mathrm{d}\hat{\rho}_t = -\mathrm{i}\hbar^{-1}[\hat{H}, \hat{\rho}_t]\mathrm{d}t + \tfrac{1}{8}\sigma^2 \left(2\hat{H}\hat{\rho}_t\hat{H} - \hat{H}^2\hat{\rho}_t - \hat{\rho}_t\hat{H}^2\right)\mathrm{d}t$$
$$+ \tfrac{1}{2}\sigma\left((\hat{H} - H_t)\hat{\rho}_t + \hat{\rho}_t(\hat{H} - H_t)\right)\mathrm{d}W_t, \tag{4.6}$$

where $H_t = \mathrm{tr}\, \hat{\rho}_t \hat{H}$.

We take a moment to spell out some of the mathematical ideas implicit in the dynamics. In accordance with the well-established Kolmogorovian outlook, we introduce a probability space $(\Omega, \mathcal{F}, \mathbb{P})$ as the basis of the theory. We do not necessarily say in detail what the structure of this space is, but we assume that it is endowed with sufficient richness to support the various structures that we wish to consider. Thus Ω is a set on which we introduce a σ-algebra $\mathcal{F}$ (no relation to the σ above) and a probability measure $\mathbb{P}$. By an *algebra* we mean a collection of subsets of Ω such that $\Omega \in \mathcal{F}$, $A \in \mathcal{F}$ implies $\Omega \backslash A \in \mathcal{F}$, $A \in \mathcal{F}$ and $B \in \mathcal{F}$ implies $A \cup B \in \mathcal{F}$. If for any countable collection of elements $A_i \in \mathcal{F}$, $i \in \mathbb{N}$, it holds that $\cup_{i\in\mathbb{N}} A_i \in \mathcal{F}$, then we say that $\mathcal{F}$ is a σ-*algebra*. The pair $(\Omega, \mathcal{F})$ is called a *measurable space*. By a *probability measure* on $(\Omega, \mathcal{F})$ we mean a function $\mathbb{P} : \mathcal{F} \to [0, 1]$ satisfying $\mathbb{P}[\Omega] = 1$, and $\mathbb{P}[\cup_{i\in\mathbb{N}} A_i] = \sum_{i\in\mathbb{N}} \mathbb{P}[A_i]$ for any countable collection of elements $A_i \in \mathcal{F}$, $i \in \mathbb{N}$, such that $A_i \cap A_j = \emptyset$ if $i \neq j$. A measurable space endowed with a probability measure defines a *probability space*. A function $X : \Omega \to \mathbb{R}$ is said to be $\mathcal{F}$-*measurable*, or measurable on $(\Omega, \mathcal{F})$, if for all $A \in \mathcal{B}_\mathbb{R}$, where $\mathcal{B}_\mathbb{R}$ is the Borel σ-algebra on $\mathbb{R}$, it holds that $\{\omega : X(\omega) \in A\} \in \mathcal{F}$. Thus for each $A \in \mathcal{B}_\mathbb{R}$ we require $X^{-1}(A) \in \mathcal{F}$. If X is a measurable function on a probability space $(\Omega, \mathcal{F}, \mathbb{P})$, we say that X is a *random variable*, and the associated distribution function is defined for $x \in \mathbb{R}$ by $F_X(x) = \mathbb{P}[X < x]$, where $\mathbb{P}[X < x]$ denotes the measure of the subset $\{\omega \in \Omega : X(\omega) < x\}$.

By a *random process* on $(\Omega, \mathcal{F}, \mathbb{P})$, we mean a family of random variables $\{X_t\}_{t\geq 0}$ parametrized by time. To formulate a theory of random processes some additional structure is required. First we need the idea of a complete probability space. A σ-algebra $\mathcal{F}^P$ is said to be an *augmentation* of the σ-algebra $\mathcal{F}$ with respect to

ℙ if $\mathcal{F}^{\mathbb{P}}$ contains all subsets $B \subset \Omega$ for which there exist elements $A, C \in \mathcal{F}$ satisfying $A \subseteq B \subseteq C$ and $\mathbb{P}[C \setminus A] = 0$. If $\mathcal{F}^{\mathbb{P}} = \mathcal{F}$, we say that $(\Omega, \mathcal{F}, \mathbb{P})$ is *complete*. Next we need the idea of a *filtration* on $(\Omega, \mathcal{F}, \mathbb{P})$, by which we mean a nondecreasing family $\mathbb{F} = \{\mathcal{F}_t\}_{t \geq 0}$ of sub-σ-algebras of $\mathcal{F}$. We say that a filtration $\mathbb{F}$ is *right continuous* if for all $t \geq 0$ it holds that $\mathcal{F}_t = \mathcal{F}_{t^+}$ where $\mathcal{F}_{t^+} = \cap_{u > t} \mathcal{F}_u$. If additionally we assume, as we do, that for any $A \in \mathcal{F}$ such that $\mathbb{P}[A] = 0$ it holds that $A \in \mathcal{F}_0$, then we say that the filtered probability space $(\Omega, \mathcal{F}, \mathbb{P}, \mathbb{F})$ satisfies *the usual conditions*. A random process $\{X_t\}$ is said to be *adapted* to $\mathbb{F}$ if the random variable X_t is $\mathcal{F}_t$-measureable for all $t \geq 0$. We say that $\{X_t\}$ is right continuous if the sample paths $\{X_t(\omega)\}_{t \geq 0}$ are right continuous for almost all $\omega \in \Omega$. By a *standard Brownian motion* or *Wiener process* on a filtered probability space $(\Omega, \mathcal{F}, \mathbb{P}, \mathbb{F})$ we mean a continuous, adapted process $\{W_t\}_{t \geq 0}$ such that (a) $W_0 = 0$ almost surely, (b) $W_t - W_s$ is normally distributed with mean 0 and variance $t - s$ for $t > s \geq 0$, and (c) $W_t - W_s$ is independent of $\mathcal{F}_s$ for $t > s$. The filtration $\mathbb{F}$ may be strictly larger than that generated by the Brownian motion itself. The existence of processes satisfying these conditions is guaranteed by the following (Hida 1980, Karatzas and Shreve 1986). Let $\Omega = C[0, \infty)$ be the space of continuous functions from $\mathbb{R}^+$ to $\mathbb{R}$. Each point $\omega \in \Omega$ corresponds to a continuous function $\{W_t(\omega)\}_{t \geq 0}$, and we write $\mathcal{F} = \sigma[\{W_t\}_{t \geq 0}]$ for the σ-algebra generated by $\{W_t\}_{t \geq 0}$. The σ-algebra generated by a collection $\mathcal{C}$ of functions $X : \Omega \to \mathbb{R}$ is defined to be the smallest σ-algebra Ξ on Ω such that each function $X \in \mathcal{C}$ is Ξ-measurable. Then there exists a unique measure $\mathbb{P}$ on the $(\Omega, \mathcal{F})$, called *Wiener measure*, such that properties (a), (b), and (c) hold, and we take $\mathbb{F}$ to be the filtration $\{\mathcal{F}_t\}_{t \geq 0}$ generated by $\{W_t\}_{t \geq 0}$, defined by $\mathcal{F}_t = \sigma[\{W_s\}_{0 \leq s \leq t}]$ for each $t \geq 0$.

In what follows, we assume that $(\Omega, \mathcal{F}, \mathbb{P}, \mathbb{F})$ satisfies the usual conditions. Equalities and inequalities for random variables are understood to hold ℙ-almost-surely. One checks by the use of Ito calculus that if $\hat{\rho}_t$ takes the form of a pure projection operator

$$\hat{\rho}_t = \frac{|\psi_t\rangle \langle \psi_t|}{\langle \psi_t | \psi_t \rangle}, \tag{4.7}$$

then the stochastic Schrödinger equation (4.1) for the state vector implies that the pure density matrix (4.7) satisfies the stochastic master equation (4.6). The relevant calculation is shown, for example, in sections 6.1–6.2 of Adler (2004). Since (4.6) is a nonlinear stochastic differential equation, it does not immediately follow that (4.6) should be applicable to general states rather than merely to pure states. Nevertheless, this is what we propose, and, as we shall see, the theory that follows from Definition 1 has many desirable properties, both physical and mathematical. For some purposes, it is useful if we write equation (4.6) in integral form, incorporating the initial condition explicitly. In that case we have

$$\hat{\rho}_t = \hat{\rho}_0 - i\hbar^{-1} \int_0^t [\hat{H}, \hat{\rho}_s] ds + \tfrac{1}{8}\sigma^2 \int_0^t \left(2\hat{H}\hat{\rho}_s\hat{H} - \hat{H}^2\hat{\rho}_s - \hat{\rho}_s\hat{H}^2 \right) ds$$
$$+ \tfrac{1}{2}\sigma \int_0^t \left((\hat{H} - H_s)\hat{\rho}_s + \hat{\rho}_s(\hat{H} - H_s) \right) dW_s. \quad (4.8)$$

Then it follows, by taking the expectation of each side, which eliminates the term involving the stochastic integral, that the mean state of the system satisfies

$$\langle \hat{\rho}_t \rangle = \hat{\rho}_0 - i\hbar^{-1} \int_0^t [\hat{H}, \langle \hat{\rho}_s \rangle] ds + \tfrac{1}{8}\sigma^2 \int_0^t \left(2\hat{H}\langle \hat{\rho}_s \rangle \hat{H} - \hat{H}^2 \langle \hat{\rho}_s \rangle - \langle \hat{\rho}_s \rangle \hat{H}^2 \right) ds.$$
(4.9)

Here $\langle \hat{\rho}_t \rangle = \mathbb{E}[\hat{\rho}_t]$, where $\mathbb{E}[\,\cdot\,]$ denotes expectation under $\mathbb{P}$. One recognizes (4.9) as the integral form of a master equation of the type derived by Lindblad (1976), Gorini et al. (1976), and, in a different context, Banks et al. (1984), and we have the following:

Proposition 1 *If the state of a quantum system satisfies the energy-driven stochastic master equation, then the mean state of the system satisfies a linear master equation of the form*

$$\frac{d\langle \hat{\rho}_t \rangle}{dt} = -i\hbar^{-1}[\hat{H}, \langle \hat{\rho}_t \rangle] + \tfrac{1}{8}\sigma^2 \left(2\hat{H}\langle \hat{\rho}_t \rangle \hat{H} - \hat{H}^2 \langle \hat{\rho}_t \rangle - \langle \hat{\rho}_t \rangle \hat{H}^2 \right). \quad (4.10)$$

In the pure case, it is well known (see, for example, Gisin 1989) that if $|\psi_t\rangle$ satisfies (4.1) then the expectation of the corresponding pure density matrix, given by

$$\langle \hat{\rho}_t \rangle = \mathbb{E}\left[\frac{|\psi_t\rangle \langle \psi_t|}{\langle \psi_t | \psi_t \rangle} \right], \quad (4.11)$$

satisfies the autonomous stochastic differential equation (4.10). This is not so obvious if one works directly with the dynamics of a state vector, but if one takes the stochastic master equation as the starting point then the linearity of the dynamics of $\langle \hat{\rho}_t \rangle$ is immediate. Proposition 1 shows that in the generic situation where the density matrix is of rank greater than unity and follows the general nonlinear stochastic dynamics given by (4.6), the associated mean density matrix $\langle \hat{\rho}_t \rangle$ still satisfies (4.10).

We are thus led to postulate that the energy-driven stochastic master equation presented in Definition 1, with a prescribed initial state $\hat{\rho}_0$, characterizes the stochastic evolution of the state of a quantum system as reduction proceeds. In saying that we take the initial state as prescribed, we avoid for the moment entering into a discussion about how that can be achieved. Likewise, we avoid asking how one can determine what the initial state of the system is. It is meaningful to ask such questions, but we separate the problem of working out the consequences of the evolution of the state from the problem of working out what the state of the system is in the first place or how to create a system in a given state. In Sections 4.2, 4.3, and 4.4, we work out properties of the energy-driven stochastic master equation. A number of the results obtained are generalizations of corresponding results known to hold in the case when the state is pure. In Proposition 2 we show that the expectation of the variance of the energy goes to zero in the limit as t grows large. In Proposition 3, we show that there exists a random variable $H_\infty = \lim_{t\to\infty} H_t$ taking values in the spectrum of the Hamiltonian such that we have $\mathbb{E}[H_\infty] = \text{tr}\,\hat{\rho}_0 \hat{H}$ and $\text{Var}[H_\infty] = \text{tr}\,\hat{\rho}_0 \hat{H}^2 - (\text{tr}\,\hat{\rho}_0 \hat{H})^2$. The proofs of Propositions 2 and 3 generalize arguments appearing in Hughston (1996). In Section 4.5, we present a derivation of the Born rule for general states, summarized in Proposition 4, extending arguments of Ghiradi et al. (1990), Adler and Horwitz (2000), and Adler et al. (2001). In the case of a degenerate Hamiltonian, the reduction leads for a given outcome to the associated Lüders state. Then in Sections 4.6 and 4.7, we proceed to construct a general solution of the energy-driven stochastic master equation using techniques of nonlinear filtering theory. Here we extend results known for the dynamics of pure states (Brody and Hughston 2002). The solution, which takes the form of a completely positive stochastic map, is obtained by the introduction of a so-called *information process* $\{\xi_t\}_{t\geq 0}$ defined by $\xi_t = \sigma t H + B_t$ where the random variable H takes values in the spectrum of the Hamiltonian operator, and $\{B_t\}_{t\geq 0}$ is an independent Brownian motion. We show that it is possible to construct the processes $\{\hat{\rho}_t\}_{t\geq 0}$ and $\{W_t\}_{t\geq 0}$ in terms of $\{\xi_t\}_{t\geq 0}$ in such a way that $\{\hat{\rho}_t\}_{t\geq 0}$ satisfies the energy-driven stochastic master equation and $\{W_t\}$ is a standard Brownian motion on $(\Omega, \mathcal{F}, \mathbb{P}, \mathbb{F})$, where $\mathbb{F}$ is the filtration generated by $\{\xi_t\}$. The results are summarized in Propositions 5 and 6. Then we introduce the notion of a *potential*, and in Propositions 7 and 8 we show that the decoherence of the density matrix can be characterized in a rather natural way by the fact that its off-diagonal terms are potentials. Section 4.8 concludes.

4.2 Dynamic Properties of the Energy Variance

We proceed to show that many of the important properties of the pure state dynamics (4.1) carry forward to the general state dynamics (4.6). First, one can check that the trace of $\hat{\rho}_t$ is preserved under (4.6). Thus, if $\text{tr}\,\hat{\rho}_0 = 1$ then equation (4.8) implies that $\text{tr}\,\hat{\rho}_t = 1$ for all $t > 0$. Next, one can check that the energy expectation process $\{H_t\}_{t\geq 0}$ defined by $H_t = \text{tr}\,\hat{\rho}_t \hat{H}$ is a *martingale*. In fact, even in the pure case, the result can be obtained rather more directly by use of (4.6) than (4.1), for if we transvect each side of (4.6) with $\hat{H}$ and take the trace, we are immediately led to the following dynamical equation for the energy:

$$dH_t = \sigma V_t \, dW_t. \tag{4.12}$$

Here we have written $V_t = \text{tr}\,\hat{\rho}_t\,(\hat{H} - H_t)^2$ for the variance of the energy. Thus we have

$$H_t = H_0 + \sigma \int_0^t V_s\,dW_s. \tag{4.13}$$

Since the expectation and the variance of the energy are bounded random variables, it follows from (4.13) that $\{H_t\}_{t\geq 0}$ is a martingale. Letting $\mathbb{E}_t[\,\cdot\,] = \mathbb{E}[\,\cdot\,|\,\mathcal{F}_t]$ denote conditional expectation with respect to $\mathcal{F}_t$, we have $\mathbb{E}_s[H_t] = H_s$ for $0 \leq s \leq t$. The martingale property represents conservation of energy in a conditional sense. This property is known to be satisfied by the energy expectation process in the case of a pure state, and we see that the martingale property holds more generally in the case of a mixed state governed by the energy-driven stochastic master equation. A further calculation shows that

$$dV_t = -\sigma^2 V_t^2\,dt + \sigma\beta_t\,dW_t, \tag{4.14}$$

where $\{\beta_t\}_{t\geq 0}$ denotes the so-called energy skewness process, defined by

$$\beta_t = \text{tr}\,\hat{\rho}_t\,(\hat{H} - H_t)^3. \tag{4.15}$$

The dynamical equation (4.14) can be obtained as follows. Write the variance in the form

$$V_t = \text{tr}\,\hat{\rho}_t\,\hat{H}^2 - H_t^2. \tag{4.16}$$

The dynamics of the term $\text{tr}\,\hat{\rho}_t\,\hat{H}^2$ can be worked out by transvecting each side of equation (4.6) with $\hat{H}^2$. The dynamics of the second term can be deduced by applying Ito's lemma to H_t^2 and using (4.12). The two results combined give (4.14).

The stochastic equation satisfied by the variance of the Hamiltonian in the case of a general state has the same form that it has in the pure case. In the pure case, (4.14) implies that the variance tends to zero asymptotically and thus that the state evolves to an energy eigenstate. We shall show that the argument carries through to the case of a general initial state. That is to say, for any initial state the result of the evolution given by (4.6) is an energy eigenstate. By an energy eigenstate with energy E we mean a state $\hat{\rho}$ such that $\hat{H}\hat{\rho} = E\hat{\rho}$. If the Hamiltonian is nondegenerate, then the energy eigenstates are pure states. In the case of a degenerate Hamiltonian, the situation is more complicated. If the outcome of the collapse is an eigenstate with energy E_r, then it can be shown that the state that results is the so-called Lüders state given by outcome of the Lüders (1951) projection postulate associated with that energy and the given initial state (Adler et al. 2001).

Definition 2 *Let $\hat{P}_r$ denote the projection operator onto the Hilbert subspace $\mathcal{H}_r$ consisting of state vectors that are eigenstates of $\hat{H}$ with eigenvalue E_r. Then for any initial state $\hat{\rho}_0$ the associated Lüders state $\hat{L}_r$ is defined by*

$$\hat{L}_r = \frac{\hat{P}_r\,\hat{\rho}_0\,\hat{P}_r}{\text{tr}\,\hat{\rho}_0\,\hat{P}_r}. \tag{4.17}$$

If the Hamiltonian is degenerate, and if the initial state is pure, then the final state will be pure. On the other hand, if the initial state is impure, then the final state need not be pure and in general will be impure.

To show that collapse to an energy eigenstate occurs as a consequence of (4.14) for a general initial state, we establish the following, which is known to hold for pure states:

Proposition 2 *Let $\{\rho_t\}$ satisfy the energy-driven stochastic master equation. Then the expectation of the variance of the Hamiltonian vanishes asymptotically:*

$$\lim_{t \to \infty} \mathbb{E}[V_t] = 0. \tag{4.18}$$

Proof We integrate (4.14) to obtain

$$V_t = V_0 - \sigma^2 \int_0^t V_s^2 \, ds + \sigma \int_0^t \beta_s \, dW_s. \tag{4.19}$$

The integrals are defined since the variance and the skewness are bounded. Since the drift in (4.14) is negative, we see that $\mathbb{E}_s[V_t] \leq V_s$ for $0 \leq s \leq t$ and hence that $\{V_t\}_{t \geq 0}$ is a *supermartingale*. Taking the unconditional expectation on each side of (4.19), we have

$$\mathbb{E}[V_t] = V_0 - \sigma^2 \mathbb{E}\left[\int_0^t V_s^2 \, ds\right], \tag{4.20}$$

which shows that $\mathbb{E}[V_t]$ decreases as t increases, and hence that $\lim_{t \to \infty} \mathbb{E}[V_t]$ exists. We say that an $\mathbb{R}$-valued random process $\{X_t\}_{t \geq 0}$ on a probability space $(\Omega, \mathcal{F}, \mathbb{P})$ is *measurable* if for all $A \in \mathcal{B}_\mathbb{R}$, where $\mathcal{B}_\mathbb{R}$ is the Borel σ-algebra on $\mathbb{R}$, it holds that

$$\{(\omega, t) : X_t(\omega) \in A\} \in \mathcal{F} \times \mathcal{B}_{\mathbb{R}^+}, \tag{4.21}$$

where $\mathcal{B}_{\mathbb{R}^+}$ denotes the Borel σ-algebra on the positive "time axis" $\mathbb{R}^+ = [0, \infty)$. A sufficient condition for a process to be measurable is that it should be right continuous. Then one has the following (Liptser and Shiryaev 1975):

Fubini's theorem. *If a process $\{X_t\}_{t \geq 0}$ is measurable and $\int_S \mathbb{E}[|X_t|] \, dt < \infty$ for some $S \in \mathcal{B}_{\mathbb{R}^+}$, then $\int_S |X_t| \, dt < \infty$ almost surely and*

$$\mathbb{E}\left[\int_S X_t \, dt\right] = \int_S \mathbb{E}[X_t] \, dt. \tag{4.22}$$

As a consequence of Fubini's theorem, we can interchange the order of the expectation and the integration on the right side of (4.20) to obtain

$$\mathbb{E}[V_t] = V_0 - \sigma^2 \int_0^t \mathbb{E}[V_s^2] \, ds, \tag{4.23}$$

from which it follows that

$$\frac{d\mathbb{E}[V_t]}{dt} = -\sigma^2\,\mathbb{E}\left[V_t^2\right]. \qquad (4.24)$$

Thus we can write

$$\frac{d\mathbb{E}[V_t]}{dt} = -\sigma^2\,\mathbb{E}[V_t]^2(1+\alpha_t), \qquad (4.25)$$

where

$$\alpha_t = \frac{1}{\mathbb{E}[V_t]^2}\mathbb{E}[(V_t - \mathbb{E}[V_t])^2], \qquad (4.26)$$

and we note that α_t is nonnegative. If we set $\gamma_t = \int_0^t \alpha_s ds$, we can integrate (4.25) to obtain

$$\mathbb{E}[V_t] = \frac{V_0}{1 + V_0\,\sigma^2(t+\gamma_t)}. \qquad (4.27)$$

Since γ_t is nonnegative, we have

$$\mathbb{E}[V_t] \le \frac{V_0}{1 + V_0\,\sigma^2 t}, \qquad (4.28)$$

and this gives (4.18). $\square$

4.3 Asymptotic Properties of the Variance

As a consequence of (4.18), one deduces that the energy variance vanishes as t goes to infinity. More precisely, it holds that $V_\infty = 0$ almost surely. To see this, we need to show that the limit $V_\infty = \lim_{t\to\infty} V_t$ exists, in an appropriate sense, and then we need to show that the order of the limit and the expectation in (4.18) can be interchanged. If both of these conditions hold, then we conclude from (4.18) that $V_\infty = 0$. Now, when we ask whether a limit exists, we are not asking whether the result is finite. Limits, if they exist, are allowed to be infinite. The question is one of convergence. Moreover, even if a random process converges, that does not imply that the resulting function on Ω to which the process converges is a random variable (that is to say, a measurable function). So the question is whether there exists a random variable V_∞ to which the variance process converges for large t with probability one. If the answer is yes, then one can ask whether the interchange of limit and expectation is valid, and if so then we are able to conclude that the result of the collapse process is a state of zero energy variance and hence an energy eigenstate.

To show that (4.18) implies $V_\infty = 0$ almost surely, we use the *martingale convergence theorem*. There are various versions of this theorem, and it will be sufficient to have at hand the version that follows (Protter 2003). First we introduce some additional terminology. We fix a probability space and let $p \in \mathbb{R}$ satisfy $p \geq 1$. A random process $\{X_t\}_{t \geq 0}$ is said to be bounded in $\mathcal{L}^p$ if

$$\sup_{0 \leq t < \infty} \mathbb{E}[\,|X_t|^p\,] < \infty. \tag{4.29}$$

As usual, by the supremum we mean the least upper bound. A random process $\{X_t\}_{t \geq 0}$ is said to be right-continuous if it holds almost surely that $\lim_{\epsilon \to 0} X_{t+\epsilon} = X_t$ for all $t \geq 0$. Then we have:

Martingale convergence theorem. *If a right-continuous supermartingale $\{X_t\}_{t \geq 0}$ is bounded in $\mathcal{L}^1$ then $\lim_{t \to \infty} X_t$ exists almost surely and defines a random variable X_∞ satisfying $\mathbb{E}[|X_\infty|] < \infty$.*

Note that in asserting that $\lim_{t \to \infty} X_t$ exists almost surely we mean that $\limsup_{t \to \infty} X_t(\omega) = \liminf_{t \to \infty} X_t(\omega)$ for all $\omega \in \Omega'$ for some set $\Omega' \in \mathcal{F}$ such that $\mathbb{P}[\Omega'] = 1$ and that there exists a random variable X_∞ such that $X_\infty(\omega) = \lim_{t \to \infty} X_t(\omega)$ for all $\omega \in \Omega$ apart from a set of measure zero.

As we shall see, the martingale convergence theorem is just the tool one needs in order to show that the energy variance process converges to zero. In particular, since the energy variance is bounded for all $t \geq 0$, we have $\sup_{0 \leq t < \infty} \mathbb{E}[\,|V_t|^p\,] < \infty$ for all $p \geq 1$. It follows by the martingale convergence theorem that $V_\infty = \lim_{t \to \infty} V_t$ exists almost surely and that $\mathbb{E}[V_\infty] < \infty$. To proceed further, we make use of the following (see, e.g., Williams 1991):

Fatou's lemma. *Let $\{Y_k\}_{k \in \mathbb{N}}$ be a countable sequence of nonnegative integrable random variables. Then $\mathbb{E}\left[\liminf_{k \to \infty} Y_k\right] \leq \liminf_{k \to \infty} \mathbb{E}[Y_k]$.*

If $\{t_k\}_{k \in \mathbb{N}}$ is a countable sequence of times such that $\lim_{k \to \infty} t_k = \infty$, then for any process $\{X_t\}_{t \geq 0}$ such that $X_\infty = \lim_{t \to \infty} X_t$ exists it holds that $\lim_{k \to \infty} X_{t_k} = X_\infty$. Thus, in our case we have $\lim_{k \to \infty} \mathbb{E}[V_{t_k}] = 0$ and $\lim_{k \to \infty} V_{t_k} = V_\infty$. We know that if $\lim_{k \to \infty} Y_k$ exists then it is equal to $\liminf_{k \to \infty} Y_k$. Then by Fatou's lemma we have $\mathbb{E}\left[\lim_{k \to \infty} V_{t_k}\right] \leq \lim_{k \to \infty} \mathbb{E}[V_{t_k}]$. It follows that $\mathbb{E}[V_\infty] = 0$ and hence $V_\infty = 0$ almost surely, since the variance is nonnegative.

4.4 Terminal Value of the Energy

Let $\text{Spec}[\hat{H}]$ denote the spectrum of the Hamiltonian. Then we have the following result, which shows that H_t and V_t are given at each time $t \geq 0$ respectively by the conditional mean and the conditional variance of the terminal value of the energy:

Proposition 3 *There exists a random variable H_∞ on $(\Omega, \mathcal{F}, \mathbb{P})$ taking values in $\text{Spec}[\hat{H}]$ such that $H_t = \mathbb{E}_t[H_\infty]$ and $V_t = \mathbb{E}_t[(H_\infty - \mathbb{E}_t[H_\infty])^2]$.*

Proof Since $\{H_t\}_{t\geq 0}$ is bounded by the highest and lowest eigenvalues of $\hat{H}$, we have $\sup_{0\leq t<\infty} \mathbb{E}[\,|H_t|\,] < \infty$, and hence by the martingale convergence theorem, the random variable $H_\infty = \lim_{t\to\infty} H_t$ exists and $\mathbb{E}[H_\infty] < \infty$. A process $\{X_t\}_{t\geq 0}$ on a probability space $(\Omega, \mathcal{F}, \mathbb{P})$ is said to be uniformly integrable if, given any $\epsilon > 0$, there exists a δ such that

$$\mathbb{E}[\,|X_t|\,\mathbb{1}(\,|X_t| > \delta\,)] < \epsilon \tag{4.30}$$

for all $t \geq 0$, where $\mathbb{1}(\,\cdot\,)$ is the indicator function. Let $\{M_t\}_{t\geq 0}$ be a right-continuous martingale on a probability space $(\Omega, \mathcal{F}, \mathbb{P})$ with filtration $\{\mathcal{F}_t\}_{t\geq 0}$. Then it is known that the following conditions are equivalent: (i) there exists a random variable M_∞ such that $\lim_{t\to\infty} \mathbb{E}[|M_t - M_\infty|] = 0$; (ii) there exists a random variable M_∞ satisfying $\mathbb{E}[M_\infty] < \infty$ such that $M_t = \mathbb{E}_t[M_\infty]$ for all $t \geq 0$; (iii) $\{M_t\}_{t\geq 0}$ is uniformly integrable. Clearly, any bounded martingale is uniformly integrable. Since $\{H_t\}_{t\geq 0}$ is bounded, we have

$$H_t = \mathbb{E}_t[H_\infty], \tag{4.31}$$

as claimed. We turn now to the variance, in connection with which we use the following:

Monotone convergence theorem. *For any increasing sequence $\{Y_k\}_{k\in\mathbb{N}}$ of nonnegative integrable random variables such that $\lim_{k\to\infty} Y_k = Y_\infty$, where Y_∞ is not necessarily integrable, it holds that $\lim_{k\to\infty} \mathbb{E}[Y_k] = \mathbb{E}[Y_\infty]$.*

By use of (4.18) and (4.20), together with the monotone convergence theorem, we deduce that

$$\mathbb{E}\left[\int_0^\infty V_s^2\,\mathrm{d}s\right] < \infty. \tag{4.32}$$

Hence, it follows from (4.19) that

$$V_0 + \sigma \int_0^\infty \beta_s\,\mathrm{d}W_s = \sigma^2 \int_0^\infty V_s^2\,\mathrm{d}s. \tag{4.33}$$

If we take a conditional expectation, we obtain

$$V_0 + \sigma \int_0^t \beta_s\,\mathrm{d}W_s = \sigma^2\,\mathbb{E}_t \int_0^\infty V_s^2\,\mathrm{d}s. \tag{4.34}$$

Combining this relation with (4.19), we deduce that

$$V_t = \sigma^2\,\mathbb{E}_t \int_t^\infty V_s^2\,\mathrm{d}s. \tag{4.35}$$

Quantum State Reduction

Next we observe that as a consequence of (4.13), we have

$$H_\infty - H_t = \sigma \int_t^\infty V_s \, dW_s. \tag{4.36}$$

Taking the square of each side of this equation, forming the conditional expectation, and using the Ito isometry, we obtain

$$\mathbb{E}_t (H_\infty - H_t)^2 = \sigma^2 \mathbb{E}_t \int_t^\infty V_s^2 \, ds, \tag{4.37}$$

and therefore

$$V_t = \mathbb{E}_t (H_\infty - \mathbb{E}_t H_\infty)^2, \tag{4.38}$$

as claimed. □

The significance of this result is that the conventional expectation value H_0 of the observable $\hat{H}$ with respect to the initial state $\hat{\rho}_0$ is equal to the expectation of the terminal value of the energy on the completion of the reduction process. This may seem like a tautology, but it is not, since the statistical interpretation of the expectation value of an observable in quantum mechanics is an assumption, not a conclusion, of the theory.

Likewise, we see that the conventional squared uncertainty V_0 is the variance of the terminal value of the energy on the completion of the reduction process. Again, the statistical interpretation of the squared uncertainty is an assumption in quantum mechanics, not a conclusion of the theory. But under the dynamics of the stochastic master equation these properties are deduced rather than assumed.

The methods used in the proof of Proposition 3 can be used to give an alternative derivation of the fact that $\lim_{t \to \infty} \mathbb{E}[V_t] = 0$ implies that $V_\infty = 0$ almost surely. We have already seen that this follows as a consequence of Fatou's lemma, but the same result can be obtained by use of the martingale convergence theorem. The proof is as follows. We observe that by the definition of the variance process, we have

$$V_t = \operatorname{tr} \hat{\rho}_t \hat{H}^2 - H_t^2, \tag{4.39}$$

where $H_t = \operatorname{tr} \hat{\rho}_t \hat{H}$. Writing $U_t = \operatorname{tr} \hat{\rho}_t \hat{H}^2$, we see that $\{U_t\}_{t \geq 0}$ is a bounded martingale. It follows by the martingale convergence theorem that $\{U_t\} \to U_\infty$, and as a consequence we have $\{V_t\} \to U_\infty - H_\infty^2$, from which it follows that

$$V_\infty = U_\infty - \left(H_0 + \sigma \int_0^\infty V_s \, dW_s\right)^2. \tag{4.40}$$

Since $\mathbb{E}[U_\infty] = U_0$, it follows by use of the Ito isometry that

$$\mathbb{E}[V_\infty] = U_0 - H_0^2 - \sigma^2 \mathbb{E} \int_0^\infty V_s^2 \, ds. \tag{4.41}$$

On the other hand, on account of (4.33), we have

$$V_0 = \sigma^2 \mathbb{E} \int_0^\infty V_s^2 \, ds. \tag{4.42}$$

Since $U_0 - H_0^2 = V_0$, it follows that $\mathbb{E}[V_\infty] = 0$, and therefore $V_\infty = 0$ almost surely.

4.5 Derivation of the Born Rule

The foregoing arguments show that the dynamic approach to reduction extends to the situation in which the initial state of the system need not be pure. The Born rule is another example of an assumption of quantum mechanics that can be derived from the stochastic master equation. As before, for the given Hamiltonian let $\hat{P}_r$ denote the projection operator on to the Hilbert subspace of energy E_r. Let the number of distinct energy levels be D.

Proposition 4 *Under the dynamics of the energy-driven stochastic master equation, with initial state ρ_0, the probability that the outcome will be a state with energy E_r is given by*

$$\mathbb{P}[H_\infty = E_r] = \operatorname{tr} \hat{\rho}_0 \hat{P}_r. \tag{4.43}$$

Proof It is straightforward to check that the process $\{\pi_{rt}\}_{t \geq 0}$ defined for each value of $r = 1, 2, \ldots, D$ by $\pi_{rt} = \operatorname{tr} \hat{\rho}_t \hat{P}_r$ is a martingale. Thus we have $\pi_{rt} = \mathbb{E}_t[\pi_{r\infty}]$ and hence

$$\operatorname{tr} \hat{\rho}_0 \hat{P}_r = \mathbb{E}[\operatorname{tr} \hat{\rho}_\infty \hat{P}_r]. \tag{4.44}$$

On the other hand, because the state reduces asymptotically to a random energy eigenstate, we know that

$$\operatorname{tr} \hat{\rho}_\infty \hat{P}_r = \mathbb{1}(H_\infty = E_r), \tag{4.45}$$

and since

$$\mathbb{E}[\mathbb{1}(H_\infty = E_r)] = \mathbb{P}[H_\infty = E_r], \tag{4.46}$$

we are led to the Born rule (4.43). $\square$

It may seem tautological to assert that the probability of the outcome E_r is given by the trace of the product of the initial density matrix and the projection operator $\hat{P}_r$, but it is not. In quantum mechanics, the Born rule is an assumption, part of the statistical interpretation of the theory. Physicists are on the whole quite comfortable with this assumption, but that does not change the fact that there is no generally accepted "derivation" of the Born rule as a probability law arising from

Quantum State Reduction   63

within quantum theory itself. Indeed, it is one of the features of the energy-driven stochastic reduction model that a mathematically satisfactory explanation for this otherwise baffling aspect of quantum theory emerges.

4.6 Solution to Stochastic Master Equation

A solution to the energy-driven stochastic master equation (4.6) can be written down as follows. We start afresh and consider a finite-dimensional quantum system for which the Hamiltonian (possibly degenerate) is $\hat{H}$ and the initial state (which we regard as prescribed) is $\hat{\rho}_0$. Let a probability space $(\Omega, \mathcal{F}, \mathbb{P})$ be given, upon which we introduce a standard Brownian motion $\{B_t\}_{t\geq 0}$ and an independent random variable H, taking values in Spec[$\hat{H}$] with the distribution $\mathbb{P}[H = E_r] = \text{tr}\,\hat{\rho}_0 \hat{P}_r$, where $\hat{P}_r$ denotes the projection operator onto the Hilbert subspace of energy E_r. Then we introduce a so-called *information process* on $(\Omega, \mathcal{F}, \mathbb{P})$ denoted $\{\xi_t\}_{t\geq 0}$, defined by

$$\xi_t = \sigma t H + B_t. \tag{4.47}$$

Thus $\{\xi_t\}$ takes the form of a Brownian motion with a random drift, the rate of drift being determined by the random variable H and the parameter σ. Processes of this type arise in the theory of stochastic filtering (Wonham 1965, Liptser and Shiryaev 2000). In the language of filtering theory, one refers to H as the *signal*, B_t as the *noise*, and ξ_t as the *observation*. Of course, the notion of observation as it is understood in the context of filtering theory has no immediate connection with the notion of observation as it is usually understood in quantum mechanics. Nevertheless, the ideas that have been developed in filtering theory are rather suggestive, so it is worth keeping the associated terminology in mind as we proceed. Loosely speaking, one can think of "that which has been observed" in the context of filtering theory as equivalent to "that which has irreversibly manifested itself in the world" in the context of a physical theory. Now, let $\{\mathcal{F}_t\}_{t\geq 0}$ denote the filtration generated by $\{\xi_t\}_{t\geq 0}$. We have the following:

Proposition 5 *Let the operator-valued process $\{\hat{K}_t\}_{t\geq 0}$ be defined by*

$$\hat{K}_t = \exp\left[-i\hbar^{-1}\hat{H}t + \tfrac{1}{2}\sigma\hat{H}\xi_t - \tfrac{1}{4}\sigma^2\hat{H}^2 t\right]. \tag{4.48}$$

Then the process $\{\hat{\rho}_t\}_{t\geq 0}$ defined by

$$\hat{\rho}_t = \frac{\hat{K}_t \hat{\rho}_0 \hat{K}_t^*}{\text{tr}[\hat{K}_t \hat{\rho}_0 \hat{K}_t^*]} \tag{4.49}$$

has trace unity, is nonnegative definite, and satisfies a stochastic master equation of the form

$$d\hat{\rho}_t = -i\hbar^{-1}[\hat{H}, \hat{\rho}_t]dt + \tfrac{1}{8}\sigma^2 \left(2\hat{H}\hat{\rho}_t\hat{H} - \hat{H}^2\hat{\rho}_t - \hat{\rho}_t\hat{H}^2\right) dt$$
$$+ \tfrac{1}{2}\sigma \left((\hat{H} - H_t)\hat{\rho}_t + \hat{\rho}_t(\hat{H} - H_t)\right) dW_t, \tag{4.50}$$

where $H_t = \operatorname{tr} \hat{\rho}_t \hat{H}$ and the process $\{W_t\}_{t \geq 0}$ defined by

$$W_t = \xi_t - \sigma \int_0^t H_s \, ds \tag{4.51}$$

is an $\{\mathcal{F}_t\}$-Brownian motion.

Remark Here we look at the stochastic master equation from a new point of view. Instead of regarding $\{W_t\}_{t \geq 0}$ as an "input" to the model, we regard $\{\xi_t\}_{t \geq 0}$ as the input. Then both $\{\hat{\rho}_t\}_{t \geq 0}$ and $\{W_t\}_{t \geq 0}$ are defined in terms of $\{\xi_t\}_{t \geq 0}$, and together they satisfy equation (4.50).

Proof Let us set $\Lambda_t = \operatorname{tr}[\hat{K}_t \hat{\rho}_0 \hat{K}_t^*]$. From the cyclic property of the trace, we obtain

$$\Lambda_t = \operatorname{tr} \hat{\rho}_0 \exp\left(\sigma \hat{H} \xi_t - \tfrac{1}{2}\sigma^2 \hat{H}^2 t\right). \tag{4.52}$$

By Ito's lemma, along with $(d\xi_t)^2 = dt$, which follows from (4.47), we have

$$d\Lambda_t = \sigma \operatorname{tr} \hat{\rho}_0 \hat{H} \exp\left(\sigma \hat{H} \xi_t - \tfrac{1}{2}\sigma^2 \hat{H}^2 t\right) d\xi_t, \tag{4.53}$$

and therefore $d\Lambda_t = \sigma H_t \Lambda_t d\xi_t$, since

$$H_t = \frac{\operatorname{tr} \hat{\rho}_0 \hat{H} \exp\left(\sigma \hat{H} \xi_t - \tfrac{1}{2}\sigma^2 \hat{H}^2 t\right)}{\operatorname{tr} \hat{\rho}_0 \exp\left(\sigma \hat{H} \xi_t - \tfrac{1}{2}\sigma^2 \hat{H}^2 t\right)}. \tag{4.54}$$

If we write (4.49) in the form

$$\hat{\rho}_t = \frac{1}{\Lambda_t} \hat{K}_t \hat{\rho}_0 \hat{K}_t^*, \tag{4.55}$$

a straightforward calculation using the Ito quotient rule then gives (4.50). To establish that the process $\{W_t\}_{t \geq 0}$ defined by (4.51) is an $\{\mathcal{F}_t\}$-Brownian motion under $\mathbb{P}$, we use the so-called *Lévy criterion*. We need to show (i) that $(dW_t)^2 = dt$ and (ii) that $\{W_t\}_{t \geq 0}$ is an $\{\mathcal{F}_t\}$-martingale under $\mathbb{P}$.

The first property follows immediately as a consequence of the Ito multiplication rules applied to (4.47) and (4.51). To check that the second property holds, we need to verify for $s \leq t$ that $\mathbb{E}[W_t \mid \mathcal{F}_s] = W_s$. Let $\mathcal{G}_t$ denote the σ-algebra generated by H and $\{\xi_u\}_{0 \leq u \leq t}$. Then $\mathcal{F}_t$, which is generated by $\{\xi_u\}_{0 \leq u \leq t}$ alone, is a sub-σ-algebra of $\mathcal{G}_t$, and for all $t \geq 0$ we have the tower property of conditional expectation:

$$\mathbb{E}[\mathbb{E}[\,\cdot\mid \mathcal{G}_t] \mid \mathcal{F}_t] = \mathbb{E}[\,\cdot\mid \mathcal{F}_t]. \tag{4.56}$$

Now, by (4.51) it holds that

$$\mathbb{E}[W_t \mid \mathcal{F}_s] = \mathbb{E}[\xi_t \mid \mathcal{F}_s] - \sigma \int_0^t \mathbb{E}[H_u \mid \mathcal{F}_s] \, du. \tag{4.57}$$

As for the first term on the right side of (4.57), it follows from (4.47) that

$$\begin{aligned}
\mathbb{E}[\xi_t \mid \mathcal{F}_s] &= \mathbb{E}[B_t \mid \mathcal{F}_s] + \sigma t \, \mathbb{E}[H \mid \mathcal{F}_s] \\
&= \mathbb{E}[B_t \mid \mathcal{F}_s] + \sigma t \, H_s \\
&= \mathbb{E}\left[\mathbb{E}[B_t \mid \mathcal{G}_s] \mid \mathcal{F}_s\right] + \sigma t \, H_s \\
&= \mathbb{E}[B_s \mid \mathcal{F}_s] + \sigma t \, H_s \\
&= \xi_s + \sigma(t - s) H_s,
\end{aligned} \tag{4.58}$$

where we use the tower property to go from the second to the third line. In the second term on the right side of (4.57), we use the fact that $\{H_t\}$ is a martingale to deduce that

$$\int_0^t \mathbb{E}[H_u \mid \mathcal{F}_s] \, du = \int_0^s H_u \, du + \int_s^t H_s \, du = \int_0^s H_u \, du + (t-s)H_s. \tag{4.59}$$

Thus putting together the results for the two terms on the right side of (4.57), we have

$$\mathbb{E}[W_t \mid \mathcal{F}_s] = \xi_s - \sigma \int_0^s H_u \, du = W_s, \tag{4.60}$$

which is what we wished to show. □

4.7 Information Filtration

The collapse property in the case of a general state admits a remarkable interpretation in the language of stochastic filtering. As before, let us write $\hat{P}_r$ ($r = 1, \ldots, D$) for the projection operator onto the Hilbert subspace $\mathcal{H}_r$ consisting of state vectors with eigenvalue E_r. For any element $|a\rangle \in \mathcal{H}_r$, we have $\hat{H}|a\rangle = E_r |a\rangle$, and for the Hamiltonian we can write

$$\hat{H} = \sum_{r=1}^D E_r \hat{P}_r. \tag{4.61}$$

Therefore, if we set $\hat{R}_{nmt} = \hat{P}_n \hat{\rho}_t \hat{P}_m$, then for the diagonal terms we have

$$\hat{R}_{nnt} = \frac{\hat{P}_n \hat{\rho}_0 \hat{P}_n \exp\left[\sigma E_n \xi_t - \frac{1}{2}\sigma^2 E_n^2 t\right]}{\sum_{r=1}^D p_r \exp\left[\sigma E_r \xi_t - \frac{1}{2}\sigma^2 E_r^2 t\right]}, \tag{4.62}$$

where $p_r = \operatorname{tr} \hat{\rho}_0 \hat{P}_r$.

Proposition 6 *For each n the process $\{\hat{R}_{nnt}\}_{t\geq 0}$ is a uniformly integrable martingale, given by*

$$\hat{R}_{nnt} = \mathbb{E}\left[\mathbb{1}(H = E_n) \mid \mathcal{F}_t\right] \frac{\hat{P}_n \hat{\rho}_0 \hat{P}_n}{\operatorname{tr} \hat{\rho}_0 \hat{P}_n}, \qquad (4.63)$$

where

$$\mathbb{E}\left[\mathbb{1}(H = E_n) \mid \mathcal{F}_t\right] = \frac{p_n \exp\left[\sigma E_n \xi_t - \tfrac{1}{2}\sigma^2 E_n^2 t\right]}{\sum_{r=1}^{D} p_r \exp\left[\sigma E_r \xi_t - \tfrac{1}{2}\sigma^2 E_r^2 t\right]}. \qquad (4.64)$$

Thus, $\{\xi_t\}$ carries *partial information* about the value of the random variable H, which is revealed as time progresses, leading asymptotically to the outcome

$$\hat{R}_{nn\infty} = \mathbb{1}(H = E_n) \frac{\hat{P}_n \hat{\rho}_0 \hat{P}_n}{\operatorname{tr} \hat{\rho}_0 \hat{P}_n}, \qquad (4.65)$$

which is the Lüders state that results under the projection postulate in the standard theory as a consequence of an energy measurement, with the outcome E_n, given that the initial state is $\hat{\rho}_0$. In the present context, there is no measurement as such. Nevertheless, the final state of the reduction process is a Lüders state. For each value of n, the corresponding diagonal element of the density matrix at time t is given by the conditional expectation of the indicator function $\mathbb{1}(H = E_n)$ given the value of ξ_t.

Next we present a probabilistic formulation of the fact that the state decoheres as reduction proceeds. For any operator $\hat{O}$ let us write $|\hat{O}| = (\operatorname{tr} \hat{O}\hat{O}^\dagger)^{1/2}$.

Definition 3 *By a potential on a filtered probability space $(\Omega, \mathcal{F}, \mathbb{P}, \mathbb{F})$, we mean a strictly positive right-continuous supermartingale $\{\pi_t\}_{t\geq 0}$ with the property that $\lim_{t\to\infty} \mathbb{E}[\pi_t] = 0$.*

Then we have the following:

Proposition 7 *For each n, m such that $n \neq m$ the process $\{|\hat{R}_{nmt}|\}_{t\geq 0}$ is a potential.*

Proof Let n, m be such that $n \neq m$. The off-diagonal matrix elements of the state then take the form

$$\hat{R}_{nmt} = \hat{P}_n \hat{\rho}_0 \hat{P}_m \frac{\exp\left[-i\hbar^{-1}(E_n - E_m)t + \tfrac{1}{2}\sigma(E_n + E_m)\xi_t - \tfrac{1}{4}\sigma^2(E_n^2 + E_m^2)t\right]}{\sum_{r=1}^{D} p_r \exp\left[\sigma E_r \xi_t - \tfrac{1}{2}\sigma^2 E_r^2 t\right]}. \qquad (4.66)$$

Thus we have

$$\hat{R}_{nmt} = \hat{P}_n \hat{\rho}_0 \hat{P}_m \exp\left[-i\hbar^{-1}(E_n - E_m)t\right] \Phi_{nmt}, \qquad (4.67)$$

where

$$\Phi_{nmt} = \frac{\exp\left[\frac{1}{2}\sigma(E_n + E_m)\xi_t - \frac{1}{4}\sigma^2(E_n^2 + E_m^2)t\right]}{\sum_{r=1}^{D} p_r \exp\left[\sigma E_r \xi_t - \frac{1}{2}\sigma^2 E_r^2 t\right]}, \tag{4.68}$$

and a calculation shows that

$$\Phi_{nmt} = \Pi_{nmt} \exp\left[-\frac{1}{8}\sigma^2(E_n - E_m)^2 t\right], \tag{4.69}$$

where

$$\Pi_{nmt} = \frac{\exp\left[\frac{1}{2}\sigma(E_n + E_m)\xi_t - \frac{1}{8}\sigma^2(E_n + E_m)^2 t\right]}{\sum_{r=1}^{D} p_r \exp\left[\sigma E_r \xi_t - \frac{1}{2}\sigma^2 E_r^2 t\right]}. \tag{4.70}$$

We claim that for any $\lambda \in \mathbb{R}$ the process $\{\mu_t\}_{t \geq 0}$ defined by

$$\mu_t = \frac{\exp\left[\lambda \xi_t - \frac{1}{2}\lambda^2 t\right]}{\sum_{r=1}^{D} p_r \exp\left[\sigma E_r \xi_t - \frac{1}{2}\sigma^2 E_r^2 t\right]} \tag{4.71}$$

is a martingale. To see that this is so, note that by use of Ito's lemma, together with the relation $d\xi_t = \sigma H_t dt + dW_t$, we have $d\mu_t = (\lambda - \sigma H_t)\mu_t dW_t$, and thus

$$\mu_t = \exp\left[\int_0^t (\lambda - \sigma H_s)\, dW_s - \frac{1}{2}\int_0^t (\lambda - \sigma H_s)^2\, ds\right]. \tag{4.72}$$

Since $\{H_t\}_{t \geq 0}$ is bounded, we deduce that $\{\mu_t\}_{t \geq 0}$ is a martingale. Therefore, $\{\Pi_{nmt}\}_{t \geq 0}$ is a martingale and $\{\Phi_{nmt}\}_{t \geq 0}$ is a supermartingale. By (4.69) one sees that

$$\mathbb{E}[\Phi_{nmt}] = \exp\left[-\frac{1}{8}\sigma^2(E_n - E_m)^2 t\right], \tag{4.73}$$

and hence

$$\lim_{t \to \infty} \mathbb{E}[\Phi_{nmt}] = 0, \tag{4.74}$$

so $\{\Phi_{nmt}\}_{t \geq 0}$ is a potential. Finally, we observe that

$$\left|\hat{R}_{nmt}\right| = \left|\hat{P}_n \hat{\rho}_0 \hat{P}_m\right| \Phi_{nmt}, \tag{4.75}$$

from which we obtain

$$\lim_{t \to \infty} \mathbb{E}\left[\left|\hat{R}_{nmt}\right|\right] = 0 \tag{4.76}$$

for $n \neq m$, which is what we wished to prove. □

Thus, the potential property of the off-diagonal terms of the density matrix in the energy representation captures the essence of what is meant by decoherence. We see that the decay of the off-diagonal terms of the density matrix is exponential in time and that the decay rate for any particular such term is proportional to the square of the difference of the associated energy levels. In fact, we can take the representation of $\{\Phi_{nmt}\}_{t\geq 0}$ as a potential a step further. A calculation making use of the Ito quotient rule shows that

$$d\Phi_{nmt} = -\tfrac{1}{8}\sigma^2(E_n - E_m)^2\, \Phi_{nmt}\, dt + \tfrac{1}{2}\sigma(E_n + E_m - 2H_t)\, \Phi_{nmt}\, dW_t. \quad (4.77)$$

As a consequence, for each n, m such that $n \neq m$ we have

$$\Phi_{nmt} = 1 - \tfrac{1}{8}\sigma^2(E_n - E_m)^2 \int_0^t \Phi_{nms}\, ds + \tfrac{1}{2}\sigma \int_0^t (E_n + E_m - 2H_s)\, \Phi_{nms}\, dW_s. \quad (4.78)$$

Taking the limit as t goes to infinity and using the fact that $\Phi_{nm\infty} = 0$ almost surely, we deduce that

$$1 + \tfrac{1}{2}\sigma \int_0^\infty (E_n + E_m - 2H_s)\, \Phi_{nms}\, dW_s = \tfrac{1}{8}\sigma^2(E_n - E_m)^2 \int_0^\infty \Phi_{nms}\, ds. \quad (4.79)$$

Then by taking a conditional expectation we obtain

$$1 + \tfrac{1}{2}\sigma \int_0^t (E_n + E_m - 2H_s)\, \Phi_{nms}\, dW_s = \tfrac{1}{8}\sigma^2(E_n - E_m)^2\, \mathbb{E}_t\left[\int_0^\infty \Phi_{nms}\, ds\right], \quad (4.80)$$

from which it follows by use of (4.78) that

$$\Phi_{nmt} = \tfrac{1}{8}\sigma^2(E_n - E_m)^2\, \mathbb{E}_t\left[\int_t^\infty \Phi_{nms}\, ds\right]. \quad (4.81)$$

This identity gives us a representation of $\{\Phi_{nmt}\}_{t\geq 0}$ as a so-called type-D potential. More explicitly, if we define an increasing process $\{A_{nmt}\}_{t\geq 0}$ by setting

$$A_{nmt} = \tfrac{1}{8}\sigma^2(E_n - E_m)^2 \int_0^t \Phi_{nms}\, ds, \quad (4.82)$$

then we have

$$\Phi_{nmt} = \mathbb{E}_t[A_{nm\infty}] - A_{nmt}, \quad (4.83)$$

which is the canonical form for a potential of type D (Meyer 1966). Thus we arrive at the following:

Proposition 8 *The state process under energy-driven stochastic reduction is of the form*

$$\hat{\rho}_t = \sum_{n=1}^{D} \mathbb{E}_t \left[\mathbb{1}(H = E_n) \right] \frac{\hat{P}_n \hat{\rho}_0 \hat{P}_n}{\operatorname{tr} \hat{\rho}_0 \hat{P}_n}$$
$$+ \sum_{n,m=1}^{D} \mathbb{1}_{n \neq m} \hat{P}_n \hat{\rho}_0 \hat{P}_m \exp\left[-i\hbar^{-1}(E_n - E_m) t \right] \Phi_{nmt}. \qquad (4.84)$$

The conditional expectation in the first term is given by (4.64) and the potential in the second term is given by (4.69). At time zero, the two terms combine to give the initial density matrix $\hat{\rho}_0$. As the collapse proceeds, the first term converges to the Lüders state associated with the selected energy eigenvalue E_n, and the second term tails off to zero. It should be emphasized that if the initial state is impure, and if the Hamiltonian is degenerate, then the final state will in general also be impure.

4.8 Conclusion

In our development of the dynamic reduction program we have taken the view that the state of a single system can be described by a density matrix that may or may not be pure. The initial state $\hat{\rho}_0$ is prescribed, and its value at time t is given by the random density matrix $\hat{\rho}_t$. The model is understood as describing an "objective" reduction process, so there are no observers in the theory in the usual sense. All the same, one can ask what is known at time t, in the sense of what has manifested itself in the world (or, let's say, in the experimenter's laboratory) at that time. For this purpose it seems reasonable to adopt the view that the standard interpretation of the filtration $\{\mathcal{F}_t\}$ gives an adequate answer. This means that for any overall outcome of chance $\omega \in \Omega$, the value of any $\mathcal{F}_t$-measureable random variable X_t will be "known" or will have manifested itself at (or before) time t. In particular, the value of $\hat{\rho}_t$ itself will be known at time t, as will the value of the information process ξ_t. Now, it is not quite meaningful to ask how ξ_t can be measured in a theory in which there are no measurements. Nevertheless, we are forced to the conclusion that the theory only makes sense if ξ_t is known (in the sense of having manifested itself) at time t. In fact, Diósi (2015) has arrived at what we believe to be in essence a similar conclusion, that stochastic reduction models only really make sense if the $\{\xi_t\}$ process can in some appropriate sense be monitored in real time. This is not the same thing as saying that the quantum system is being *actively* monitored (in the sense of Diósi 1988, Barchielli and Belavkin 1991, Barchielli 1993, Wiseman 1996, Wiseman and Diósi 2001, Barchielli and Gregoratti 2009), since the monitoring that takes place in such considerations is within a framework of standard quantum dynamics, and some form of *ad hoc* collapse is required as an additional assumption to make the infinitesimal collapses occur in response to the monitoring. But it may be that in a laboratory situation it is possible to monitor $\{\xi_t\}$, or equivalently $\{H_t\}$, in the passive sense implicit in the structure of the information filtration of the models that we have here described. The class of information-based models that can be developed by use of the filtering techniques discussed in Sections 4.6 and 4.7 can be extended to a wider set of models, in which the underlying noise is not Brownian motion but rather a general Lévy process. Such processes, like Brownian motion, have the property of being stationary with independent increments, but are not generally Gaussian and can be discontinuous. Providing that a condition is satisfied ensuring the existence of exponential moments, Lévy trajectories are suitable for characterizing a wide and extraordinarily diverse family of noise processes (Brody, Hughston and Yang 2013). The development of relativistic analogues of the models considered here remains an open problem, though it seems reasonable to conjecture that in the relativistic case the reduction process should lead to states for which the total mass and spin take definite values.

Acknowledgments

This work was carried out in part at the Perimeter Institute for Theoretical Physics, which is supported by the government of Canada through Innovation, Science and Economic Development Canada and by the province of Ontario through the Ministry of Research, Innovation and Science, and at the Aspen Center for Physics, which is supported by National Science Foundation grant PHY-1066293. DCB acknowledges support from the Russian Science Foundation (project 16-11-10218). We are grateful to S. L. Adler, T. Benoist, I. Egusquiza, S. Gao, and B. K. Meister for helpful comments and discussions.

References

Adler, S. L. (2002) Environmental influence on the measurement process in stochastic reduction models. *J. Phys. A: Math Gen.* **35**, 841–858.

Adler, S. L. (2003a) Why decoherence has not solved the measurement problem: a response to P. W. Anderson. *Studies Hist. Phil. Mod. Phys.* **34**, 135–142.

Adler, S. L. (2003b) Weisskopf-Wigner decay theory for the energy-driven stochastic Schrödinger equation. *Phys. Rev. D* **67**, 025077.

Adler, S. L. (2004) *Quantum Theory as an Emergent Phenomenon* (Cambridge University Press).

Adler, S. L., Brody, D. C., Brun, T. A. and Hughston, L. P. (2001) Martingale models for quantum state reduction. *J. Phys. A: Math. Gen.* **34**, 8795–8820.

Adler, S. L. and Horwitz, L. P. (2000) Structure and properties of Hughston's stochastic extension of the Schrödinger equation. *J. Math. Phys.* **41**, 2485–2499.

Adler, S. L. and Mitra, I. L. (2000) Equilibrium distribution of gas molecules adsorbed on an active surface. *Phys. Rev. E* **2**, 4386–4388.

Banks, T., Susskind, L. and Peskin, M. E. (1984) Difficulties for the evolution of pure states into mixed states. *Nucl. Phys. B* **244**, 125–134.

Barchielli, A. and Belavkin, V. P. (1991) Measurements continuous in time and *a posteriori* states in quantum mechanics. *J. Phys. A: Math. Gen.* **24**, 1495–1514.

Barchielli, A. (1993) On the quantum theory of measurements continuous in time. *Rep. Math. Phys.* **33**, 21–34.

Barchielli, A. and Gregoratti, M. (2009) *Quantum Trajectories and Measurements in Continuous Time*, Lecture Notes in Physics **782** (Berlin: Springer-Verlag).

Bassi, A. and Ghirardi, G. C. (2003) Dynamical reduction models. *Phys. Rep.* **379**, 257–426.

Bassi, A., Ippoliti, E., and Vacchini, B. (2005) On the energy increase in space-collapse models. *J. Phys. A: Math. Gen.* **38**, 8017–8038.

Bassi, A. (2007) Dynamical reduction models: present status and future developments. *J. Phys. Conf. Series* **67**, 012013.

Bassi, A., Lochan, K., Satin, S., Singh, T. P. and Ulbricht, H. C. (2013) Models of wavefunction collapse, underlying theories, and experimental tests. *Rev. Mod. Phys.* **85**, 471–527.

Bassi, A., Dürr, D. and Henriks, G. (2013) Uniqueness of the equation for quantum state vector collapse. *Phys. Rev. Lett.* **111**, 210401.

Bell, J. S. (1987) Are there quantum jumps? In: *Schrödinger, Centenary of a Polymath* (Cambridge University Press). Reprinted in: J. S. Bell (1987) *Speakable and Unspeakable in Quantum Mechanics* (Cambridge University Press).

Brody, D. C. and Hughston, L. P. (2002a) Efficient simulation of quantum state reduction. *J. Math. Phys.* **43**, 5254–5261.

Brody, D. C. and Hughston, L. P. (2002b) Stochastic reduction in nonlinear quantum mechanics. *Proc. R. Soc. Lond. A* **458**, 1117–1127.

Brody, D. C. and Hughston, L. P. and Syroka J. (2003) Relaxation of quantum states under energy perturbations. *Proc. R. Soc. Lond. A* **459**, 2297–2316.

Brody, D. C. and Hughston, L. P. (2005) Finite-time stochastic reduction models. *J. Math. Phys.* **46**, 082101.

Brody, D. C. and Hughston, L. P. (2006) Quantum noise and stochastic reduction. *J. Phys. A: Math. Gen.* **39**, 833–876.

Brody, D. C., Constantinou, I. C., Dear, J. D. C. and Hughston, L. P. (2006) Exactly solvable quantum state reduction models with time-dependent coupling. *J. Phys. A: Math. Gen.* **39**, 11029–11051.

Brody, D. C., Hughston, L. P. and Yang, X. (2013) Signal processing with Lévy information. *Proc. Roy. Soc. Lond. A* **469**, 20120433.

Davies, E. B. (1976) *Quantum Theory of Open Systems* (London: Academic Press).

Diosi, L. (1988) Continuous quantum measurement and Ito formalism. *Phys. Lett. A* **129**, 419–423.

Diosi, L. (1989) Models for universal reduction of macroscopic fluctuations. *Phys. Rev. A* **40**, 1165–1174.

Diosi, L. (2015) Is spontaneous wave function collapse testable at all? *J. Phys. Conf. Series* **626**, 012008.

Gao, S. (2013) A discrete model of energy-conserved wave function collapse. *Proc. R. Soc. Lond. A* **469**, 20120526.

Ghirardi, G. C., Rimini, A. and Weber, T. (1986) Unified dynamics for microscopic and macroscopic systems. *Phys. Rev. D* **34**, 470–491.

Ghirardi, G. C., Pearle, P. and Rimini, A. (1990) Markov processes in Hilbert space and continuous spontaneous localization of systems of identical particles. *Phys. Rev. A* **42**, 78–89.

Ghirardi, G. C. (2000) Beyond conventional quantum mechanics. In: *Quantum Reflections*, J. Ellis and D. Amati, eds., 79–116 (Cambridge University Press).

Ghirardi, G. C. (2016) Collapse theories. In: *Stanford Encyclopedia of Philosophy*, Spring 2016 edition, E. N. Zalta, ed., https://plato.stanford.edu/archives/spr2016/entries/qm-collapse/.

Gisin, N. (1989) Stochastic quantum dynamics and relativity. *Helv. Phys. Acta* **62** 363–371.

Gorini, V., Kossakowski, A. and Sudarshan, E. C. G. (1976) Completely positive dynamical symmetry groups of N-level systems. *J. Math. Phys.* **17**, 821–825.

Haag, R. and Kastler, D. (1964) An algebraic approach to quantum field theory. *J. Math. Phys.* **7**, 848–861.

Hida, T. (1980) *Brownian Motion* (New York: Springer-Verlag).

Hughston, L. P. (1996) Geometry of stochastic state vector reduction. *Proc. R. Soc. Lond. A* **452**, 953–979.

Isham, C. J. (1995) *Lectures on Quantum Theory* (London: Imperial College Press).

Karatzas, I. and Shreve, S. E. (1991) *Brownian Motion and Stochastic Calculus*, second edition (Berlin: Springer).

Karolyhazy, F. (1966) Gravitation and quantum mechanics of macroscopic objects. *Nuovo Cimento* **42**, 390–402.

Karolyhazy, F., Frenkel, A. and Lukács, B. (1986) On the possible role of gravity in the reduction of the wave function. In: *Quantum Concepts in Space and Time*, R. Penrose and C. J. Isham, eds. (Oxford University Press).

Kolmolgorov, A. (1933) *Grundbegriffe der Wahrscheinlichkeitsrechnung* (Berlin: Springer). English translation (1956) *Foundations of the Probability Theory* (New York: Chelsea).

Lindblad, G. (1976) On the generators of quantum dynamical semigroups. *Commun. Math. Phys.* **48**, 119–130.

Liptser, R. S. and Shiryaev, A. N. (2000) *Statistics of Random Processes*, Vols. I and II, second edition (Berlin: Springer).

Lüders, G. (1951) Über die Zustandsänderung durch den Messprozess. *Ann. Physik* **8**, 322–328.

Mengütürk, L. A. (2016) Stochastic Schrödinger evolution over piecewise enlarged filtrations. *J. Math. Phys.* **57**, 032106.

Meyer, P. A. (1966) *Probability and Potentials* (Waltham, Massachusetts: Blaisdell).

Mielnik, B. (1974) Generalized quantum mechanics. *Commun. Math. Phys.* **37**, 221–256.

von Mises, R. (1919) Grundlagen der Wahrscheinlichkeitsrechnung. *Mathematische Zeitschrift* **5**, 52–99.

von Mises, R. (1928) *Wahrscheinlichkeit, Statistik und Wahreheit* (Wien: Julius Springer). English translation by J. Neyman, D. Sholl and E. Rabinowitsch (1939) *Probability, Statistics and Truth* (London: William Hodge & Company, Limited).

von Neumann, J. (1932) *Mathematische Grundlagen der Quantenmechanik* (Berlin: Springer). English translation by R. T. Beyer (1971) *Mathematical Foundations of Quantum Mechanics* (Princeton University Press).

Pearle, P. (1976) Reduction of the state vector by a nonlinear Schrödinger equation. *Phys. Rev. D* **13**, 857–868.

Pearle, P. (1989) Combining stochastic dynamical state-vector reduction with spontaneous localization. *Phys. Rev. D* **39**, 2277–2289.

Pearle, P. (1999) Collapse models. In: *Open Systems and Measurement in Relativistic Quantum Theory*, H. P. Breuer and F. Petruccione, eds. (Heidelberg: Springer).

Pearle, P. (2000) Wavefunction collapse and conservation laws. *Found. Phys.* **30**, 1145–1160.

Pearle, P. (2004) Problems and aspects of energy-driven wave-function collapse models. *Phys. Rev. A* **69** 042106.

Pearle, P. (2007) How stands collapse I. *J. Phys. A: Math. Gen.* **40**, 3189–3204.

Pearle, P. (2009) How stands collapse II. In: *Quantum Reality, Relativistic Causality, and Closing the Epistemic Circle: Essays in Honour of Abner Shimony*, W. C. Myrvold and J. Christian, eds., The Western Ontario Series in Philosophy of Science, Vol. **73**, 257–292 (Springer Netherlands).

Penrose, R. (1986) Gravity and state vector reduction. In: *Quantum Concepts in Space and Time*, R. Penrose and C. J. Isham, eds. (Oxford University Press).

Penrose, R. (1996) On gravity's role in quantum state reduction. *General Relativity and Gravitation* **28**, 581–600.

Penrose, R. (2016) *Fashion, Faith and Fantasy in the New Physics of the Universe* (Princeton University Press).

Percival, I. C. (1994) Primary state diffusion. *Proc. R. Soc. Lond. A* **447**, 189–209.

Percival, I. C. (1998) *Quantum State Diffusion* (Cambridge University Press).

Protter, P. (2003) *Stochastic Integration and Differential Equations: A New Approach*, second edition (Berlin: Springer).

Segal, I. E. (1947) Postulates for general quantum mechanics. *Ann. Math.* **48**, 930–948.

Shimony, A. (1990) Desiderata for modified quantum dynamics. In: *PSA 1990*, A. Fine, M. Forbes and L. Wessels, eds., Vol. **2** (East Lansing, Michigan: Philosophy of Science Association).

van Lambalgen, M. (1999) Randomness and foundations of probability: von Mises' axiomatization of random sequences. In: *Festschrift for David Blackwell*, T. Ferguson, ed., IMS Lecture Notes, Monograph Series **50**, Institute for Mathematical Statistics (California: Hayward).

Weinberg, S. (2012) Collapse of the state vector. *Phys. Rev. A* **85**, 062116.

Williams, D. (1991) *Probability with Martingales* (Cambridge University Press).

Wiseman, H. M. (1996) Quantum trajectories and quantum measurement theory. *Quantum Semiclass. Opt.* **8**, 205–222.

Wiseman, H. M. and Diosi, L. (2001) Complete parameterization, and invariance, of diffusive quantum trajectories for Markovian open systems. *Chem. Phys.* **268**, 91–104.

Wonham, W. M. (1965) Some applications of stochastic differential equations to optimal nonlinear filtering. *J. SIAM Control A* **2**, 347–369.

5
Collapse Models and Space–time Symmetries

DANIEL J. BEDINGHAM

A picture of dynamical collapse of the wave function which is relativistic and time symmetric is presented. The part of the model which exhibits these features is the set of collapse outcomes. These play the role of matter distributed in space and time. It is argued that the dynamically collapsing quantum state, which is both foliation dependent and follows a time-asymmetric dynamics, is not fundamental: it represents a state of information about the past matter distribution for the purpose of estimating the future matter distribution. It is also argued from the point of view of collapse models that both special and general relativistic considerations point towards a discrete space–time structure and that gravity may not need to be quantised to give a theory that is consistent with quantum matter.

5.1 Introduction

5.1.1 Destination

Consider the following description of the physical world:

The world is described by a definite-valued distribution of matter in space and time. The matter densities assigned to points in the future with respect to some time slice cannot be predicted with certainty. The best that can be done is to estimate these values on the basis of information about matter densities at points in the past and to use *standard methods of Bayesian inference* to update our estimates as new information about the distribution of matter is acquired. We find good agreement with quantum theory. Furthermore, our description is grounded in definite local features—the distribution of matter in space and time. The uncertainty which characterises quantum behaviour results from the uncertain future values of matter density given information about the way that matter was distributed in the past.

This may not sound like a model of objective collapse of the wave function, yet we aim to show that this is the natural way to understand the meaning of collapse models in light of the symmetries that we expect of their dynamics under transformations of the space–time coordinates. In order to get there we must tell the story from the beginning.

5.1.2 Journey

In the orthodox picture of quantum theory our only way to access the quantum world is via the results of measurements. It is only via the results of measurements that we can infer the form of the wave function which cannot be observed directly. Despite this lack of direct experience, the ability of the wave function to undergo interference effects which appear in the precisely determined distribution of measurement results leads many to believe that the wave function is a real physical object (see also Pusey et al. [1]).

The orthodox quantum rules determine that the wave function should be updated following a measurement in correspondence with the observed outcome. Treating this as a real physical collapse process acting on a real physical wave function raises the question of what should count as a measurement capable of triggering it—this is the infamous measurement problem [2, 3, 4]. The original idea behind the solution to this problem offered by dynamical collapse models [5, 6] was to give the collapse process a firm foundation by having it obey a well-defined law. The essential feature is that it happens not in response to the occurrence of a measurement but spontaneously as a result of a mathematically expressed rule. For any sensible dynamical collapse model, it should be such that it reproduces the overall effect of the collapse of the wave function in situations that are unambiguously quantum measurements, that it has little effect on the usual Schrödinger dynamics of micro systems, and that, since it is supposed to be a description of matter on all scales, it reduces to a classical description of macro objects.

The fact that it is possible to find such a model is remarkable [7, 8, 9]. It offers the chance to solidify the foundations of quantum theory and resolve the measurement problem in a way that adheres to orthodox intuitions. But it may not persuade everyone. Many believe that the collapse process should properly be viewed as a process of inference for which it is natural to regard the wave function as a state of information and that this should be taken into account in the foundations. Indeed, the point will be stressed in this chapter that collapse models can be viewed in this way, where the wave function is demoted from real physical object to a device for encoding information about objective facts. Furthermore, this is a view which is strongly motivated by concerns about the symmetries of the collapse dynamics under transformations of the space–time coordinates.

In collapse models, as will be explicitly described, when a collapse takes place, it is formally equivalent in terms of its effect on the wave function to the effect of a quantum measurement. Associated to the collapse is a random variable which plays the role of a measurement outcome. Just as in the orthodox picture where the measurement outcomes give our only direct physical access to the quantum world, it is natural to treat the collapse outcomes as the part of the model which corresponds to real physical events at definite points in space and time. This view with primitive status given to the collapse outcomes rather than the wave function is close to the way in which collapse models were understood by Bell [10]. The chance of any particular outcome is determined by the wave function, but the wave function itself is determined by the past collapse outcomes. This means that the wave function can be seen as a useful intermediary whose job is to process information about collapses that have occurred in the past in order to estimate future collapse outcomes. Something that we aim to clarify here is that the wave function plays the role of a probability distribution for future collapse outcomes in a surprisingly conventional way.

In the terminology of Goldstein [11, 12] and Allori [13, 14, 15], the collapse outcomes are to be regarded as the *primitive ontology* (PO). The PO is the part of a theory which provides a picture of the distribution of matter in space and time, allowing for a straightforward comparison between the theory and the world of our experience. In the context of collapse models (and in particular the Ghirardi–Rimini–Weber (GRW) model [7]—see what follows), treating the collapse outcomes as the PO has come to be known as the flash ontology.

The generic structure of collapse models will be presented in Section 5.2 and it will be demonstrated that a collapse of the wave function is nothing more than a Bayesian updating in response to information about the present distribution of matter. It will also be shown that, under given conditions, this structure has time reversal symmetry at the level of collapse outcomes. This is to be contrasted with the quantum state description, which is clearly time asymmetric since it is shaped by collapses in the past but not in the future. In Section 5.3 the structure of specially relativistic dynamical collapse is presented where it is shown how the probability of a given set of collapse outcomes in a given space–time region can be Lorentz invariant and independent of space–time foliation. A quantum state history, by contrast, must be defined with reference to a space–time foliation. If the quantum state were the basis of local physical properties this feature would be troublesome (local features should not depend on an arbitrary choice of time slice). However, with primitive status given to collapse outcomes and the quantum state representing maximal information about past collapses the hypersurface dependence of the quantum state is natural, since a hypersurface is needed to demarcate the past-future boundary. It is also argued that considerations of energy conservation point towards a discrete space–time structure. In Section 5.4 this view is strengthened by the possibility that it opens a way to consistently describe classical gravity and quantum matter.

5.2 Generic Non-relativistic Structure

For all the variation in proposed dynamical collapse models there is a common underlying structure which we present here [16, 17]. There is a quantum state or wave function $|\psi_t\rangle$ which, if there is no collapse event between times s and $t > s$, evolves in time, satisfying the Schrödinger equation with solution

$$|\psi_t\rangle = \hat{U}(t-s)|\psi_s\rangle, \tag{5.1}$$

where $\hat{U}(t) = e^{-i\hat{H}t}$ is the usual unitary time evolution operator for a Hamiltonian $\hat{H}$. At certain times governed by the model there is a collapse event (we assume for now that these times are discrete; they may be randomly distributed or periodic and deterministic). The collapse events involve a sudden change to the wave function. A collapse at time t is described by

$$|\psi_t\rangle \to |\psi_{t+}\rangle = \hat{J}(z_t)|\psi_t\rangle, \tag{5.2}$$

where $\hat{J}$ is the collapse operator (J for "jump") and z_t is a random variable upon which it depends. The chance of realising a particular z_t is given by the probability distribution

$$P(z_t|\psi_t) = \frac{\langle \psi_{t+}|\psi_{t+}\rangle}{\langle \psi_t|\psi_t\rangle} = \frac{\langle \psi_t|\hat{J}^\dagger(z_t)\hat{J}(z_t)|\psi_t\rangle}{\langle \psi_t|\psi_t\rangle}. \tag{5.3}$$

This means that the state immediately prior to the collapse event determines the chance of z_t. Finally, in order for the probability distribution to be normalised we require the completeness condition

$$\int dz \hat{J}^\dagger(z)\hat{J}(z) = \hat{1}. \tag{5.4}$$

The last four equations are the essence of the idea of dynamical collapse. A specific model corresponds to a particular choice for $\hat{J}$ and for the sequence of times at which the collapses occur. Continuous models described by stochastic differential equations can be understood as the limit case of this construction in which the collapses happen at each moment (this requires a suitable weakening of the effect of each collapse as the rate of collapses is increased). The choice of collapse operator commits to a definite basis (or set of bases) within which collapse takes place.

The two most well-known models are the Ghirardi–Rimini–Weber (GRW) model [7] and the Continuous Spontaneous Localisation (CSL) model [8, 9]. For the GRW model we have

$$\hat{J}(\mathbf{z}_t) = \left(\frac{\alpha}{\pi}\right)^{3/4} \exp\left\{-\frac{\alpha}{2}(\hat{\mathbf{x}} - \mathbf{z}_t)^2\right\}, \tag{5.5}$$

where $\hat{\mathbf{x}}$ is the position operator for an individual constituent particle and α is a constant. The effect of the collapse operator is to localise the state of that particle about the random position z_t. The particles are assumed to be distinguishable, and each particle experiences independent collapses at independent Poisson-distributed random times with rate λ. The parameter α sets a length scale for the localisation.

So why is it that we only see collapse during measurements if the collapses are perpetually affecting all the particles? The reason is that during a quantum measurement, a large number of particles in the measuring device (assumed to be something like a rigid pointer) become entangled with the state of the system of interest, and it only requires that one constituent particle undergoes a collapse event to cause the entire collapse process. This means that the rate of collapse per particle λ can be made sufficiently small that collapses rarely affect small systems, but given the large number of particles in the device, the chance of a constituent particle undergoing a collapse in a small time can be sufficiently high.

For the CSL model we have [9] (Sec. IVC)

$$\hat{J}[z_t] = W^{1/2} \exp\left\{-\frac{\beta}{2} \int d\mathbf{x} \left[\hat{N}(\mathbf{x}) - z_t(\mathbf{x})\right]^2\right\}, \qquad (5.6)$$

where W is a normalisation factor, $\hat{N}(\mathbf{x})$ is the (smeared) particle number density operator about point $\mathbf{x}$, and β is a collapse strength parameter. These collapses occur with Poisson-distributed random times with rate μ. Here $z_t(\mathbf{x})$ is a real random function. The effect of this collapse operator is to sharpen the state of number density at each point in space. This is not a precise projection of the $N(\mathbf{x})$-state; there is some spread determined by the parameter β and the timescale of interest. For systems of small number density this spread can be large enough to leave spatial superposition states untroubled; but for systems with large number density (such as measuring devices) a Schrödinger cat state of, for example, different macro pointer positions with a large quantum variance in number density would be rapidly suppressed. This collapse operator also has a localisation effect which can be easily understood in the case of a single particle, since the only state of well-defined number density (i.e., zero quantum variance) is one that is perfectly localised. Strictly, the CSL model as it is usually presented is the continuous limit of this model which involves taking the rate μ to infinity whilst letting $\beta \to 0$ in such a way that $\mu\beta = 2\gamma$ is a constant, and where γ is the usual CSL parameter [9] (Sec. IVC). For the sake of austerity we shall continue to call this the CSL model.

At this stage we might be tempted to argue that the wave function represents the state of the world. This seems reasonable in light of the fact that having incorporated collapses into the dynamics, the wave function has been purged of unwanted macro superposition states. With collapse, the dynamical process accurately updates the wave function to reflect previous measurement events.

Ultimately we require that our theory explain the world of our experience which involves spatial arrangements of matter which change with time or, more generally, local properties in fixed regions of space–time [18]. But the wave function is not an object that belongs in the 3+1 dimensional space–time of our experience. It belongs to a large (or perhaps infinite) dimensional Hilbert space. If the wave function represents the state of the world there is some work to do in converting it into definite local properties of the world at fixed points in space and time. One approach is to determine the precise density of matter in space as the quantum expectation of a suitable matter density operator (this is often called the matter density ontology). A problem with this kind of scheme is that it is possible to come up with other characteristics that can be defined precisely in expectation but which would not be directly observable (a simple example being the expected momentum of a well-localised particle). For a precise correspondence between the wave function and local properties in space–time it would seem that we must choose some preferred operator(s) to act as directly experienced features. This might not seem too bad given that we have already chosen a preferred basis for collapse, but the idea that even though the wave function describes the world, we only have direct access to a seemingly limited feature of it, seems arbitrary.

In practice in the laboratory, the wave function only reflects the state of a system insofar as it accurately predicts measurement statistics. For an experimenter it is the measurement results that are the experienced occurrences. Noting that the description of a generic collapse model presented in equations (5.1) through (5.4) is formally equivalent in its effect on the wave function to a generalised quantum measurement (see Nielsen and Chuang [19]), and that in this correspondence the collapse outcomes z play the role of measurement outcomes, leads to the suggestion that the collapse outcomes (rather than operator expectations) should be treated as local properties in space–time [10]. This is a far more natural option given that they are a component part of the model, integral to the dynamical process, and that they belong in space–time. (See Esfeld and Gisin [20], Esfeld [21], Egg and Esfeld [22] for further discussion.)

In the GRW model the collapse outcomes are points in space–time representing a sample of individual particle locations. A physical object then corresponds to a collection of collapses in a localised space–time region. A massive object (i.e., a dense collection of collapses in a local region) will persist in time, since the particles (of the wave function description) will continue to undergo collapses.

In the CSL model the collapse outcomes are assignments to points throughout space and time of values z representing the local number density of particles. This is a much more straightforward ontology—the collapse outcomes are a dense sample of local matter density throughout space–time. However, it has been pointed out by Bacciagaluppi [23] that in the continuous version, the variance in z is effectively infinite so that it is difficult to see how a smooth distribution of matter is recovered. This is also the case for the discrete-in-time version (5.6), but the problem can easily be rectified by having the collapses occur at discrete points in space as well as time (see Figure 5.1). With a view to forming a relativistic model of collapse we can suppose that the CSL collapses are randomly located with uniform distribution in *space–time* volume (up to Poisson-type fluctuations expected to result from the random sprinkling) with density μ and that each collapse takes the form

$$\hat{J}(z_x) = \left(\frac{\beta}{\pi}\right)^{1/4} \exp\left\{-\frac{\beta}{2}\left[\hat{N}(\mathbf{x}) - z_x\right]^2\right\}, \qquad (5.7)$$

where $x = (\mathbf{x}, t)$. This collapse occurs at time t and leads to a narrowing of the (smeared) number density state at position $\mathbf{x}$. The variance in each z_x is finite, and the usual CSL model is recovered by letting μ tend to infinity whilst $\beta \to 0$ with $\mu\beta = 2\gamma$ (recall γ is the CSL parameter).

In both these examples, when we gather together many of the realised z values, we find that they provide a representation of where matter is in space–time. It thus makes sense to regard these zs as actually representing the definite matter distribution in space–time. Since the collapse outcomes correspond to the outcomes of generalised measurements, empirical predictions of the distribution of matter in the world in terms of z will agree with those of the standard quantum formalism. Small differences are to be expected since the collapse process leads to loss of coherence and heating, offering a way to experimentally test for collapse (see Bassi et al. [6] and references therein).

However, this construction sees the matter distribution as a kind of by-product of the collapse dynamics of the wave function. It is natural to ask whether it is possible to begin with a description in terms of physical events $\{z\}$ in which the role of the wave function is simply to determine the probability distribution of zs in the future given information about zs in the past. In the next subsection we outline how this can be achieved. In light of this it seems wrong to call $\{z\}$ the collapse outcomes; we should simply regard them as the distribution of matter.

5.2.1 Collapse as Inference

It may seem that the distribution of matter (defined in terms of the collapse outcomes) has more of an intertwined relationship with the wave function than suggested by the claim that the wave function is simply a state which makes probabilistic interpretations about future variables given known past ones. After all, the collapse outcomes are generated by the physical collapse process of the wave function. But in fact, as we will show here, one can simply regard the wave function as defining a probability distribution and think of a collapse event described by Eq. (5.2), not as a physical process, but as representing an inference.

In order to make this point consider the following (see also related work in Bedingham [24], Tumulka [25]). We assume that there is some c-number valued time-dependent quantity $A(t)$ which is not directly observable. In fact the only access to this quantity is via discrete noisy information which takes the form

$$z_t = A(t) + B_t, \tag{5.8}$$

where B_t is a noise variable assumed to have a Gaussian distribution with standard deviation σ so that the distribution of z_t is centred about $A(t)$ with width of order σ. The variable $A(t)$ can be thought of as an attempt to capture the behaviour of the mean value of z_t, the value once it has been stripped of random fluctuations. We further assume that A is 'quantum' in the sense that the probability distribution for $A(t) = A$ at time t is given by

$$P_t(A) = \frac{|\langle A|\psi_t\rangle|^2}{\langle \psi_t|\psi_t\rangle}, \tag{5.9}$$

where $|\psi_t\rangle$ is a Schrödinger-picture wave function satisfying the Schrödinger equation and where possible values of A are represented by an operator $\hat{A}$ with eigenstates $|A\rangle$ satisfying $\hat{A}|A\rangle = A|A\rangle$. Now let us ask how we should update this probability distribution on the basis of the noisy information z_t at time t. Using Bayes' theorem we have

$$P_t(A|z_t) = \frac{P_t(A) P_t(z_t|A)}{P_t(z_t)}. \tag{5.10}$$

The probability distribution for z_t given A is a Gaussian of width σ centred on A so that

$$P_t(z_t|A) = \frac{1}{(2\pi\sigma^2)^{1/2}} e^{-(A-z_t)^2/2\sigma^2}, \tag{5.11}$$

and the probability distribution for $z_t = A + B_t$ is a convolution of the probability distributions of A and B_t,

$$P_t(z_t) = \int dA \frac{1}{(2\pi\sigma^2)^{1/2}} e^{-(A-z_t)^2/2\sigma^2} \frac{|\langle A|\psi_t\rangle|^2}{\langle \psi_t|\psi_t\rangle}. \tag{5.12}$$

This means that we can write

$$P_t(A|z_t) = \frac{|\langle A|\psi_{t+}\rangle|^2}{\langle \psi_{t+}|\psi_{t+}\rangle}, \tag{5.13}$$

where

$$|\psi_{t+}\rangle = \frac{1}{(2\pi\sigma^2)^{1/4}} e^{-(\hat{A}-z_t)^2/4\sigma^2} |\psi_t\rangle. \tag{5.14}$$

This is the standard dynamical collapse process for a jump operator of the form

$$\hat{J}(z_t) = \frac{1}{(2\pi\sigma^2)^{1/4}} e^{-(\hat{A}-z_t)^2/4\sigma^2}, \qquad (5.15)$$

and this is the form taken in the GRW model (with an $\hat{A}$ operator for each component of the position operator for each particle). The collapse process is mathematically equivalent to a Bayesian updating of the probability distribution on the basis of noisy information about the precise value of the variable A. Furthermore, the unconditioned probability for the variable z_t given in Eq. (5.12) is precisely of the form of the standard collapse model probability $P_t(z_t) = \langle \psi_{t+} | \psi_{t+} \rangle / \langle \psi_t | \psi_t \rangle$.

The CSL model of Eq. (5.6) can be treated in a similar way, and we arrive at the same conclusion by choosing a noisy information field of the form

$$z_t(\mathbf{x}) = A(\mathbf{x}) + B_t(\mathbf{x}), \qquad (5.16)$$

where the field variable $A(\mathbf{x})$ represents the (unsmeared) number density of particles, and $B_t(\mathbf{x})$ is a field of Gaussian noise with spatial correlations described by a covariance function which is related to the smearing function of the CSL model. The noise field expressed here as $z_t(\mathbf{x})$ is a linear transformation of that of the CSL model of Eq. (5.6).

The CSL model of Eq. (5.7) is treated in the same way as the GRW model where now A represents the smeared number density at each randomly sprinkled point $x = (\mathbf{x}, t)$ (see Figure 5.1).

Note that a choice of initial wave function at the beginning of the Universe is essentially a choice of the initial probability distribution for A. Supposing that this

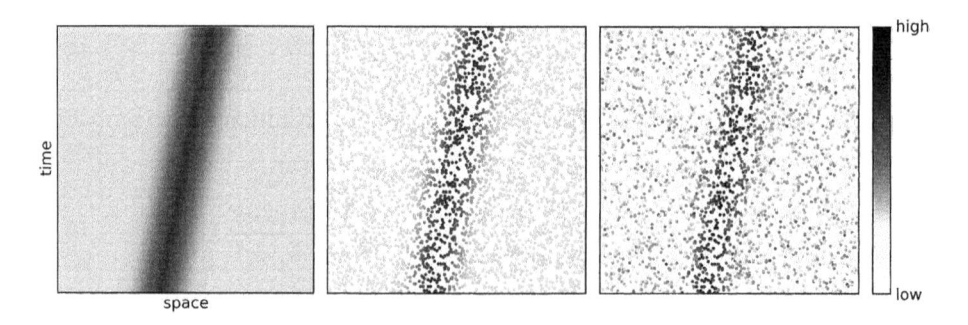

Figure 5.1 A comparison of pictures of matter density in space–time. The left panel displays an object with Gaussian distribution of matter density in space moving with constant velocity to the right. The central panel shows a pointillistic view of the same matter density for a Poisson-distributed sample of points in space–time. The right panel shows the same points but where the matter density values now include a Gaussian noise component—this picture is the natural ontology of the CSL model as expressed in Eq. (5.7).

is an objective chance (perhaps obtained via a best-system analysis) makes the wave function determinate (up to an overall phase) at every time slice provided that we always update for past zs. The unitary dynamics of the wave function defines the probability distribution for A at future times, and it is here that quantum behaviour enters.

In practice we do not know the initial state of the Universe, and we are unable to keep up to date with every single past value of z. However, if we prepare a wave function for a small system on the basis of a suitable measurement we can provide a good approximation to the wave function for performing quantum mechanics in practice. A wave function for which we have ignored the results of previous collapse outcomes (most likely we can only perform the update approximately when we are able to amplify the effects in a quantum measurement) is still a perfectly valid tool for estimating the value of A; it simply rests on incomplete information. Furthermore, there are two clear examples of when a collapse can be ignored: (i) when it happens in a distant region where the quantum state is uncorrelated with the system of interest; and (ii) for small systems where the width of the noise distribution σ (equivalently the width of the collapse operator) is much greater than the uncertainty in A determined by the wave function.

5.2.2 Time Reversal Symmetry

The picture of dynamical collapse presented earlier appears to be inherently time asymmetric. More generally the idea of objective collapse of the wave function is at odds with time symmetry since the wave function at any point in time is shaped by those collapse events that have happened in the past but not in the future (see also Penrose [26] ch. 30.3). Here we will show that there is an underlying time-reversal-symmetric structure [16, 17].

Consider the entire history of the Universe from the initial state $\hat{\rho}_I$ (which we generalise to a density matrix state) at the beginning of time in the past, all the way to a final condition $\hat{\rho}_F$ at the end of time in the future (we treat the final condition as a POVM on the final state; there is no demand that the final state be precisely of the form $\hat{\rho}_F$, and we can have $\hat{\rho}_F \propto \hat{1}$ expressing no constraint). By repeatedly using the probability rule (5.3) we can determine the probability of a given set of time-ordered collapse outcomes $\{z_1, z_2, \ldots, z_n\}$ (of which we assume that there are n of them) occurring at a given set of time intervals $\{\Delta t_0, \Delta t_1, \ldots, \Delta t_n\}$ (Δt_0 is the time interval between the start of the Universe and the first collapse event; Δt_1 is the time interval between the first and second collapse events, etc.; Δt_n is the time interval between the last collapse event and the end of the Universe). We further condition this probability on the initial and final states of the Universe. Given some simple constraints (see what follows), it turns out that this probability is exactly the same as the probability of the reverse sequence of collapse outcomes $\{z_n, \ldots, z_2, z_1\}$ with the reversed set of time intervals $\{\Delta t_n, \ldots, \Delta t_1, \Delta t_0\}$ given the initial condition $\hat{\rho}_F^*$ and final condition $\hat{\rho}_I^*$ [17], or

$$P(\{z_1, z_2, \ldots, z_n\} | \{\Delta t_0, \Delta t_1, \ldots, \Delta t_n\}; \hat{\rho}_I; \hat{\rho}_F)$$
$$= P(\{z_n, \ldots, z_2, z_1\} | \{\Delta t_n, \ldots, \Delta t_1, \Delta t_0\}; \hat{\rho}_F^*; \hat{\rho}_I^*). \quad (5.17)$$

A sufficient condition for this to be the case is that there exists a basis $\{|\phi_i\rangle\}$ in which

$$\langle \phi_i | \hat{U}(t) | \phi_j \rangle^* = \langle \phi_i | \hat{U}(-t) | \phi_j \rangle;$$
$$\langle \phi_i | \hat{J}(z) | \phi_j \rangle^* = \langle \phi_i | \hat{J}^\dagger(z) | \phi_j \rangle. \quad (5.18)$$

For a time-independent Hermitian Hamiltonian this means that there exists a basis in which both $\hat{U}$ and $\hat{J}$ are symmetric (this can be shown to be the case for both the GRW and CSL models where the basis is that to which the collapse occurs [17]). The existence of such a basis allows us to define $\hat{\rho}^*$ by

$$\langle \phi_i | \hat{\rho}^* | \phi_j \rangle = \langle \phi_i | \hat{\rho} | \phi_j \rangle^*. \quad (5.19)$$

So at the level of collapse outcomes $\{z_i\}$ there is time symmetry. The stochastic laws determining the probability of a given complete set of collapse outcomes can be used in either time direction, and in both cases they give the same probability. From a dynamical point of view it doesn't matter which direction of time we use. They both give the same result. Assuming that the collapse outcomes alone (without the wave function) give an empirically adequate description of the world then we can conclude that collapse models are time symmetric.

In terms of the wave function things indeed look time asymmetric. But this is simply because the wave function is a way of encoding and conditioning on all the past collapse events. For example, given the initial condition $\hat{\rho}_I$ and the set of all past collapse outcomes $\{z_i | i < j\}$ at time t_j, then the quantum state is determined and is given by

$$\hat{\rho}_j = \hat{U}(\Delta t_j)\hat{J}(z_j)\hat{U}(\Delta t_{j-1}) \cdots \hat{U}(\Delta t_1)\hat{J}(z_1)\hat{U}(\Delta t_0)\hat{\rho}_I$$
$$\hat{U}^\dagger(\Delta t_0)\hat{J}^\dagger(z_1)\hat{U}^\dagger(\Delta t_1) \cdots \hat{U}^\dagger(\Delta t_{j-1})\hat{J}^\dagger(z_j)\hat{U}^\dagger(\Delta t_j). \quad (5.20)$$

This encapsulates the maximal information about the past necessary to predict future collapse events. The wave function is time asymmetric for the simple reason that it is defined in a time asymmetric way.

We can just as easily define a backwards-in-time quantum state from the final condition of the Universe and all collapse outcomes to the future. This would be suitable for estimating past collapse outcomes. We naturally make use of the forward-in-time picture for the reason that we can keep records of the past but not the future. This can be put down to the special initial condition of the Universe (one that is essentially a clean slate upon which to chalk). Indeed it is the imposition of time-asymmetric initial and final conditions of the Universe that must be responsible for observed asymmetries predicted by collapse models—these include average energy increase.

The argument presented in this section supports the conclusion that we should treat the collapse outcomes as the primitive elements of collapse models. At the level of collapse outcomes the dynamics are fundamentally time symmetric. At the wave function level it is natural that the description is time asymmetric since the wave function is an asymmetrically derived concept representing a state of information about past collapse events.

5.3 Relativistic Structure

The non-relativistic structure presented in the previous section can easily be generalised to one that is explicitly relativistic [27, 28, 29]. Earlier, our use of a time parameter t defined a specific time slice on which the state was specified. In general we would like to be able to consistently specify our state with reference to a time slice t' in a different frame or to specify the state on an arbitrary spacelike hypersurface Σ. The interaction picture [30] allows us to do this: to unitarily evolve the state from a surface Σ to a new surface Σ' (such that $\Sigma \prec \Sigma'$, meaning that no points in Σ' are to the past of Σ; see Fig. 5.2) we use

$$|\Psi_{\Sigma'}\rangle = \hat{U}[\Sigma', \Sigma]|\Psi_{\Sigma}\rangle, \qquad (5.21)$$

where the unitary operator is

$$\hat{U}[\Sigma', \Sigma] = T \exp\left[-i \int_{\Sigma}^{\Sigma'} \hat{\mathcal{H}}_{\text{int}}(x) dV\right], \qquad (5.22)$$

and where $\hat{\mathcal{H}}_{\text{int}}$ is the interaction Hamiltonian (a scalar operator, ensuring covariance), V is the space–time volume measure, and T is the time-ordering operator. There are many different foliations of space–time that include both Σ and Σ', and to ensure that the final result does not depend to the particular choice we use to get from Σ to Σ', we must impose the microcausality condition: $[\hat{\mathcal{H}}_{\text{int}}(x), \hat{\mathcal{H}}_{\text{int}}(y)] = 0$ for spacelike separated x and y. An easy way to see why this must be true is to note that if we advance the hypersurface first at x and then at y we should get the same result as when we do this in the opposite order. The microcausality condition ensures that this works out.

Now we add the collapse events. These are associated to space–time points. By randomly sprinkling the locations of the events in proportion to space–time volume

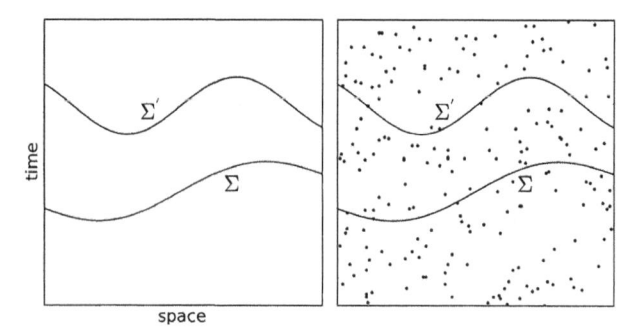

Figure 5.2 Left panel: space–time diagram indicating spacelike hypersurfaces Σ and Σ' with $\Sigma \prec \Sigma'$. Right panel: includes collapse events randomly sprinkled in proportion to space–time volume using a Poisson process.

using a Poisson process (see Figure 5.2) we ensure that the events are distributed in this way in all frames of reference [31]. The construction is therefore Lorentz invariant. We assume that as the hypersurface Σ crosses a point x to which a collapse event is associated the state changes spontaneously according to

$$|\Psi_\Sigma\rangle \to |\Psi_{\Sigma+}\rangle = \hat{J}_x(z_x)|\Psi_\Sigma\rangle, \qquad (5.23)$$

where $\hat{J}_x$ is a scalar operator (for covariance). The random variable z_x has the probability distribution

$$P(z_x|\Psi_\Sigma) = \frac{\langle\Psi_{\Sigma+}|\Psi_{\Sigma+}\rangle}{\langle\Psi_\Sigma|\Psi_\Sigma\rangle} = \frac{\langle\Psi_\Sigma|\hat{J}_x^\dagger(z_x)\hat{J}_x(z_x)|\Psi_\Sigma\rangle}{\langle\Psi_\Sigma|\Psi_\Sigma\rangle}, \qquad (5.24)$$

and the collapse operator $\hat{J}_x$ should satisfy the completeness property

$$\int dz \hat{J}_x^\dagger(z)\hat{J}_x(z) = \hat{1}. \qquad (5.25)$$

Then, as with the non-relativistic case in Section 5.2, the construction is analogous to a generalised quantum measurement ($\hat{J}_x$ corresponds to a generalised measurement operator, z_x corresponds to the outcome). Furthermore, the results of Sections 5.2.1 (collapse process understood as Bayesian inference) and 5.2.2 (time reversal symmetry of collapse outcomes) are straightforwardly generalised to this relativistic scheme. In the relativistic setting, for the same reason that $\hat{\mathcal{H}}_{\text{int}}$ must satisfy the microcausality condition, the collapse operators must also satisfy further microcausality conditions

$$[\hat{J}_x(z_x), \hat{J}_y(z_y)] = 0 \quad \text{and} \quad [\hat{J}_x(z_x), \hat{\mathcal{H}}_{\text{int}}(y)] = 0, \qquad (5.26)$$

for spacelike separated x and y. As demonstrated in detail in Bedingham et al. [29], with these conditions, it follows that: (i) given the set of collapse outcomes

$\{z_x\}$ for all collapse events at locations $\{x|\Sigma \prec x \prec \Sigma'\}$ the final state on Σ' is unambiguously defined by the dynamical process given the initial state on Σ, in particular it is independent of the foliation used to get from Σ to Σ'; and (ii) the probability of a given set of collapse outcomes at given locations $\{x|\Sigma \prec x \prec \Sigma'\}$ is independent of the foliation used to get from Σ to Σ'.

Therefore, the procedure by which we predict the set of collapse outcomes in a space–time region is covariant and does not depend on a particular frame or foliation. The probabilities for collapse events in a given region as detailed in the previous paragraph may depend on the hypersurfaces Σ and Σ'; however, Σ is necessary in order to specify the initial state upon which the probabilities are conditioned, and Σ' is used simply to specify the region of space–time which is of interest. By contrast, in order to invoke a state history it is necessary to choose a particular foliation through which the state evolves. The state histories for two different foliations might be very different. In particular if the states of spatially separated regions are entangled, the local state description for one of these regions can depend significantly on the particular leaf on the foliation passing through it. This does not alter the fact that collapse events are consistent and predictable in a way that is independent of foliation.

It is the collapse outcomes which are invariantly specified, and it is the collapse outcomes that tie together the different state histories on different foliations such that they are really just different descriptions of the same events [32]. Furthermore, treating the quantum state as a state of information about past collapses makes it perfectly natural that it should be dependent on a choice of hypersurface since this is needed to demarcate the past that we are talking about. Relativistic considerations therefore further boost our conjecture that it is the collapse outcomes that are to be seen as primitive and the quantum state as a derivative concept.

It remains only to find a form for $\hat{J}_x$ which satisfies these constraints (scalar + microcausality). For a complex scalar field, for example, a well-motivated choice is [33]

$$\hat{J}_x(z_x) = \left(\frac{\beta}{\pi}\right)^{1/4} \exp\left\{-\frac{\beta}{2}(\hat{\phi}^\dagger(x)\hat{\phi}(x) - z_x)^2\right\}, \qquad (5.27)$$

[cf. Eq. (5.7)]. The effect of this collapse operator is to focus the quantum amplitude associated to the modulus squared of the field operator about the randomly chosen value z_x. Without the unitary dynamics we would expect many such collapse events to gradually bring the state to an eigenstate of $\hat{\phi}^\dagger(x)\hat{\phi}(x)$. In fact the effect of the free unitary dynamics is to disperse the state in competition with the collapse so that typically we would find a stable state with a non-zero variance in the field value.

It is well known that the choice of collapse operator (5.27) is problematic [33]. It is to be expected that collapse models lead to an increase in average energy: for realistic models the collapses lead to localisation, and by the uncertainty principle this results in momentum dispersion and therefore average energy increase. The problem with the present model is that the rate of increase in the energy density is infinite. If $\hat{H}$ is the standard free complex scalar field Hamiltonian, then the expected change in energy for a collapse event occurring at x is

$$\Delta E = \int dz \frac{\langle \Psi_\Sigma | \hat{J}_x(z)[\hat{H}, \hat{J}_x(z)] | \Psi_\Sigma \rangle}{\langle \Psi_\Sigma | \Psi_\Sigma \rangle} = \delta^3(0) \frac{\beta}{2} \langle \hat{\phi}^\dagger(x) \hat{\phi}(x) \rangle, \qquad (5.28)$$

(in the limit in which β is small). A simple resolution follows from the assumption that space–time is fundamentally discrete. Even for classical space–times there are approaches to doing this in a Lorentz-invariant manner [34]. The point is that the discreteness sets a fundamental length scale, a, which means that $\delta^3(0) \sim a^{-3}$, rendering the energy increase finite. In remains to set the parameters of the model in such a way that energy increase is small yet collapse effects are significant enough to provide an explanation of the macro world of our experience. Furthermore, we are not necessarily constrained by the fact that a might be the Planck scale. If for the collapse model to work a larger length scale is necessary, then within a discrete space–time it is conceptually unproblematic to consider a quantum field which only exists on a coarse subset of the space–time lattice. If only this field experiences collapses, then other fields would collapse if they had direct interaction with this field. (We also note that there is an alternative resolution to the problem of infinite energy increase making use of an unusual 'off-mass-shell' quantum field to mediate the collapse process [27, 28, 29].)

For multiple quantum fields we can replace $\hat{\phi}^\dagger(x)\hat{\phi}(x)$ in the collapse operator (5.27) by a sum of equivalent 'mass' terms for each of the fields. Another option could be that each individual field undergoes its own individual collapse process or else that only one field undergoes collapse and others collapse as a result of their unitary interaction and subsequent entanglement with this field.

In summary, the lessons for collapse models from relativity are a consolidation of the idea that it is the collapse outcomes that form the primitive ontology and further that space–time has a discrete structure. In the next section we will try to strengthen the case for these features by arguing that the problems of incorporating gravity might also be resolved by taking this view.

5.4 Including Gravity

A remarkable feature of dynamical collapse, and in particular the picture of collapse presented here with primitive status given to the collapse outcomes, is that it undermines the main arguments for the need for a quantum theory of gravity and that it may enable the formation of a self-consistent theory of classical gravity and quantum matter (see also Tilloy and Diósi [37]). The starting point for any attempt to do this is to try to make sense of Einstein's equation

$$G_{\mu\nu} = 8\pi T_{\mu\nu}, \qquad (5.29)$$

where, given that the left side must be a c-number-valued tensor, the question is what to use for the right side. We cannot simply use the quantum operator for the energy momentum tensor, and so the usual approach is to use the expectation of the quantum-matter energy momentum operator. For collapse models the more natural choice is to build the energy momentum tensor from the collapse outcomes which represent the distribution and flow of matter through space–time. This will clearly be a challenge, and we offer no clear route to doing this here. However, we will analyse some objections to the semiclassical approach to treating gravity and point out how they are avoided by this picture.

The first two issues are due to Eppley and Hannah [38]. The first arises when it is assumed that a gravitational wave can interact with quantum matter without causing the state to collapse. In a simplified version which captures the essential idea, a particle is prepared in a superposition of two spatially separated localised states, each in its own box [39]. Then, by monitoring one of the boxes by scattering gravitational waves from the particle, we can, if the gravitational wave does not collapse the particle state, observe the particle as belonging to a superposition state and detect a change if a measurement is performed on the other box to determine if the particle is located there. This enables signals to be sent at speeds greater than the speed of light.

The reason why this argument does not work for gravity sourced by the collapse outcomes is due to the fact that in order for there to be any influence from matter on gravity, there must be collapses. Only those collapses afflicting the box being monitored will influence the gravitational waves being used to monitor it. The no-signalling theorem can then be used to show that any measurement on one box cannot influence the collapse outcome statistics for the other box and therefore cannot influence the scattering amplitude of a gravitational wave.

The second issue of Eppley and Hannah [38] concerns cases when interaction between a gravitational wave and some matter requires the collapse of the matter state. A gravitational wave could in principle be used to measure the location of some matter prepared in a well-defined momentum state. Since the gravitational wave could have arbitrarily small momentum so that the momentum state of the matter is undisturbed, then both the position and momentum could be known with accuracy.

This argument works on the premise the momentum is conserved. This is simply not true for realistic collapse models, and if the state becomes localised by any means (including interaction with a gravitational wave), then by its wavelike nature, its momentum becomes uncertain.

Further arguments are made by Page and Geilker [35], who point out that if there is no collapse of the wave function, and if gravity is sourced by the expectation of quantum matter energy momentum, then the theory actually contradicts experimental results since it predicts gravitational effects sourced by quantum matter on branches of the wave function other than ours. Their experiment involves quantum mechanical decision and amplification to test gravitational effects from other branches: depending on the decision outcome a large mass is moved and its gravitational field observed.

This issue is clearly not relevant for collapse models where the energy momentum of the quantum state is continually representative of the distribution of energy and matter that we experience, but, Page and Geilker [35] point out, any collapse model of the wave function leads to a contradiction of the Einstein equation since the left side is automatically conserved, $\nabla^\mu G_{\mu\nu} = 0$ (the contracted Bianchi identity), yet the right side will not be conserved under a collapse process. It is a generic feature of realistic collapse models that they do not conserve energy momentum. Collapse models are by nature stochastic, and as pointed out earlier, the spontaneous localising feature is accompanied by jumps in momentum and energy. Furthermore if we are to consider using the noisy collapse outcomes as the basis for the energy momentum source the problem is compounded. The Einstein equation simply won't work with a stochastic energy momentum source.

We have already seen in Section 5.3 that discrete space–time structure offers a possible resolution to the problem of infinite energy increase in relativistic versions of collapse models; there is also an indication that it could resolve this inconsistency of the Einstein equation.

An example of a space–time with discrete geometry is Regge calculus [36], in which a manifold is built from elementary simplexes (e.g. a 2-dimensional simplex is a triangle; a 3-dimensional simplex is a tetrahedron, etc.). Within each simplex the continuous space is flat, and we can define coordinates for any two adjacent cells which are flat throughout the two. The result is a Riemannian structure on all but the boundary of the joints of several simplexes. This is where the curvature is concentrated.

We will only note that the Regge analogue of the contracted Bianchi identity is not precise. In fact only in the limit that the space becomes smooth is it the case that the analogue of the Einstein tensor is conserved [36]. This opens the door to a consistent equation for discrete space–time geometry sourced by stochastic and non-conserved matter.

A promising approach to describing discrete space–time with Lorentz invariance is as a causal set [34], which is a set of points with a causal order relation $\prec$ such that if $x \prec y$ then the point y is to the causal future of x. Discreteness is expressed by the fact that the set is locally finite meaning that between any two causally ordered points $x \prec y$ there is a finite number of points z such that $x \prec z \prec x$. Comparison with a Lorentzian manifold is made using the notion of faithful embedding: a causal set can be faithfully embedded in a Lorentzian manifold if the points of the set can be mapped into the manifold in such a way that the order relation matches the causal ordering of the manifold and the number of set elements mapped into a region is on average proportional to the volume of the region. It is a central conjecture that a given set cannot be faithfully embedded into two manifolds that are not similar on scales greater than the discretisation scale.

The fundamental structure of causal sets does not demand that they can be embedded in any manifold, and to the extent that a causal set is well approximated by a manifold, it need not satisfy the Bianchi identities precisely. Given a causal set that can be faithfully embedded in a Lorentzian manifold one can always add and take away points and order relations [provided that new order relations do not violate the fundamental axioms: antisymmetry ($x \prec y \prec x \implies x = y$), and transitivity ($x \prec y \prec z \implies x \prec z$)], allowing for stochastic deviations of the causal structure and local volume measure with respect to the manifold. The problem for causal sets is rather how the correspondence with a 4-dimensional manifold should arise from all possibilities.

5.5 Summary

We have argued for an understanding of collapse models in which the substance of the world is definite-valued matter density distributed in space and time and where the wave function describes the patterns which relate matter densities at different points. We have shown that the wave function can be understood as a probability distribution for the future matter density conditional on matter density in the past and in which the collapse process corresponds to a Bayesian update on the basis of new matter density information.

We have seen that the distribution of matter density can be explained in a way that is structurally time symmetric and that the time asymmetry of the wave function dynamics is naturally understood as resulting from an asymmetry in the way that the wave function is defined as a conditional probability distribution. Similarly the probability of a given distribution of matter density can be specified with Lorentz invariance and the foliation dependence of a given state history is a natural consequence of the need to demarcate a past–future boundary for the purpose of conditioning on past data.

Finally we have argued that this picture of quantum matter undermines some key arguments in favour of the need for a quantum theory of gravity, raising the possibility of a consistent theory of quantum matter and classical gravity.

Acknowledgements

I would like to thank Chris Timpson, Owen Maroney, and Harvey Brown for useful discussions and comments on an earlier draft. This work was funded by a grant from the Templeton World Charity Foundation.

References

[1] Pusey, M. F., Barrett, J. and Rudolph, T. [2012], 'On the reality of the quantum state', *Nature Physics* **8**(6), 475–478.

[2] Albert, D. Z. [2009], *Quantum Mechanics and Experience*, Harvard University Press.

[3] Bell, J. [1981], Quantum mechanics for cosmologists, *in* 'Quantum Gravity II', Oxford University Press, p. 611.

[4] Bell, J. [1990], 'Against "measurement"', *Physics World* **3**(8), 33.

[5] Bassi, A. and Ghirardi, G. [2003], 'Dynamical reduction models', *Physics Reports* **379**(5), 257–426.

[6] Bassi, A., Lochan, K., Satin, S., Singh, T. P. and Ulbricht, H. [2013], 'Models of wave-function collapse, underlying theories, and experimental tests', *Reviews of Modern Physics* **85**(2), 471.

[7] Ghirardi, G. C., Rimini, A. and Weber, T. [1986], 'Unified dynamics for microscopic and macroscopic systems', *Physical Review D* **34**(2), 470.

[8] Pearle, P. [1989], 'Combining stochastic dynamical state-vector reduction with spontaneous localization', *Physical Review A* **39**(5), 2277.

[9] Ghirardi, G. C., Pearle, P. and Rimini, A. [1990], 'Markov processes in Hilbert space and continuous spontaneous localization of systems of identical particles', *Physical Review A* **42**(1), 78.

[10] Bell, J. S. [2001], 'Are there quantum jumps?', *John S. Bell on the Foundations of Quantum Mechanics*, p. 172.

[11] Goldstein, S. [1998*a*], 'Quantum theory without observers', *Physics Today* **51**(3), 42–47.

[12] Goldstein, S. [1998*b*], 'Quantum theory without observers: Part two', *Physics Today* **51**(4), 38–42.

[13] Allori, V. [2013*a*], On the metaphysics of quantum mechanics, *in* S. LeBihan, ed., 'Precis de la philosophie de la physique', Vuibert, p. 116.

[14] Allori, V. [2013*b*], 'Primitive ontology and the structure of fundamental physical theories', *The Wave Function: Essays on the Metaphysics of Quantum Mechanics* pp. 58–75.

[15] Allori, V. [2015], 'Primitive ontology in a nutshell', *International Journal of Quantum Foundations* **1**(3), 107–122.

[16] Bedingham, D. and Maroney, O. [2015], 'Time reversal symmetry and collapse models', *arXiv preprint arXiv:1502.06830*.

[17] Bedingham, D. and Maroney, O. [2016], 'Time symmetry in wave function collapse', *arXiv preprint arXiv:1607.01940*.

[18] Bell, J. S. [1976], 'The theory of local beables', *Epistemological Letters* **9**(11).

[19] Nielsen, M. A. and Chuang, I. L. [2010], *Quantum Computation and Quantum Information*, Cambridge University Press.

[20] Esfeld, M. and Gisin, N. [2014], 'The GRW flash theory: a relativistic quantum ontology of matter in space–time?', *Philosophy of Science* **81**(2), 248–264.

[21] Esfeld, M. [2014], 'The primitive ontology of quantum physics: guidelines for an assessment of the proposals', *Studies in History and Philosophy of Science Part B: Studies in History and Philosophy of Modern Physics* **47**, 99–106.

[22] Egg, M. and Esfeld, M. [2015], 'Primitive ontology and quantum state in the GRW matter density theory', *Synthese* **192**(10), 3229–3245.

[23] Bacciagaluppi, G. [2010], 'Collapse theories as beable theories', *Manuscrito* **33**(1), 19–54.

[24] Bedingham, D. J. [2011b], 'Hidden variable interpretation of spontaneous localization theory', *Journal of Physics A: Mathematical and Theoretical* **44**(27), 275303.

[25] Tumulka, R. [2011], 'Comment on "hidden variable interpretation of spontaneous localization theory"', *Journal of Physics A: Mathematical and Theoretical* **44**(47), 478001.

[26] Penrose, R. [2004], *The Road to Reality: A Complete Guide to the Physical Universe*, Jonathan Cape.

[27] Bedingham, D. J. [2011c], 'Relativistic state reduction dynamics', *Foundations of Physics* **41**(4), 686–704.

[28] Bedingham, D. [2011a], Relativistic state reduction model, *in* 'Journal of Physics: Conference Series', Vol. 306, IOP Publishing, p. 012034.

[29] Bedingham, D., Modak, S. K. and Sudarsky, D. [2016], 'Relativistic collapse dynamics and black hole information loss', *Phys. Rev. D* **94**, 045009.

[30] Tomonaga, S.-I. [1946], 'On a relativistically invariant formulation of the quantum theory of wave fields', *Progress of Theoretical Physics* **1**(2), 27–42.

[31] Bombelli, L., Henson, J. and Sorkin, R. D. [2009], 'Discreteness without symmetry breaking: a theorem', *Modern Physics Letters A* **24**(32), 2579–2587.

[32] Myrvold, W. C. [2002], 'On peaceful coexistence: is the collapse postulate incompatible with relativity?', *Studies in History and Philosophy of Science Part B: Studies in History and Philosophy of Modern Physics* **33**(3), 435–466.

[33] Ghirardi, G. C., Grassi, R. and Pearle, P. [1990], 'Relativistic dynamical reduction models: general framework and examples', *Foundations of Physics* **20**(11), 1271–1316.

[34] Bombelli, L., Lee, J., Meyer, D. and Sorkin, R. D. [1987], 'space–time as a causal set', *Physical Review Letters* **59**(5), 521.

[35] Page, D. N. and Geilker, C. [1981], 'Indirect evidence for quantum gravity', *Physical Review Letters* **47**(14), 979.

[36] Regge, T. [1961], 'General relativity without coordinates', *Il Nuovo Cimento (1955–1965)* **19**(3), 558–571.

[37] Tilloy, A. and Diósi, L. [2016], 'Sourcing semiclassical gravity from spontaneously localized quantum matter', *Physical Review D* **93**(2), 024026.

[38] Eppley, K. and Hannah, E. [1977], 'The necessity of quantizing the gravitational field', *Foundations of Physics* **7**(1–2), 51–68.

[39] Carlip, S. [2008], 'Is quantum gravity necessary?', *Classical and Quantum Gravity* **25**(15), 154010.

Part II
Ontology

6

Ontology for Collapse Theories

WAYNE C. MYRVOLD

In this chapter, I will discuss what it takes for a dynamical collapse theory to provide a reasonable description of the actual world. I will start with discussions of what is required, in general, of the ontology of a physical theory and then apply these considerations to the quantum case. One issue of interest is whether a collapse theory can be a quantum state monist theory, adding nothing to the quantum state and changing only its dynamics. Although this was one of the motivations for advancing such theories, its viability has been questioned, and it has been argued that, in order to provide an account of the world, a collapse theory must supplement the quantum state with additional ontology, making such theories more like hidden-variables theories than would first appear. I will make a case for quantum state monism as an adequate ontology and, indeed, the only sensible ontology for collapse theories. This will involve taking dynamical variables to possess not sharp values, as in classical physics, but *distributions* of values.

6.1 Introduction

The issue of how to interpret dynamical collapse theories is one that has been much discussed in the literature. As originally construed, such theories were taken to be ones according to which the wave function is everything (it would be better to say: physical reality is exhausted by whatever it is in the physical world that the wave function represents). The viability of this has been questioned, and it has been argued that dynamical collapse theories, in order to afford a description of a world anything like the one we live in, must be supplemented by additional ontology, making such theories more akin to the de Broglie-Bohm pilot wave theory than they might originally have seemed.[1] If one accepts such an argument, then one is left with a choice to make of primitive ontology, resulting in distinct but empirically equivalent theories, depending on choice of ontology.

---

[1] Allori et al. [1] is the *locus classicus*.

This chapter will not attempt to survey these discussions. Rather, I will try to make a case that the appropriate way to think of such theories—indeed, I will argue, the only sensible way—is still quantum state monism. On a quantum state monist approach, there is nothing over and above the quantum state. There must, however, be physical objects such as tables and chairs and laboratory equipment. If the theory is to describe a world in which there are such things, a quantum state monist must adopt the proposal, made early on in these discussions, of revising the way we attribute physical properties to systems, from the strict eigenstate-eigenvalue link to a slightly relaxed criterion on which we can talk of a system possessing a property if the state is close enough to an eigenstate of the operator corresponding to that property. These possessed properties do not, however, exhaust the ontology; a full account of the world, according to these theories, is one on which dynamical quantities almost never have precise values but rather have *distributions* of values, an account that has been developed and advocated by Philip Pearle [2].

We will begin with some general considerations about what it takes for a physical theory to give an account of the world. I will first argue for the need for what Bell called "local beables" and then proceed with an analysis, which will draw heavily on Newton's own thoughts on the matter, of what is required in order for a theory to represent a world of physical objects located in space.

Many of the discussions about these matters have invoked conformity to common sense or to intuition as a means to judge the acceptability of proposals concerning ontology. This, it seems to me, is an inadequate basis for making such judgments. We have no reason to expect either common sense or intuition to be reliable guides to what the world is like or to what the world could be like. Our common sense and intuitions are likely to be shaped by contact with aspects of the world in which quantum phenomena do not reveal themselves and things behave quasi-classically. Further, it should be uncontroversial that quantum mechanics is counterintuitive in *some* aspect or another. And even when it comes to classical physics, our intuitions can lead us astray. Some readers will, I expect, take the Newton-inspired account of physical objects offered in Section 6.3, to violate some of their intuitions about what things are. This should not be taken as an objection; as always, when we find ourselves in the grip of an intuition, we should ask of it whether it can survive scrutiny and be turned into a considered judgments. Any intuitions one might have that are at odds with the account given, I claim, fail to survive such scrutiny.

For an argument for quantum state monism, along different lines but to a conclusion that I take to be compatible with and complementary to the one advanced in this chapter, see Peter J. Lewis's contribution to this volume [3]. For a contrary view, see Roderich Tumulka's contribution [4].

6.2 The Need for Things to Be Met With in Space

What is required for a sensible ontology for a physical theory? We must certainly not presuppose that the world must conform to our prejudices about the way things are. If there is one thing that we have learned from the history of science, it is that science can teach us that the world differs in some fairly drastic ways from what we might have otherwise thought.

We are open to learning that the world is radically different from what everyday experience or the science of ages past might suggest. Does this entail an "anything goes" attitude toward the ontology of physical theory? Not quite. The conclusions of science, radical as they may be, gain their credibility and their right to be taken seriously by virtue of being supported by empirical evidence. The process of gathering such evidence involves manipulation of experimental and observational apparatus and recording of the results of experiments and observations, which are then used as input for theory construction and theory testing (and, of course, this is a dialectical process, as theory informs and guides our experimentation and observation). If, via this process, one arrived at a theory with no room for any such things as experimental apparatus in its ontology, one would find oneself in an awkward position, as the theory would undermine its own empirical base. This would be a peculiar situation. Though there is no *logical contradiction* involved in the supposition that everything that we have taken to be an experimental result is in fact an illusion, we should bear in mind that what we are pursuing is not merely a logically consistent account of the world but one that we have some reason to believe is at least approximately correct in its broad outlines. This cannot be if the theory is empirically self-undermining. To say that a theory is empirically self-undermining does not entail that it is false, but it *does* entail that we do not have good reason to believe in its truth.

Our empirical evidence is based on experience. As Tim Maudlin [5] has emphasized, it does not follow that an adequate physical theory entail facts about experiences of conscious beings. Physics proceeds while the mind–body problem remains a vexed question. Our evidence is based on manipulations and observations of experimental apparatus and, if we are able to represent this apparatus within the theory, at least in a schematic way, so that we are in a position to compare how the apparatus is observed to behave with what the theory would lead one to expect of it, then we are in a position to test the theory empirically and gain evidence for or against the theory.

This is nothing new. Consider, for example, the case of Newtonian gravitational theory, applied to the solar system. The evidence base is observations of planetary positions from various observatories. The theoretical treatment consists of positing a mass distribution—which consists, for example, of representations of the planets and the sun by means of points with which are associated appropriate masses—and studying the evolution of this mass distribution under gravitational interactions. Once this is done, there is still the matter of connecting the theory with observations. There is a superb discussion of this by Howard Stein in a paper entitled "Some Reflections on the Structure of Our Knowledge in Physics."

> Let me underscore the point that there can be no thought of deducing observations within that framework. To do so in the strict sense, one would need to have a physical theory of the actual observer, and to incorporate it into the Newtonian framework. I certainly do not want to say that there is a reason "in principle" why such a thing can never be done, for any possible (future) physical framework; but everyone knows that Newton could not do it, and that we—in the best versions of our own physics—cannot do it. Even waiving the theory of the observer, it is clear that all astronomical observations are intermediated by light; therefore, to deduce anything like observations, one would have to include the theory of light within the framework [6, p. 649].

What we do, instead, is represent the observer (or, more accurately, the observatory) schematically within the theory.

> One represents the observer within the spatio-temporal framework by a world-line (or a system of world-lines). Putting in—for the gravitational theory of the solar system—the world-lines of the planets and satellites, as calculated from suitable initial data, one can then determine at each instant all the relevant angles between lines drawn from the observer to the bodies of the system (including, if the theory is properly handled, with the earth represented as an extended body and its rotation treated systematically, lines from the observer to terrestrial landmarks). As a first approximation, such lines are treated as lines of sight. With more sophistication on the observational side, the results are turned over to the experts in observational astronomy, who will take such account as they are able to of atmospheric refraction, of aberration of starlight, and so on (p. 650).

A few comments about this. First of all, there can be no question of trying to ascertain, from the mathematical structure of the theory alone, what the components of the mathematical structure are supposed to represent. One knows, from the beginning, which world-line is meant to be the world-line of the Sun, which of Jupiter, of the Earth, *etc*. The formalism is from the outset a physically interpreted formalism. Some philosophers have imagined a set of explicit "correspondence rules" providing the link between the formalism employed in physical theory and the physical systems to be represented. There is something right about this, but also something misleading. Physicists' treatments of physical systems typically include verbal descriptions of what it is that the variables introduced into the analysis are meant to represent. But, as Stein himself rightly emphasizes, the connection of the theoretical apparatus to observational practice is never wholly explicit; it is something learned in practice and, for the case of laboratory experiments, in the lab, via actual experience with the relevant instruments.

Second, Newtonian gravitational theory is a theory only of a limited aspect of physical phenomena. As Stein emphasizes, it does not treat of the behavior of light or of the interactions of ponderable matter with light, and, as such, does not treat of the key attribute of the Sun and the planets that permits us to gain information about where they are. But it is an intelligible project to extend the theory so as to include these aspects and to embed Newtonian gravitational theory within a theory that also treats of light and its interactions with matter. Moreover, any theory with any pretensions of being in principle a comprehensive account of the physical world would *have* to include this extension.

At the close of the paper, Stein adds some remarks about quantum mechanics. "In this theory, we just do not know how to 'schematize' the observer and the observation" (p. 653). It might be better to say: we do not know how to schematize experimental apparatus and experimental results. That this is not a straightforward matter within the usual framework of the theory is precisely what gives rise to the so-called measurement problem. More on this in Section 6.4.

6.3 What Is It to Be a Material Body?

6.3.1 The Newtonian Conception of Bodies

It must, therefore, be possible to represent, at least schematically, within a theory that has the potential to be a comprehensive one, the sorts of things that our experimental devices are. Let's pause for a moment and ask: what does it take to be such a thing?

On this matter, there is an interesting discussion, at the very birth of modern physics, by Isaac Newton, in a manuscript entitled "*De gravitatione et æquipondio fluidorum,*" unpublished until 1962. In this manuscript, Newton attacks, among other things, the Cartesian doctrine that the essential attribute of body is *extension*. Newton distinguishes body from extension and, indeed, accepts that there are extensions—that is, extended regions of space—that are not the locations of bodies. What, then, we may ask Newton, is the difference between extended regions that are the locations of bodies and those that are not?

To explain his answer to this question, Newton presents a sort of creation myth.

it must be agreed that God, by the sole action of thinking and willing, can prevent a body from penetrating any space defined by certain limits.

If he should exercise this power, and cause some space projecting about the Earth, like a mountain or any other body, to be impervious to bodies and thus stop or reflect light and all impinging things, it seems impossible that we should not consider this space to be truly body from the evidence of our senses (which constitute our sole judge on this matter); for it will be tangible on account of its impenetrability, and visible, opaque and coloured on account of the reflection of light, and it will resonate when struck because the adjacent air will be moved by the blow.

Thus we may imagine that there are empty spaces scattered through the world, one of which, defined by certain limits, happens by divine power to be impervious to bodies, and *ex hypothesi* it is manifest that this would resist the motions of bodies and perhaps reflect them, and assume all the properties of a corporeal particle, except that it will be motionless. If we may further imagine that that impenetrability is not always maintained in the same part of space but can be transferred hither and thither according to certain laws, yet so that the amount and shape of that impenetrable space are not changed, there will be no property of body which this does not possess. It would have shape, be tangible and mobile, and be capable of reflecting and being reflected, and no less constitute a part of the structure of things than any other corpuscle, and I do not see that it would not equally operate on our minds and in turn be operated upon ...

...In the same way if several spaces of this kind should be impervious to bodies and to each other, they would all sustain the vicissitudes of corpuscles and exhibit the same phænomena. And so if all this world were constituted of this kind of beings, it would hardly seem to be any different in character. And hence these beings will be either bodies or like bodies. If they are bodies, then bodies can be defined to be *determinate quantities of extension which the omnipresent God affects with certain conditions*: these are (1) that they be mobile (and therefore I have not declared them to be numerical parts of space, which are strictly immobile, but only definite quantities which may be transferred from space to space). (2) That two such be unable in any part to coincide, or that they be impenetrable and so when by their motions they meet they obstruct one another and are reflected in accordance with certain laws. (3) That they be able to excite various perceptions of the senses and the fancy in created minds, and in turn to be moved by the latter .... [7, pp. 139–140].

Some readers, familiar with the all-too-common attribution of space–time substantivalism to Newton, will be tempted to gloss Newton's ontology as: the basic ontology is space, with the properties that are characteristic of bodies, impenetrability, opacity, and the like, being attributes of some regions of space. As if anticipating this reading, Newton takes pains to make it clear that he rejects it.

Perhaps now it may be expected that I should define extension as substance or accident or else nothing at all. But by no means, for it has its own manner of existence which fits neither substances nor accidents. It is not substance...

...although philosophers do not define substance as an entity that can act upon things, yet all tacitly understand this of substances, as follows from the fact that they would readily allow extension to be substance in the manner of body if only it were capable of motion and of sharing in the action of body. And on the contrary they would hardly allow that body is substance if it could not move nor excite in the mind any sensation or perception whatever [7, pp. 131–132].

Space itself is not a substance. Where the requisite dynamical conditions hold, there is a material body, and Newton explicitly denies that those conditions must inhere in some substantival substrate, itself devoid of properties, which Newton rejects as unintelligible.

for the existence of these beings it is not necessary that we suppose some unintelligible substance to exist in which as subject there may be an inherent substantial form ....

Between extension and its impressed form there is almost the same analogy that the Aristotelians posit between the *materia prima* and substantial forms, namely when they say that the same matter is capable of assuming all forms, and borrows the denomination of numerical body from its form. ... They differ, however, in that extension (since it is *what* and *how constituted* and *how much*) has more reality than *materia prima*, and also in that it can be understood, in the same way as the form that I assigned to bodies [7, pp. 140–141].

A few comments about this "creation myth" of Newton's are in order. First, however central the agency of God was to Newton's thought about the physical world, reference to divine decree is not essential to the account of body that he presents. What matters is that it be a matter of physical law that the requisite conditions hold in the relevant regions, whether or not one takes those laws to issue from divine decree. Second, though Newton talks as if there is an absolute distinction between rest and motion, nothing in the account requires this, which is equally at home in Galilean or Minkowski or some other relativistic space–time.

Third, the various dynamical properties, such as opacity, reflectivity, and impenetrability, can be separated. For example, a body need not be impermeable, or may be permeable to some sorts of things and not others, or permeable to various degrees to various sorts of things. Though Newton took it for granted Jupiter and Saturn are impenetrable and seemed to assume that his readers would, too,[2] we now know that they are giant balls of gas. They nonetheless share with other extended bodies the properties of reflectivity and mobility.

Fourth, as illustrated by the cases of Jupiter and Saturn, the region occupied by a body need not have sharp boundaries. Sharp as the visual boundary of Jupiter may seem to be through your favorite telescope, the upper atmosphere does not a precise upper boundary, and any precise delimitation of it would have to involve an element of convention. One might, perhaps, imagine that the atmosphere is composed of smaller parts, atoms or molecules, which themselves have precise regions that they occupy, and that any ambiguity in the boundary of Jupiter has to do only with which of these well-located particles to include as its parts. But there is no necessity for bodies to be composed of parts like this. All of the dynamical properties that make a body a body are a matter of degree, and it would not detract at all from their status as bodies if the intensity of these properties did not go abruptly to zero outside a precisely defined region but, rather, dropped off sharply, quickly becoming negligible outside a region that is only specifiable within some tolerable limits of fuzziness.

Fifth, though Newton expresses some doubt about whether the bodies that make up our world are the kind of being he describes, he rightly says that, if this world were constituted of this kind of being, it could hardly be said to be any different in

---

[2] See the remarks to the Third Rule for the Study of Natural Philosophy in the *Principia*: "We find those bodies that we handle to be impenetrable, and hence we conclude that impenetrability is a property of all bodies universally" [8, p. 795]. This is explicitly extended to the heavenly bodies in the last paragraph of the remarks on Rule 3.

character. Mobile quantities of extension endowed with the right sorts of dynamical properties *would be* bodies, whatever else might be true of them, and, without the right sort of dynamical properties, nothing could be said to be a body or to have a location in space.

With these qualifications, there seems to be something deeply right about Newton's account of body. *To be a body is to have a certain sort of place in a network of dynamical relations.* A dynamically inert substrate for these properties is neither needed nor intelligible. Moreover, these considerations are independent of the details of physical theory and may justifiably be taken as a constraint on any physical theory that has any pretense to include in its ontology things to be met with in space.

One consequence of the Newtonian view of a physical body is worth mentioning. One might be tempted to consider, as a skeptical hypothesis, a supposition that there is, in fact, no table in front of me, merely a region of empty space that, by God's decree, looks and feels in all ways like a table. Our Newton-inspired conception of physical body declares this to be nonsensical. There is nothing more to being a table than to act in every way like a table.

6.3.2 Is Talk of "Primitive Ontology" Helpful?

In an influential and in many ways insightful paper, Allori et al. [1] argue that the ontology of any physical theory can be partitioned into *primitive ontology*, which is meant to be what the theory is primarily about, and nonprimitive ontology, whose role it is to provide dynamics for the primitive ontology.

I understand the motivation for talk of this sort in the context of Bohmian mechanics. This theory is primarily about the motion of the Bohmian corpuscles, and the role of the wave function is to guide the motion of these corpuscles. It is less clear to me that all physical theories fall naturally into this mode. And there seem to me to be some risks associated with this way of thinking. One risk is to fall into thinking that the local beables that make up ordinary objects must be part of the fundamental ontology of the theory. This is not a condition placed on the ontology in Allori et al. [1], but it is a natural extension of this way of thinking, and it *is* taken as a requirement in Allori [9]. Another risk is regarding the non-primitive ontology as unreal. Again, this is not a step by taken in Allori et al. [1], but it *is* taken in Allori et al. [10, p. 332].

The most serious danger, though, it seems to me, is the danger of sliding into thinking that there is a two-step process. *First*, one posits an ontology, with no dynamical assumptions, that is, no assumptions about how it behaves, and *then* one posits a dynamics for it. This suggests that it could make sense to speaking of something being present that is dynamically inert. This, again, is not a pitfall that Allori et al. [1] fall into, though they come close when they say, "The wave function in each of these theories, which has the role of generating the dynamics for the PO [primitive ontology], has a nomological character utterly absent in the PO" [1, p. 363]. This is, however, a step taken by Esfeld [11] and by Esfeld et al. [12], who posit "primitive stuff, *materia prima*, having no physical properties at all" (p. 135), matter points possessing only metrical relations to each other.

The problem with this is that, until we have said something about how the purported ontology acts, we haven't yet given sense to the claim that it is there at all. What it is for an object to occupy a region of space or, indeed, to have any sort of spatial relations to anything is for it to do something there—exclude other objects or reflect light or something of the sort.

There is a potentially misleading analogy with classical planetary physics. Eudoxus, Ptolemy, Copernicus, Tycho, Kepler took the existence of the planets for granted and sought to learn how they move, and it was Newton who finally gave us dynamical laws for their motions. One is tempted to proclaim, *existence precedes dynamics!*

The reason this is misleading is obvious from consideration of how locations and motions were attributed to the planets in the first place. It was taken for granted by all those researchers that the planets were located in the regions from which they reflected light. That is, certain aspects of the way that these bodies interact with things were used to attribute to them spatial relations to other things, and then these spatial relations were used to investigate the ways in which their locations (or, rather, relative locations) changed with time. A dynamically inert object could not be said to be located anywhere or to have distances from other dynamically inert things.

6.4 Schematizing Experimentation in Quantum Mechanics

Now let us take a look at how one might go about schematizing experiments and their results, within quantum mechanics, with an eye to what goes wrong.

First, a few words about how one sets up a quantum representation of a system, say, a hydrogen atom, to invoke the theory's first serious physical application.

One first identifies some dynamical variables of the system of interest, for example, position and momentum of the electron and of the proton that make up a hydrogen atom. One then associates with these variables operators—Dirac called them "q-numbers"—on which are imposed the appropriate commutation relations. One then constructs a Hilbert space for the operators to act upon. A (pure) state of the system in which it has a given value of a dynamical variable is represented by a vector that is an eigenvector of the corresponding operator. This is readily extended to mixed states; a mixture of pure states that are represented by eigenvectors of an operator with a common eigenvalue is a state in which the system has the corresponding value of the dynamical variable. This interpretational rule, which is half of the *eigenstate–eigenvalue link*, is the primary means by which a Hilbert space vector can be taken to indicate anything at all about a physical system. Thus, for example, if a hydrogen atom is in a state in which its total energy is E, that fact is indicated by taking the vector that represents its state to be an eigenvector, with eigenvalue E, of the operator associated with the total energy of the system.[3]

It is useful to distinguish between the two halves of the eigenstate–eigenvalue link. Let us call the conditional, "If the state of a system is an eigenstate of an operator corresponding to some dynamical quantity, the system possesses the corresponding eigenvalue" the *positive eigenstate–eigenvalue link*, and its converse, obtained by replacing "if" with "only if," the *negative eigenstate–eigenvalue link*. The negative eigenstate–eigenvalue link leads to trouble; the positive link is indispensable for interpreting the quantum formalism as any sort of physical theory at all.

Some experiments double as preparation procedures. These are the *repeatable* experiments, those that, if repeated on the same system, are sure to yield the same result. It is natural to suppose that, after such an experiment, the system indeed possesses the corresponding value of the variable in question. And, indeed, this was introduced as an explicit postulate, the collapse postulate, in von Neumann's *Mathematische Grundlagen der Quantenmechanik* (1932), whence it made its way into the second edition (1935) of Dirac's *Principles of Quantum Mechanics* (it does not appear in the first edition of 1931). In Dirac's formulation, "a measurement always causes the system to jump into an eigenstate of the dynamical variable that is being measured, the eigenvalue this eigenstate belongs to being equal to the result of the measurement" [16, p. 36].

How are we to schematize the process of performing an experiment in quantum theory? We schematize the experimental apparatus as a quantum system with at least as many distinguishable states as there are distinguishable outcomes of the experiment. Among its dynamical degrees of freedom there is a "pointer observable" whose role it is to indicate the outcome. We assume an interaction between the system of interest and the apparatus that is such that, if the initial state of the system of interest is an eigenstate of the quantity to be "measured," then the apparatus will end up indicating it, which is taken to mean: the apparatus will end up in the appropriate eigenstate of the pointer observable. And therein lies the problem. Unless we assume a collapse postulate, the interaction between the system and

---

[3] David Wallace [13, p. 215] has claimed that "the E-E link has nothing much to do with quantum theory" and that "it seems to be purely an invention of philosophers" [14, p. 4580]. Gilton [15] has decisively shown, with ample documentation of the relevant texts, that this is false.

apparatus, when the initial state of the system is *not* an eigenstate of the quantity to be "measured," will lead to an entangled state of system and apparatus that is not an eigenstate of the pointer observable. The eigenstate–eigenvalue link leaves it somewhat mysterious what we are to make of such a state. The prescription that, if the state of the apparatus is an eigenstate of the pointer observable, the value of the pointer observable is the corresponding eigenvalue, tells us nothing about what to say when the state is not an eigenstate. If we add the converse, "only if," then we conclude that the pointer is not pointing to any result and that experiments don't have outcomes, which is hard to make sense of. We can, of course, deny that the pointer observable has a definite value *only if* the apparatus's quantum state is an eigenstate of the pointer observable; this leads us to posit structure beyond the quantum state.

The quantum state monist, if she is to hold on to the usual assumption that experiments have unique definite outcomes, must modify the dynamics so as to avoid states in which the pointer fails to point. This is where the collapse postulate comes in. A dynamical collapse theory, taken as a candidate for a solution to the measurement problem, is meant to be a substitute for the problematic versions of the postulate found in textbooks, a substitute that avoids the textbook postulate's invocation of "measurement" or "experiment" as a primitive concept and avoids also the problematic assumption of two distinct sorts of processes, those that transpire during a measurement and those that transpire at all other times. That is, a theory of this sort is meant to provide, to borrow from GRW's title [17], unified dynamics for microscopic and macroscopic systems.

6.5 Ontology for Ideal Collapse Theories

Let us entertain, for the duration of this section, the fiction that we could construct a dynamical collapse theory that succeeded in producing and maintaining eigenstates of some physical quantity. Which variables should these be in order for the theory to be capable of yielding a world full of things to be met with in space? And, if we could choose an appropriate quantity, would this be sufficient, or would there still be something missing from the world?

Ghirardi et al. [18] argue that a smeared mass density fills the bill. To avoid complications arising from relativity, suppose we have a nonrelativistic quantum field theory (that is, it permits particle creation and annihilation), containing particle types that may be indicated with an index k (associated with each type is a characteristic mass, charge, spin, *etc.*). Let $\hat{\psi}_k^{\dagger}(\mathbf{x}, t)$, $\hat{\psi}_k(\mathbf{x}, t)$ be creation and annihilation operators for particles of type k, satisfying the usual bosonic or fermionic commutation or anticommutation relations, and define particle number density operators by

$$\hat{N}^{(k)}(\mathbf{x}, t) = \hat{\psi}_k^\dagger(\mathbf{x}, t)\, \hat{\psi}_k(\mathbf{x}, t). \tag{6.1}$$

We can use these to define smeared number operators. Given a nonnegative, real-valued function g with

$$\int_{-\infty}^{\infty} d^3\mathbf{x}\, g(\mathbf{x}) = 1, \tag{6.2}$$

we define

$$\hat{N}_g^{(k)}(\mathbf{r}) = \int_{-\infty}^{\infty} d^3\mathbf{x}\, g(\mathbf{x} - \mathbf{r}) \hat{N}_k(x). \tag{6.3}$$

To get a feel for this, if $g(\mathbf{x})$ is taken to be constant within a sphere of radius a centred on the origin and zero outside it, then $\hat{N}_g^{(k)}(\mathbf{r})$ would be the operator corresponding to type-k particle number in a sphere of radius a. That is, its expectation value, in any quantum state, would be the expectation value of the result of an experiment that counts the number of k-type particles in that sphere. An eigenstate of this operator would be one in which there is a definite number of k-type particles in the sphere.

Given a smearing function g, one can define smeared mass density operators by weighting particle numbers with their masses and summing over all particle types:

$$\hat{M}_g(\mathbf{r}) = \sum_k m_k \hat{N}_g^{(k)}(\mathbf{r}). \tag{6.4}$$

There are principled reasons for choosing variables of this sort as the ones that a collapse theory will tend to make definite; see Ghirardi et al. [18] and Bassi and Ghirardi [19] for discussion. The mass density should be smeared over distances small on a macroscopic scale but large compare to atomic dimensions to avoid excessive narrowing of wave packets to keep the unavoidable energy increase that accompanies collapse within acceptable limits.

For a toy version of a smearing of this sort, partition all of space into disjoint cubes with sides of a length a, a length that is small on the macroscopic scale but large compared to atomic dimensions—say, 10^{-5} cm, which is small on the human scale of things but is nevertheless 10,000 times the Bohr radius, and, for each of these cubes, define a smeared mass density operator via a function that is uniform over that cube. Let us imagine a dynamical collapse theory whose stable states are eigenstates of mass density smeared uniformly over these cubes and tolerate superpositions of distinct values of these smeared mass densities for time intervals that are minuscule on the human time scale. Could this theory describe a world of macroscopic objects anything like our own?

The answer, I think, is yes, but some argumentation is required to reach this. There is a temptation to think the answer is simple. The theory will include stable states in which there is a table-shaped region in this room of mass density higher than its surroundings and might be tempted to say that we can simply declare that, according to the theory, that is what it is to be a table. But this is too quick, because we need assurance that this table-shaped region has the right sort of dynamical properties to make it a table. Does it reflect light? Of what frequencies? What happens if I place my mug-shaped region of high mass density on the upper boundary of the table-shaped region?

These questions have some urgency because the dynamical properties of a table—its stability, its capacities to reflect light and to support mugs, and the like—are chiefly a matter of its electromagnetic interactions with other things and to specify a mass density, and to say nothing more is to say nothing about electromagnetic properties. One could consider also smeared charge densities, but, since our cell size is large compared to atomic dimensions, for ordinary matter, charge densities smeared over such regions will be zero or near zero. Nonetheless, the electromagnetic properties of ordinary objects are crucial to their being what they are. A table-shaped region of high mass density composed entirely of electrically neutral matter would not hold its shape except momentarily, nor would it reflect light or support a mug, which would pass right through it without resistance.

An analysis that pays heed to the details of the quantum state yields the correct answer. A table is not merely a mass density located in a table-shaped region, but a structure composed of nucleons and electrons in bound states, forming atoms, which are bound together in a more-or-less rigid configuration. Our quantum state, which, by supposition, localizes a mass corresponding to definite integral multiples of particle masses in each of these cubes, will be one such that includes many-body correlations between the positions of the various constituents. Even if the location of a single nucleon is somewhat indefinite within the cube, it will be correlated with the positions of the other nuclei that make up the atomic nucleus of which it is a part, and the position of the nucleus of the atom will be correlated with its electrons. Thus, even though the smeared charge density might be zero within a cube, and even if there is not a definite matter of fact about the charge distribution within the cube, it *will* be a matter of fact that the charge is not *uniformly* zero within the cube and that positive charges corresponding to atomic nuclei are surrounded by clouds of negative charge. This means that a photon that impinges on a table-region will interact primarily with outer-shell electrons and be absorbed or reflected. It also means that interactions between my mug-shaped region and the table-shaped region will be dominated by outer-shell electrons and that they will repel each other at short range.

Thus, the image we get from the theory, of regions of high mass density that behave much like we expect classical objects to behave, depends heavily on features that we regard as peculiarly quantum. In order for there to be stable atoms and molecules, let alone stable macroscopic objects, it is necessary for electron wave functions to be spread out over atomic and molecular distances. It also depends on quantum multiparticle correlations; though the nucleus of an atom might have its center-of-mass wave function spread out over a region of the size of the localization scale, there will be correlations on a finer scale between the positions of the individual nucleons and between the positions of the nucleons and their accompanying electrons.

Upshot of this: A collapse theory that yielded eigenstates of mass density smeared over regions that are small on the macroscopic scale but large enough to permit internal goings-on at the atomic scale to proceed unmolested would, indeed, solve the problem of yielding a description of a world containing things to be met with in space, but this is dependent on the physical reality of features of the quantum state that go well beyond the mass density it represents. The other features of the world represented by the quantum state are an integral part of the ontology.

6.6 Ontology for Near-Collapse Theories

6.6.1 The Fuzzy Link

In the previous section, we entertained the fiction of a collapse theory that yielded eigenstates of the mass content of the elements of a partition of space into cubes that are small on the macroscopic scale and argued that, in conjunction with other aspects of quantum theory, a theory like that can, indeed, yield an adequate description of a world containing things to be met with in space. On such a theory, there is a matter of fact about how much mass there is in each of the little cubes. The theory admits states in which there are regions of high mass density that have the requisite dynamical properties to count as bodies in the Newtonian sense.

Instead of this fictional theory, let us consider the actual Continuous Spontaneous Localization theory (CSL; Pearle 20, Ghirardi et al. 21, pp. 79–89). This differs in two ways from our fictional theory. One is that, instead of a uniform smearing function over a precise region, Ghirardi, Pearle, and Rimini employ a smooth smearing function, a spherical Gaussian, characterized by a width a. The corresponding smeared number operators, which we will call simply $\hat{N}^{(k)}(\mathbf{r})$, can be thought of as an average number of particles in an imprecisely defined region of roughly size a^3. The other difference is that the theory does not produce eigenstates of the corresponding smeared mass-density operators but, instead, tends to suppress superpositions of states in which there is a substantial spread in the value. That is, it will tend to suppress (but not completely eliminate) superpositions of states that correspond to differences in these smeared mass densities.

To get a feel for this, consider two macroscopic regions, A and B, separated by a macroscopic distance. Let $|\psi_A\rangle$ be a state in which there is a large number N of atoms mostly localized in A. That is, the many-particle wave function for this state, $\psi_1(\mathbf{x}_1,\ldots,\mathbf{x}_N)$, has nonnegligible amplitude only when all of $\mathbf{x}_1,\ldots,\mathbf{x}_N$ are in A. Let $|\psi_B\rangle$ be a state in which the same number of atoms are localized in B. The states $|\psi_A\rangle$ and $|\psi_B\rangle$ could be states that are close to eigenstates of the smeared mass density operators everywhere, but an equally weighted superposition of the two or any superposition with nonnegligible coefficients for both of these would not be close to an eigenstate of $\hat{N}(\mathbf{r})$ for $\mathbf{r}$ within the respective regions in which the states $|\psi_A\rangle$ and $|\psi_B\rangle$ localize the particles. A state like that would tend to collapse into one in which most of the mass density is in either A or B.

Most, but not all; the theory produces only approximations to eigenstates of the smeared mass density operators. The fact that collapse theories do not produce eigenstates of anything like our familiar physical quantities, but at best approximations to them, has been called the "tails problem."[4] This feature of collapse theories motivated, on the part of the developers of such theories, a modification of the eigenstate–eigenvalue link.

if one wishes to attribute objective properties to individual systems one has to accept that such an attribution is legitimate even when the mean value of the projection operator on the eigenmanifold associated to the eigenvalue corresponding to the attributed property is not exactly equal to 1, but is extremely close to it (Ghirardi et al. 26, p. 1298; see also Ghirardi et al. 27, pp. 109–123, Ghirardi and Pearle 28, pp. 109–123, Ghirardi et al. 18, pp. 5–38).

This modification has been dubbed, by Clifton and Monton [29, pp. 697–717], the *fuzzy link*.[5]

A state $|\psi\rangle$ is an eigenstate of an operator $\hat{A}$ if and only of the variance of $\hat{A}$ in that state,

$$\mathcal{V}(\hat{A}) = \langle\psi|\hat{A}^2|\psi\rangle - \langle\psi|\hat{A}|\psi\rangle^2, \qquad (6.5)$$

is equal to zero. One way to estimate closeness to being an eigenstate of some variable is to use the criterion that its variance be small. Ghirardi et al. (18, pp. 5–38) apply this criterion to mass densities. They say that the smeared mass density in the vicinity of a point $\mathbf{r}$ is *objective* if and only if the variance of $\hat{M}(\mathbf{r})$ is small compared to the square of its mean value:

---

[4] This was first flagged as a problem by Shimony [22] and Albert and Loewer [23].
[5] Peter Lewis [24, pp. 86–90] distinguishes between a *fuzzy link*, according to which there is some precise threshold p such that a system possesses the property $A = a$ if and only if $\langle\Psi|P_a|\Psi\rangle > 1 - p$, and a *vague link* according to which possession of a definite property is a matter of degree. It is hard to imagine what arguments there could be (or even what it might mean) for there to be a precise threshold. [25] argue, correctly in my opinion, that there could be no such precise threshold and that the modified link must be somewhat vague. This is what I mean by a fuzzy link.

$$\frac{\mathcal{V}(\hat{M}(\mathbf{r}))}{\langle\psi|\hat{M}(\mathbf{r}))|\psi\rangle^2} \ll 1. \tag{6.6}$$

In some later works [30, 31], the mass density is said to be "accessible" if this criterion is satisfied.

How close is close enough? Obviously, to pick a precise fixed threshold would be every bit as arbitrary as picking a precise upper limit to Jupiter's atmosphere. The dynamics of the theory should be our guide, as well as our capacities to discriminate between quantum states. If the spread of some physical quantity is too small to be discernible, then the state's behavior will not be appreciably different from that of an eigenstate, and we have license to disregard the difference. Ghirardi et al. [18] justify the criterion by arguing that, when the criterion (6.6) is satisfied, things behave as if there is a classical mass in the smearing region equal to the expectation value of the smeared mass density.

This proposal is an application of the fuzzy link to the smeared mass densities. That is, it is a proposal about when, in the context of a dynamical collapse theory, we should regard the state as one in which we should say that there is a physical quantity that is sufficiently well defined as to have, in effect, an objective value. It remains an interpretation on which "*the statevector is everything*" (Ghirardi and Grassi 30, p. 368; see also Ghirardi 31, p. 364, Ghirardi 32, p. 233).

In a world like this, crucial physical quantities, such as the amount of mass in a given region of space, fail to have precise values, though they may come very *close* to having a precise value, in the sense of satisfying the accessibility criterion. Some readers will balk at this. A proposal of this sort, it might be said, offends intuition or common sense. However, as we have said already in the introduction, we have no reason to take common sense or intuitions to be reliable guides as to how the world is. We should ask: can a case be made that, as a matter of conceptual necessity, the world *cannot* be like that?

Any attempt at such an argument is doomed to fail. It seems to me that anyone who accepts the following propositions, which strike me as incontrovertible, should accept that the world could, indeed, be like that.

- To be a physical body is nothing more and nothing less than to have a certain place in a network of dynamical and causal relations of the sort discussed in Section 6.3.
- An ideal collapse theory, if there could be one, on which sufficiently large regions of space contained perfectly definite quantities of mass, would, provided that the dynamics of the theory underwrote the right sort of dynamical behavior of these masses, be capable of delivering a sensible description of a world much like ours.
- Given the dynamics of a theory like CSL, a small difference in a state makes only a small difference to dynamical behaviour.

This is an application of a more general Principle of Continuity for metaphysics. If certain states of a physical theory are capable of bearing some metaphysical burden, then a slightly different state is capable of bearing the same burden. If this continuity principle were not accepted, it seems to me, then we would not be capable of engaging in metaphysics at all, as any physical theory we have ever formulated, and any physical theory that we are likely to formulate in the foreseeable future is at best some sort of approximation to a deeper, more fundamental theory. And, even if we were handed an exact theory on tablets of stone, we would still only know the physical states of things within a certain degree of approximation.

This perspective on ontology—that it is really how things interact with other things that counts—suggests a natural choice for the variables that the collapse theory will tend to make definite. As interactions can be represented by terms in a complex system's Hamiltonian, a natural choice would be energy density or, in a relativistic context, the stress-energy tensor. In a nonrelativistic approximation, most of a massive object's energy is in its rest mass, and so the energy density would approximate a mass density, and arguments that a mass density makes for an adequate ontology would carry over.

6.6.2 Adding Primitive Ontology

There is in the literature another option, which has become more popular in recent years, for interpreting collapse theories.

For any state $|\Psi\rangle$, whether or not it is at all close to an eigenstate of any of the operators $\hat{M}(\mathbf{r})$, one can define the *expectation value*.

$$m(\mathbf{r}) = \langle \Psi | \hat{M}(\mathbf{r}) | \Psi \rangle. \tag{6.7}$$

If one is dissatisfied with the fuzzy link, which involves vagueness at the level of basic ontology, an alternative is to postulate, *as additional ontology over and above the quantum state*, a mass density that always exists and always has the value, any time and place, given by (6.7). This can be thought of as introducing the mass density as primitive ontology.[6] This option has the advantage of being more in accord with classical intuitions. The mass density, always definite, is like a classical mass density, with the chief difference that it is subject to unclassical dynamical laws. There's no question, intuition tells us, of what we're talking about on a theory with this primitive ontology.

---

[6] As far as I can tell, this proposal first appears, explicitly, in Goldstein [33]. It is discussed extensively in Allori et al. [1]. It should be mentioned that Ghirardi has adopted this view of the status of the mass density and now regards it as "an additional element which need to be posited in order to have a complete and consistent description of the world" [34, p. 2907].

This, it seems to me, is an illusion. The reason is, essentially, precisely the reason that led Ghirardi and coauthors to distinguish between mass densities that are *objective*, or *accessible*, and those that are not.

The relation between quantum states and mass density functions is obviously many–one. A mass density function is uniquely determined by a quantum state, but the same mass-density function can correspond to many different quantum states. This is illustrated by an example discussed by Ghirardi et al. [18]. Consider two spherical regions, A and B, and consider two quantum states. The first state to be considered, $|\Psi^\oplus\rangle$, is an equally weighted superposition of two states, $|\Psi_N^A\rangle$ and $|\Psi_N^B\rangle$. In $|\Psi_N^A\rangle$, a large, even number of particles, of total mass M, are all located (within the tolerances permitted by the theory) in the region A and evenly distributed within it. In $|\Psi_N^B\rangle$, the same holds of B. The state $|\Psi^\oplus\rangle$ is,

$$|\Psi^\oplus\rangle = \frac{1}{\sqrt{2}} \left(|\Psi_N^A\rangle + |\Psi_n^B\rangle \right). \tag{6.8}$$

The second state to be considered, $|\Psi^\otimes\rangle$, is one in which each of the regions A, B contains half of the particles.

$$|\Psi^\otimes\rangle = |\Psi_{N/2}^A\rangle \otimes |\Psi_{N/2}^B\rangle. \tag{6.9}$$

The two states yield the same mass-density function, taking on a value corresponding to $\frac{1}{2}M$ in A and $\frac{1}{2}M$ in B. But there is a difference between the physical conditions of the regions in those two states. This can be seen by the behavior of a test particle projected along a line that passes directly between the two regions and which is attracted gravitationally to the masses within those regions. In state $|\Psi^\otimes\rangle$, the particle, being attracted equally by both masses, passes undeflected. In state $|\Psi^\oplus\rangle$, by contrast, even though the mass density takes on the same values that it does in state $|\Psi^\otimes\rangle$, the reaction of the test particle is very different. If initially well localized, its state will become entangled with that of the other masses, so that the state of the whole system ends up being a state that is a superposition of two terms, in one of which the test particle is deflected upward and in the other deflected downward.

In state $|\Psi^\oplus\rangle$, do each of the two regions contain a mass $\frac{1}{2}M$? It is not enough to simply declare this to be true; not even God could create masses in those regions in the absence of physical laws that ensure that they act like masses. And, in a state like $|\Psi^\oplus\rangle$, other things do not act as if there are those half masses in those regions. Something that you might be inclined to call a "mass," if it doesn't act like a mass, is not a mass. There is no sense in which, in a state like II, regions A and B each contain mass $\frac{1}{2}M$.

Only when the function defined by (6.7) acts like a mass density are we justified in calling it one. And this is only when it satisfies the criterion of "accessibility."

But this means that *nothing is added* by taking there to be additional ontology, beyond the quantum state. The mass density m acts like a mass density only when the accessibility criterion is satisfied, and, when it does, the quantum state dynamics alone ensure that everything behaves as if a mass density were present. I am for this reason in agreement with Ghirardi and Grassi when they write,[7]

> For the reader who has followed the above analysis, it should be clear that, within CSL, physical systems are fully described by the state vector which evolves according to equations embodying genuine elements of chance. However, even though *the state vector is everything* ..., since the dynamics itself tends to make objective the mass density, it turns out to be appropriate to relate the kinematical elements in Bell's sense to the mass density distribution.
>
> ... In a sense we could state that the exposed beables of the theory are the values of the mass density function and that, at the appropriate level, the dynamics allows one to take them seriously. All other properties which emerge as a consequence of this process derive from this basic dynamical feature concerning the mass density function [30, p. 368].

Though the mass density function is singled out by the dynamics as the one that is almost always "accessible," it does not exhaust what there is in the world. Other quantities, such as charge densities, could get carried along with the mass density, and, when they satisfy the criterion of accessibility, they are no less real than the mass density.[8] A region of space that acts as if it contains a certain charge *does* contain that charge; no sense can be made of a claim that only the masses are real and that other physical quantities have an inferior ontological status. Their reality may be emergent as a consequence of the dynamics, but *emergent* is not the same as *unreal*.

To some readers, it will seem as if I am smuggling in unwarranted presuppositions. The thought would go something like this. In the absence of additional ontology, a quantum state is just an abstract object, a vector in a Hilbert space, or else something like a field on a high-dimensional space with no intrinsic relation to things in three-dimensional space. It is only when additional ontology is introduced, ontology on three-dimensional space, and the quantum state placed in the role of guiding that ontology that it has some relation to our three-dimensional space.

This line of thinking—which, admittedly, has become prevalent in the literature—seems to me to misconstrue the nature of quantum theories. We are not presented with an abstract Hilbert space, with no built-in relation to anything physical, or, worse, with a mysterious function on a high-dimensional space. To construct a quantum theory, one first identifies a target system with certain

---

[7] Parallel passages, differing only in minor matters of wording, can be found in Ghirardi [31, pp. 363–364] and Ghirardi [32, p. 233].

[8] Gao [35, §6.2.1] argues, along similar lines, for the reality of charge densities when a charged system is well localized and hence the state is an eigenstate (or close to it) of charge density.

dynamical degrees of freedom. One then follows the quantization heuristic to construct a quantum theory of that system, replacing the canonical dynamical variables with operators obeying appropriate commutation relations. Insofar as the dynamical degrees of freedom are associated with certain regions of space, so are the operators that we associate with them, and the eigenvalues of those operators represent possible stable values of those quantities, and eigenstates of those operators represent states in which those values obtain in the appropriate regions of space. *All we are doing* is extending this representation, via continuity, to states that are not eigenstates.

If a quantum state were a vector in a Hilbert space that had no built-in association with dynamical quantities, there would be a serious problem with taking it to represent *anything* physical, as the intrinsic structure of such a space would not distinguish one vector from another. If a quantum state were a field on configuration space, with no built-in relation to dynamical quantities such as energy or charge density or the like, or, worse, if it were a field on a high-dimensional space with no built-in relation to our familiar three-dimensional space, then there would, indeed, be a serious problem with taking it to describe a world anything like ours. If that is what quantum states were, then we would, indeed, need to posit additional ontology above and beyond the wave function. But taking a quantum state to be something like that involves, it seems to me, an almost willful forgetting of what quantum theories are.

6.6.3 A Comment on "Flash Ontology"

Goldstein [33] and [1] adapt a suggestion of John Bell to formulate an alternate primitive ontology for the GRW theory, which they call the "flash ontology." Bell describes his proposal as follows.

> There is nothing in this theory but the wave function. It is in the wave function that we must find an image of the physical world, and in particular of the arrangements of things in ordinary three-dimensional space. But the wave-function as a whole lives in a much bigger space, of $3N$-dimensions. It has neither amplitude nor phase nor anything else until a multitude of points in ordinary three-space are specified. However, the GRW jumps (which are part of the wave-function, not something else) are well localized in ordinary space. Indeed each is centered on a particular spacetime point $(\mathbf{x}, t)$. So we can propose these events as the basis of the 'local beables' of the theory. These are the mathematical counterparts in the theory to real events at definite places and times in the real world ... A piece of matter then is a galaxy of such events. As a schematic psychophysical parallelism we can suppose that our personal experience is more or less directly of events in particular pieces of matter, our brains, which events are in turn correlated with events in our bodies as a whole, and they in turn with events in the outer world (Bell 36, in Bell 37, pp. 204–205).

Unlike Bell, for whom the jumps, or hitting events, are "part of the wave-function, not something else" (it would be better to say: are events in the evolution of the wave function), Allori *et al.* propose that these events be taken as self-subsistent primitive ontology, distinct from the wave function, whose role is to furnish a probability distribution over possible sequences of flashes. We then have two distinct physical theories with the same stochastic equation of motion for the wave function: GRWm, which is the GRW theory with mass-density ontology, and GRWf, GRW theory with the flash ontology.

As Allori *et al.* note, a table or chair will be the locus of a great many flashes per second—they estimate an order of 10^8 flashes per cubic centimeter for solid matter. On a smaller scale, the flashes would be rarer. In particular, on this theory, in a region of space that we are inclined to take to be the locus of a living cell, the interval between flashes will be of the order of 10^3 seconds, something like one flash per hour. About this, in a footnote that responds to a comment by a referee, they write, "there is presumably no cell in GRWf, though the structure of the wave function (on configuration space—even though there are no configurations) might suggest otherwise" (p. 362, fn. 5).

Of course, this does not mean that, if we look through a microscope at that region in which there is no cell, we will see nothing. On the contrary, the photons or electrons, as the case may be, employed by the microscope will become entangled with the wave function of the nonexistent cell, and the amplification process will eventually record the image in some medium involving a macroscopic number of atomic nuclei, enough so that the encoded image of a cell will be discernible in the pattern of flashes in the recording medium. We will have formed an image of something that does not exist.

Moreover, unlike the case of properties of quantum systems to which the quantum state assigns nothing like a determinate value prior to interaction with experimental apparatus—cases in which we should, indeed, say that the recorded result does not represent a fact that obtained prior to the interaction—it is possible to prepare the living cell in such a way that an observation will predictably and reliably produce an image of a cell in a given location.

We should ask: what are we saying here? Can we coherently acknowledge that is true of a given region of space that it reacts, predictably and reliably, in these ways to microscopes and other instruments, and, indeed, to anything else that might interact with it, exactly as if a cell is present, and at the same time deny that there is a cell present? The supposedly nonexistent cell would (as Newton put it) no less constitute a part of the structure of things than the bodies realized in the primitive ontology. We are being asked to draw an ontological distinction between dynamical potentialities associated with regions in which there is primitive ontology, and those located in empty space, with only the former counting as real objects. It is as if an Aristotelian, rejecting the Newtonian conception of body, declared that the only genuine bodies were those in which the dynamical properties flagged by Newton were realized in some *materia prima* and, further, declared that some of the apparent objects we interact with are of this sort, and others, mere *simulacra*. All this is, literally and without hyperbole, nonsensical.

6.7 Distributional Ontology

Considerations of situations such as the foregoing makes it clear that specifying a *mean* mass distribution (even if this were reified into primitive ontology) does not exhaust the physical facts about mass distributions. Things are very different, depending on whether the variance $\mathcal{V}(\hat{M}(\mathbf{r}))$ is appreciable or not. We must, therefore, take it that there is a matter of fact about what this variance is.

This suggests the following revision, advanced by Philip [2], to the way we attribute values to dynamical quantities. In classical physics, dynamical quantities always possess precise values. In quantum theory, there is always some imprecision, but, if the spread is small enough—that is, if the state is close enough to an eigenstate—we may treat it as one in which the quantity has a precise value. But the full reality is that associated with each dynamical variable is a *distribution of values*. This distribution, though formally like a probability distribution, is to be thought of not as a probability distribution over a precise but unknown possessed value but as reflecting a physical, ontological, lack of determinacy about what the value is.

On this proposal, *every* dynamical variable possesses some distribution. A collapse theory will tend to narrow the spread of the distributions of some of these quantities, and, when the distribution is sufficiently narrow, everything will be almost exactly as if the quantity has a precise value, and it can be treated as if it were precise. If we are seeking objects that behave like our familiar macroscopic objects, it is to those variables that we should direct our attention. But the spread-out distributions of the other variables are no less part of physical reality.

An analogy might help. For many purposes, in dealing with solar system dynamics, the planets can be treated as if they have precise, point locations. They don't, of course. Suppose, then, we wanted to develop a more nuanced concept of *location of Jupiter*. One refinement is to note that Jupiter is extended, and we could choose a certain region, an oblate spheroid, as the region in which Jupiter is located. But Jupiter's atmosphere does not have a precise upper boundary. The location of Jupiter is neither a determinate point in space nor a determinate region of space.

On this proposal, all dynamical physical quantities are like that. Take a spatial region R, and consider the quantity of mass in that region, $M(R)$. Classically, this would have a point value. On the distributional ontology, it does not have precise value but might have, instead, a distribution[9] whose density function looks

---

[9] Don't confuse this distribution, which is a distribution of values for the mass within a *fixed region of space*, with a mass distribution that attributes various mass-contents to various regions of space.

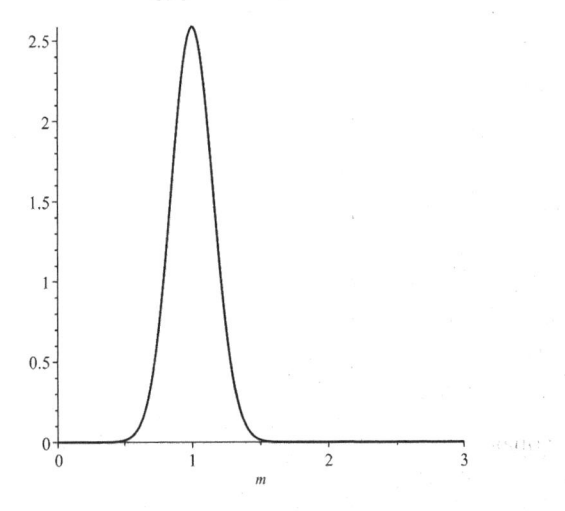

Figure 6.1 A hypothetical density function for a mass-value distribution

something like that depicted in Figure 6.1. A state in which the distribution for $M(R)$ was like that would be nearly an eigenstate of the property corresponding to $M(R)$ lying within the interval [0.5, 1.5]. It would be even more nearly an eigenstate of the property corresponding to $M(R)$ lying within the interval [0, 2]. For any interval Δ of the real line, we can consider the projection onto the subspace that, by the positive eigenstate–eigenvalue link, corresponds to $M(R)$ being in Δ. Any state vector can be written as a sum of a component within that subspace and a component orthogonal to that subspace. If the ratio of the norm of the component within that subspace to the norm of the state vector is close enough to unity, close enough for the difference to be negligible, then we are justified in asserting that the value of $M(R)$ lies in that interval.

Such assertions will, therefore, be somewhat elliptical, as they carry an implicit degree of tolerance, which can be made explicit. That is, assertions that the value of a physical quantity lies within a given interval are really what Philip Pearle [2] calls *qualified possessed value* attributions.

Consider a state $|\psi\rangle$ for which

$$\frac{\|\hat{P}(M(R) \in \Delta)|\psi\rangle\|^2}{\||\psi\rangle\|^2} \geq 1 - \varepsilon, \tag{6.10}$$

where $\hat{P}(M(R) \in \Delta)$ is the projection onto the subspace for which $M(R)$ is in the interval Δ, and ε is some small number (say 10^{-40}, as suggested by Pearle [38]). Applying the usual probability rules to outcomes of experiments, the probability in such a state that a measurement of mass in R will yield a value within Δ is at least $1 - 10^{-40}$, which means that "100 billion people making measurements every picosecond for 10 billion years can expect only one erroneous prediction"

[38, p. 151]. This seems enough to warrant the claim that we can reliably predict the result of such an experiment and that there is a fact of the matter about the region R that it will respond in the affirmative to an experiment to determine whether its mass content is in Δ.

However, in our attribution of properties to physical systems, we should not take "experiment" as a primitive term.[10] Worse, this way of speaking might be taken to presume that experimental outcomes have unqualified definite values, whereas, on this account, every physical quantity, including the pointer variables of our experimental apparatus, is distribution valued. We should be a bit careful about what it means, on this account, to say that we can reliably predict the result of an experiment.

A good experimental apparatus is one such that the dynamics of the theory will lead to near-sharp pointer values. Moreover, the sharpness of such pointer values is limited only by the resources we have available to us in building the apparatus. So, when we say that, in a state $|\psi\rangle$, we attribute a qualified possessed value to the proposition $M(R) \in \Delta$ with tolerance ε, the operational import of this is the following.

For any $\delta > 0$, no matter how small, it is possible for there to be another system, with an indicator variable Π, whose spectrum includes indicator regions Π^+ and Π^-, such that:

[1] If applied to an eigenstate of $M(R) \in \Delta$, the indicator variable will end up in Π^+ within tolerance δ.
[2] If applied to an eigenstate of $M(R) \notin \Delta$, the indicator variable will end up in Π^- within tolerance δ.
[3] If applied to the state $|\psi\rangle$, with probability at least $1 - \epsilon$, the indicator variable will end up in Π^+ within tolerance δ.

That is, we attribute properties to systems on the basis of their potential responses to other systems, and the tolerances invoked have to do with probabilities of various responses. Those responses, in turn, will be responses of distribution-valued variables and will be themselves the subjects of qualified value attributions.

Is there a threat of circularity here? There might be if we had no handle on the meaning of any of these value attributions. But they are anchored by the positive eigenstate–eigenvalue link. As the tolerance ε shrinks, a state in which a certain dynamical quantity has a given value within a tolerance ε approaches an eigenstate with that eigenvalue, and, as a consequence of that, its behavior approaches that of the eigenstate. Recall that our quantum theory was set up in the first place

---

[10] This is why I choose to say that qualified value attributions are qualified by specifying a tolerance, whereas Pearle speaks of them as qualified by the probability that the statement is false. There is nothing wrong with this way of speaking if it is understood in the way that is outlined in what follows, but it runs the risk of suggesting that we are taking experiment as an unanalyzed primitive, and not something to be schematized within the theory.

with an identification of certain dynamical variables of interest and an association of operators with those variables. We know what it means to attribute to a system a definite value of some dynamical quantity. To attribute an almost-definite value means: it behaves in almost the same way as it would if it had the definite value.

6.8 Conclusion

A distributional ontology of the sort outlined previously is, I claim, the natural way to think about a world described by a dynamical collapse theory if one takes the theory seriously, on its own terms. Attributions of values to physical quantities will always be qualified value attributions. The rationale for accepting qualified possessed value attributions is that, as ε decreases without limit, any dynamical differences between a state in which a quantity is possessed within tolerance ε and a state in which it is possessed with zero tolerance—that is, a state that, in accordance with the eigenstate–eigenvalue link, the quantity is possessed without qualification—diminish without limit. On the Newton-inspired conception of body, what it is for there to be bodies with spatial locations (or even relative spatial locations) has to do with dynamical properties associated with those locations, and there is nothing in that conception of body that requires those properties to be point valued.

This may run counter to some readers' intuitions. Those intuitions might long for an ontology in which everything is definite. They may long for an ontology in which dynamical properties are predicated of some substrate that is conceptually capable of being stripped of its dynamical properties. But neither the world nor physical theory are under any obligation to conform to such intuitions. A world like ours *could* be the way that a dynamical collapse theory says it is, and we should seriously entertain the possibility that it *is* that way.

References

[1] Allori, V., S. Goldstein, R. Tumulka, and N. Zanghì (2008). On the common structure of Bohmian mechanics and the Ghirardi-Rimini-Weber theory. *The British Journal for the Philosophy of Science 59*, 353–389.

[2] Pearle, P. (2009). How stands collapse II. In W. C. Myrvold and J. Christian (Eds.), *Quantum Reality, Relativisitic Causality, and Closing the Epistemic Circle: Essays in Honour of Abner Shimony*, pp. 257–292. New York: Springer.

[3] Lewis, P. J. (2017). On the status of primitive ontology. This volume.

[4] Tumulka, R. (2017). Paradoxes and primitive ontology in collapse theories of quantum mechanics. This volume.

[5] Maudlin, T. (2007). Completeness, supervenience and ontology. *Journal of Physics A: Mathematical and Theoretical 40*, 3151–3171.

[6] Stein, H. (1994). Some reflections on the structure of our knowledge in physics. In D. Prawitz, B. Skyrms, and D. Westerståhl (Eds.), *Logic, Methodology, and Philosophy of Science IX*, pp. 633–655. Amsterdam: Elsevier Science B.V.

[7] Newton, I. (1962). *Unpublished Scientific Papers of Isaac Newton: A Selection from the Portsmouth Collection in the University Library, Cambridge.* Cambridge: Cambridge University Press. Chosen, edited, and trasnlated by A. Rupert Hall and Marie Boas Hall.

[8] Newton, I. (1999). *The Principia: Mathematical Principles of Natural Philosophy.* Berkeley: The University of Cailfornia Press. Translated by I. Bernard Cohen and Anne Whitman.

[9] Allori, V. (2013). Primitive ontology and the structure of fundamental physical theories. See [31], pp. 58–75.

[10] Allori, V., S. Goldstein, R. Tumulka, and N. Zanghì (2014). Predictions and primitive ontology in quantum foundations: A study of examples. *The British Journal for the Philosophy of Science 65*, 323–352.

[11] Esfeld, M. (2014). Quantum Humeanism, or: Physicalism without properties. *The Philosophical Quarterly 64*, 453–470.

[12] Esfeld, M., D. Lazarovici, V. Lam, and M. Hubert (2017). The physics and metaphysics of primitive stuff. *The British Journal for the Philsophy of Science 68*, 133–161.

[13] Wallace, D. (2013). A prolegomenon to the ontology of the Everett interpretation. See [31], pp. 203–222.

[14] Wallace, D. (2012). Decoherence and its role in the modern measurement problem. *Philosophical Transactions of the Royal Society A 370*, 4576–4593.

[15] Gilton, M. J. (2016). Whence the eigenstate–eigenvalue link? *Studies in History and Philosophy of Modern Physics 55*, 92–100.

[16] Dirac, P. A. M. (1935). *Principles of Quantum Mechanics* (2nd ed.). Oxford: Oxford University Press.

[17] Ghirardi, G., A. Rimini, and T. Weber (1986). Unified dynamics for microscopic and macroscopic systems. *Physical Review D 34*, 471–491.

[18] Ghirardi, G. C., R. Grassi, and F. Benatti (1995). Describing the macroscopic world: Closing the circle within the dynamical reduction program. *Foundations of Physics 25*, 5–38.

[19] Bassi, A. and G. Ghirardi (2003). Dynamical reduction models. *Physics Reports 379*, 257–426.

[20] Pearle, P. (1989). Combining stochastic dynamical state-vector reduction with spontaneous localization. *Physical Review A 39*, 2277–2289.

[21] Ghirardi, G. C., P. Pearle, and A. Rimini (1990). Markov processes in Hilbert space and continuous spontaneous localization of systems of identical particles. *Physical Review A 42*, 78–89.

[22] Shimony, A. (1991). Desiderata for a modified quantum mechanics. In A. Fine, M. Forbes, and L. Wessels (Eds.), *PSA 1990: Proceedings of the 1990 Biennial Meeting of the Philosophy of Science Association, Volume Two: Symposia and Invited Papers*, pp. 49–59. East Lansing, MI: Philosophy of Science Association. Reprinted in Shimony [39, pp. 55–67].

[23] Albert, D. and B. Loewer (1991). Wanted dead or alive: Two attempts to solve Schrödinger's paradox. In A. Fine, M. Forbes, and L. Wessels (Eds.), *PSA 1990: Proceedings of the 1990 Biennial Meeting of the Philosophy of Science Association, Volume One: Contributed Papers*, pp. 2787–285. East Lansing, MI: Philosophy of Science Association.

[24] Lewis, P. J. (2016). *Quantum Ontology: A Guide to the Metaphysics of Quantum Mechanics*. Oxford: Oxford University Press.
[25] Albert, D. and B. Loewer (1996). Tails of Schrödinger's cat. In R. Clifton (Ed.), *Perspectives on Quantum Reality*, pp. 81–92 Dordrecht: Kluwer Academic Publishers.
[26] Ghirardi, G., R. Grassi, and P. Pearle (1990). Relativistic dynamical reduction models: General framework and examples. *Foundations of Physics 20*, 1271–1316.
[27] Ghirardi, G., R. Grassi, and P. Pearle (1991). Relativistic dynamical reduction models and nonlocality. In P. Lahti and P. Mittelstaedt (Eds.), *Symposium on the Foundations of Modern Physics 1990*, pp. 109–123. Singapore: World Scientific.
[28] Ghirardi, G. and P. Pearle (1991). Elements of physical reality, nonlocality and stochasticity in relativistic dynamical reduction models. In A. Fine, M. Forbes, and L. Wessels (Eds.), *PSA 1990: Proceedings of the 1990 Biennial Meeting of the Philosophy of Science Association, Volume Two: Symposia and Invited Papers*, pp. 35–47. East Lansing, MI: Philosophy of Science Association.
[29] Clifton, R. and B. Monton (1999). Losing your marbles in wavefunction collapse theories. *The British Journal for the Philosophy of Science 50*, 697–717.
[30] Ghirardi, G. and R. Grassi (1996). Bohm's theory versus dynamical reduction. In J. T. Cushing, A. Fine, and S. Goldstein (Eds.), *Bohmian Mechanics and Quantum Theory: An Appraisal*, pp. 353–377. Berlin: Springer.
[31] Ghirardi, G. (1997a). Quantum dynamical reduction and reality: Replacing probability densities with densities in real space. *Erkenntnis 45*, 349–365.
[32] Ghirardi, G. (1997b). Macroscopic reality and the dynamical reduction program. In M. L. D. Chiara, K. Doets, D. Mundici, and J. V. Benthem (Eds.), *Structures and Norms in Science: Volume Two of the Tenth International Congress of Logic, Methodology, and Philosophy of Science*, pp. 221–240. Dordrecht: Kluwer Academic Publishers.
[33] Goldstein, S. (1998). Quantum theory without observers—Part two. *Physics Today 51*, 38–42.
[34] Ghirardi, G. (2007). Some reflections inspired by my research activity in quantum mechanics. *Journal of Physics A: Mathematical and Theoretical 40*, 2891–2917.
[35] Gao, S. (2017). *The Meaning of the Wave Function: In Search of the Ontology of Quantum Mechanics*. Cambridge: Cambridge Univerity Press.
[36] Bell, J. S. (1987). Are there quantum jumps? In C. Kilmister (Ed.), *Schrödinger: Centenary Celebration of a Polymath*, pp. 41–52. Cambridge: Cambridge University Press. Reprinted in Bell [37, 201–212].
[37] Bell, J. S. (2004). *Speakable and Unspeakable in Quantum Mechanics* (2nd ed.). Cambridge: Cambridge University Press.
[38] Pearle, P. (1997). Tales and tails and stuff and nonsense. In R. S. Cohen, M. Horne, and J. Stachel (Eds.), *Experimental Metaphysics: Quantum Mechanical Studies for Abner Shimony, Volume One*, pp. 143–156. Dordrecht: Kluwer Academic Publishers.
[39] Shimony, A. (1993). *Search for a Naturalistic Wordlview, Volume II: Natural Science and Metaphysics*. Cambridge: Cambridge University Press.

7

Properties and the Born Rule in GRW Theory

ROMAN FRIGG

This chapter discusses the property structure of GRW theory. Taking as a starting point Lewis's (1997) so-called counting anomaly (the claim that GRW implies that arithmetic does not apply to ordinary objects), I investigate how GRW encodes properties and point out that the property structure of GRW theory is more complex than that of standard quantum mechanics because a seemingly plausible principle, the composition principle, fails.

7.1 Introduction

How are properties encoded in GRW theory? In this chapter, I discuss an influential answer to this question, the so-called *Fuzzy Link*. Lewis (1997) argued that GRW theory, when interpreted in terms of the Fuzzy Link, implies that arithmetic does not apply to ordinary objects, an argument now known as the 'counting anomaly.' I take this argument as the starting point for a discussion of the property structure of GRW theory and collapse interpretations of quantum mechanics in general. The main lesson to be drawn from the counting anomaly is that the property structure of these theories is more complex than that of standard quantum theories (and classical mechanics) because a seemingly plausible principle, the composition principle, fails.

7.2 The Fuzzy Link

The standard way to relate quantum states to physical properties is the *eigenstate–eigenvalue rule* ('E-E rule' henceforth):[1]

---

[1] A classical source for this rule is Dirac (1930, pp. 46–47).

An observable $\hat{O}$ has a well-defined value for a quantum system S in state $|\psi\rangle$ if, and only if, $|\psi\rangle$ is an eigenstate of $\hat{O}$.

States that are not eigenstates of $\hat{O}$ defy interpretation on the basis of this rule and are not assigned a property. Yet a measurement of $\hat{O}$ produces a definite outcome even if the system is not in an eigenstate of $\hat{O}$. To solve this problem standard quantum mechanics postulates that whenever a measurement is performed, the system's state instantaneously collapses into one of the eigenstates of $\hat{O}$ with a probability given by the Born rule. This leaves the system in a state that can be interpreted on the basis of the E-E rule. But the introduction of measurement-induced collapses brings a plethora of new problems with it. What defines a measurement? At what stage of the measurement process does the collapse take place (trigger problem)? And why should the properties of a physical system depend on observers?

Ghirardi, Rimini, and Weber (1986) proposed an ingenious way to overcome these difficulties.[2] Their approach is now known as 'GRW theory.' Rather than appealing to observers, GRW theory sees collapses an integral part of what happens in nature. It postulates that on average a collapse occurs every 10^{-16}s. There are, however, crucial differences between the collapses of standard quantum theory and those of GRW theory. Position is privileged in GRW theory, and the theory's mechanism induces collapses in the position basis. However, a collapse can leave a system in a proper eigenstate only if the basis is discrete and hence a system's wave function (in the position basis) cannot be arbitrarily narrow after a collapse. This fact is enshrined in GRW theory, which postulates that a collapse leaves the system not in a precise eigenstate of the position operator but in a state that is 'close' to it. Technically speaking, the original state gets multiplied by a sharply peaked Gaussian, which makes the state more localized (where the variance of the Gaussian is of the order 10^{-7}m). In the context of GRW theory it is therefore more adequate to speak of a *localization process* rather than a collapse.

However, a more localized state is still not an eigenstate of the position operator, and as far as interpreting states using the E-E rule is concerned, we're back where we started. One way around the problem is to look for an alternative to the E-E rule. Common wisdom has it that 'close enough' is good enough: for a particle to be located at x it is sufficient to say that its wave function is peaked over a narrow interval around x (see, for instance, Sakurai 1994, pp. 42–43). Albert and Loewer (1995) give an exact formulation of this idea and say that a particle with wave function $\psi(r)$ is located in the interval R iff $\int_R |\psi(r)|^2 dr \geq 1 - \epsilon$, where ϵ is a positive real number close to zero (throughout I use 'iff' for 'if and only if'). The

---

[2] The theory has been put in a particularly simple form by Bell (1987). For a comprehensive review of 'GRW type' theories, see Bassi and Ghirardi (2003). A semi-technical summary of the main ideas can be found in Frigg and Hoefer (2007). For a discussion of collapse theories in general see Gao (2017, Ch. 8).

generalization of this rule to a system with n degrees of freedom is straightforward and leads to the following definition:

Consider n objects $e_1, \ldots, e_n$. The system $E = \{e_i, \ldots, e_n\}$ consisting of these n entities with wave function $\psi(r_1, \ldots, r_n)$ has the property of being in the n-dimensional interval $R_1 \times \ldots \times R_n$ iff

$$\int_{R_1 \times \ldots \times R_n} |\psi(r_1, \ldots, r_n)|^2 d^n r \geq 1 - \epsilon. \tag{7.1}$$

Clifton and Monton (1999) call this rule the *Fuzzy Link*, and I use the label *Fuzzy Quantum Mechanics* (FQM) to refer to any interpretation of quantum mechanics (QM) that assigns properties to states using the Fuzzy Link. Regarding GRW theory as a (version of) FQM offers a natural assignment of properties to post-localization states.[3] For what follows it is convenient to let $P_{\epsilon, R_1 \times \ldots \times R_n}(e_1, \ldots, e_n)$ be the proposition stating that E has the property of being in the interval $R_1 \times \ldots \times R_n$. This proposition is true iff inequality (7.1) holds. Note that E can also consist of just one object e. In this case, the definition reduces to: the entity e with wave function $\psi(r)$ has the property of being in the interval R, i.e., $P_{\epsilon, R}(e)$ is true, iff $\int_R |\psi(r)|^2 dr \geq 1 - \epsilon$.

Peter Lewis (1997) argues that the Fuzzy Link has the undesirable conclusion that arithmetic does not apply to macroscopic objects like marbles. This argument is now known as the *counting anomaly*. Consider a marble and box (large enough for the marble to fit in). The marble can be in two states, namely $|\psi_{in}\rangle$ (the marble is inside the box) and $|\psi_{out}\rangle$ (the marble is outside the box). These states are mutually exclusive and therefore orthogonal. For the reasons we have encountered already, a localization process won't leave the marble's state in position eigenstate; the best one can expect is for the localization to leave the marble in a highly asymmetric state of the form $|\psi_m\rangle = a|\psi_{in}\rangle + b|\psi_{out}\rangle$ (or $|\psi'_m\rangle = b|\psi_{in}\rangle + a|\psi_{out}\rangle$) where $1 > |a| \gg |b| > 0$ and $|a|^2 + |b|^2 = 1$. According to the Fuzzy Link, if $|b|^2 \leq \epsilon$ then the marble is in the box: $\int_{R_{in}} |\psi_m(r)|^2 dr = |a|^2 \geq 1 - \epsilon$, where R_{in} is the region we associate with being in the box.

Now enlarge the box and put not only one but a large number n of marbles in it (assume that the box is long a slim, allowing for the marbles to be placed in it side by side so that there is no interaction between the marbles). The state of the n-marble system is $|\psi_{total}\rangle = |\psi_m\rangle_1 \ldots |\psi_m\rangle_n$. When we now interpret $|\psi_{total}\rangle$ in terms of the Fuzzy Link, we are faced with a paradox. While each individual marble

---

[3] Ghirardi and co-workers (Ghirardi et al. 1995; Bassi and Ghirardi 1999b) favour a mass-density interpretation of GRW theory. Space constraints prevent me from discussing this approach here. However, as Clifton and Monton (2000, pp. 156–161) point out, the problems I discuss in this chapter (in particular the counting anomaly) equally arise under the mass-density interpretation. So the discussion in what follows *mutatis mutandis* carries over to an interpretation of GRW theory based on the mass-density approach.

is in the box, applying the Fuzzy Link to the state of the system yields that the system is not in the box: $\int_{R_{in}\times...\times R_{in}} |\psi_{total}|^2 d^n r = |a|^{2n}$, but $|a|^{2n} \ll 1 - \epsilon$ since $|a|$ is smaller than 1.[4] So if we construct a system of n marbles each of which *individually* is in the box, we end up with an n-marble system which is not in the box. Lewis calls this the counting anomaly because ensuring (one by one) that each marble is in the box is how we count marbles. Lewis (1997, pp. 320–321) calls this the *enumeration principle*. But, as we have just seen, by doing so we end up with a state in which it is false that the system of marbles is in the box. Hence counting is impossible, and we must conclude that according to GRW theory (and indeed any version of FQM), arithmetic does not apply to macroscopic objects such as marbles.

This argument has sparked a heated debate between Ghirardi and Bassi on one side and Clifton and Monton on the other, centring around the questions whether the anomaly really arises in GRW and whether it can be observed.[5] I want to take a completely different line. I argue that the counting anomaly can be dissolved by an analysis of the property structure of FQM. I argue that the problem arises because of the seemingly innocuous but ultimately faulty assumption that the composition principle holds in FQM. On this reading, Lewis's argument performs a different function: rather than presenting a *reductio ad absurdum* of GRW theory, it highlights that the property structure of the theory is more complex than we had assumed. This restores coherence, but it comes at the cost of a violation of common sense. Everyday experience suggests that the composition principle holds true for spatial properties (an intuition which is borne out in classical mechanics as well as in versions of QM based on the E-E rule), and we have to come to terms with the realization the this is not the case in GRW theory.

7.3 The Composition Principle

The *composition principle* says that if every object e_i of an ensemble $E = \{e_1, \ldots, e_n\}$ has a certain property P, then the ensemble E itself also has property P, and vice versa. Formally, $P(e_1)\&\ldots\&P(e_n)$ iff PE, where 'PE' means that the *ensemble E itself* has property P whereas '$P(e_1)\&\ldots\&P(e_n)$' expresses the fact that *every member of E* has property P. To facilitate notation, we refer to the latter property as '$\tilde{P}$'. The composition principle then reads: $\tilde{P}E$ iff PE. This principle holds true in many cases. The concatenation of several objects of temperature T also has temperature T, and if two objects reflect light of certain wavelength, then the ensemble of both objects also reflects light of the same wavelength. However,

---

[4] There is a limit to how close $|a|$ can be to 1, and so there will always be an n so that $|a|^{2n} \ll 1 - \epsilon$.
[5] The first reply to Lewis is Ghirardi and Bassi (1999); subsequent contributions are Bassi and Ghirardi (1999, 2001), Clifton and Monton (1999, 2000), and Lewis (2003a). For discussion of the different positions, see Frigg (2003) and Lewis (2003b).

the principle does not have the status of universal law (or even a truth of logic), and it fails in certain cases. Water is wet but water molecules are not; gases have a temperature but gas molecules do not; horses have a heart but a herd of horses does not; each musician of an orchestra plays an instrument but the orchestra as a whole does not. Failures of the principle also occur in physics: the ensemble of n objects of mass m does not have mass m, and the same holds true for every additive quantity.

Examples like these highlight that asserting $\tilde{P}E$ is not the same as asserting PE. If we wish to infer $\tilde{P}E$ from PE (or *vice versa*), the composition principle has to be invoked. This principle, however, is not a universal law, and its validity in a given context needs to be justified. If we fail to provide such a justification and assume, without further argument, that the composition principle holds true, we are guilty of a *fallacy of composition*.

The problem with the marbles is a problem of composition. To see how the problem emerges, we explicitly state the composition principle as regards position. Let $e_1, \ldots, e_n$ stand for the marbles and $E = \{e_1, \ldots, e_n\}$ for the ensemble of all marbles. Let R be the property of being in the n-dimensional interval $R_1 \times \ldots \times R_n$. $\tilde{R}E$ is the statement that all marbles individually are in the relevant intervals (i.e., e_1 is in R_1, e_2 is in R_2, etc.). We then have $\tilde{R}E \equiv P_{\epsilon,R_1}(e_1) \& \ldots \& P_{\epsilon,R_n}(e_n)$ (where '$\equiv$' is the equivalence relation between propositions). This proposition is true iff $\int_{R_i} |\psi_i(r)|^2 dr \geq 1 - \epsilon, i = 1, \ldots, n$, where $|\psi_i\rangle$ is the quantum state of the ith marble and R_i the spatial interval in which it should be located. RE is the statement that the ensemble of marbles is in R and we have $RE \equiv P_{\epsilon,R_1 \times \ldots \times R_n}(e_1, \ldots, e_n)$. This statement is true iff inequality (7.1) holds. The *composition principle as regards position* (CPP) says $\tilde{R}E$ iff RE. The property of being in the box is a special case. Choose $R = R_{in} \times \ldots \times R_{in}$ and let 'B' stand for the property of being in the box. One can then define $\tilde{B}E$ and BE as before, and CPP says $\tilde{B}E$ iff BE.

Is CPP true? The answer is: it depends. Let us first consider the special case of $\epsilon = 0$, where the Fuzzy Link in effect reduces to the the E-E rule. One can show that CPP holds in this case (the proof is given in the Appendix). So under the E-E rule, spatial properties satisfy composition. The situation changes drastically if we move into FQM proper and assume $\epsilon > 0$. The implication in CPP that runs from left to right no longer holds, and CPP is false. In fact, in FQM, only the *restricted composition principle as regards position* (RCPP) holds: If RE then $\tilde{R}E$, but *not* vice versa. Applied to the marble case, RCPP says that if the ensemble of all marbles is in the box (BE), then every one of its members is in the box as well ($\tilde{B}E$). The converse, however, is false: if every member of the ensemble (i.e., every individual marble) is in the box, the same need not be true for the ensemble.

Given what we have said about composition so far, it should not come as a surprise that such an inference can fail. The surprise, however, is that the inference fails in the case of a spatial location. Being in the box or, more generally, being located within an interval seems to be a clear example of a property for which the composition should hold: if all members of E are located in $R_1 \times \ldots \times R_n$ then, so it seems, the ensemble E itself should be located within that interval as well. The counting anomaly shows that this expectation is wrong.

A critic might now respond that nothing is gained by rephrasing the counting anomaly as a failure of CPP, because any theory in which CPP fails should be rejected. So the challenge is to make plausible that one can rationally uphold a theory in which CPP fails. To meet this challenge, we need to have a closer look at the properties involved. Why is it not absurd to hold that $\tilde{B}E$ is true while BE is false?

How do we check that all marbles are in the box? Let us follow Lewis and endorse his enumeration principle. This means that we first ascertain that the first marble is in the box, then that the second marble is in the box, and so on until we reach the nth marble. If each marble turns out to be in the box, then all n marbles are in the box. Given this, it is necessary and sufficient for the n marbles to be in the box that $\tilde{B}E$ is true (which, recall, is equivalent to $P_{\epsilon,R_{in}}(e_1)\&\ldots \&P_{\epsilon,R_n}(e_{in})$).

Isn't this too strong? Does doesn't BE equally describe the state of affairs of all marbles being in the box? The crucial thing to realise is that it doesn't. If we follow the procedure given in the enumeration principle, there is just no reason to assume that BE should be true. We observe one marble after the other and make sure that it is in the box, which results in the state of all marbles *individually* being in the box but not in any property of the ensemble. One might try to resist this conclusion by arguing that it is just intuitively obvious that BE describes the state of affairs of all marbles being in the box, regardless of whether it squares with the enumeration principle. But an appeal to intuition does't cut the mustard. Properties of ensembles and of individuals are distinct, and one cannot infer one from the other without further argument. The required further argument in the current case is CPP, and we know that this principle doesn't hold in FQM. Hence, insisting that BE represents the same state of affairs as $\tilde{B}E$ is committing a fallacy of composition.

But if BE does not describe the state of affairs of all marbles being in the box, what then does it represent? We are instructed by RCPP that BE implies that all marbles are in the box but not vice versa, and so BE has 'surplus content' with respect to $\tilde{B}E$. What is this surplus? I have no answer to this question, and I think we don't need one. The interest in BE is based on the belief that it reflects the 'counting property,' but this is not the case. Furthermore, it is in general a mistake to think that everything we can define in the formalism represents something interesting in the world. Not every formal expression corresponds to a property that is physically relevant, and BE may well be one of those expressions.

There are two ways to push back against this conclusion. The first, mentioned by Lewis (1997, p. 320) and echoed by Clifton and Monton (2000, p. 160), appeals to the Born rule. If the system is in state $|\psi_{total}\rangle = (a|\psi_{in}\rangle + b|\psi_{out}\rangle)_1...(a|\psi_{in}\rangle + b|\psi_{out}\rangle)_n$, then it is unacceptable to say that all marbles are in the box. Born's rule, so the argument goes, tells us that the probability of finding the system in state $|\psi_{in}\rangle_1...|\psi_{in}\rangle_n$ is $|a|^{2n}$, and because $|a|^{2n} \ll 1$, there is only a vanishingly small probability of finding all the marbles in the box. But it makes no sense to say that the marbles are in the box if the probability of finding them there is almost negligible.

This argument is flawed, but it is flawed in an interesting way because it draws our attention to an issue that has not received much attention so far, namely how to calculate probabilities in FQM. Given that FQM alters the conditions for a property to obtain, the way of calculating probabilities has to be altered too. Consider a marble in state $|\psi_m\rangle = a|\psi_{in}\rangle + b|\psi_{out}\rangle$. What is the probability p of it being true that the marble is in the box? In standard QM, we associate the state of being in the box with $|\psi_{in}\rangle$ and using Born's rule, we get $p = |a|^2$. In FQM, however, the Fuzzy Link tells us that the marble *is* in the box if it is in state $|\psi_m\rangle$ where $|a|^2 \geq 1 - \epsilon$. Given this, it does not make sense to say that the probability of finding the marble in the box equals $|a|^2$ in $|\psi_m\rangle$. We cannot say that the proposition that the marble is in the box is true if the system is in state $|\psi_m\rangle$ and at the same time take the probability of the proposition to be smaller than 1! If we allow the proposition to be true in states like $|\psi_m\rangle$ then we have to take these *same* states when using Born's rule to calculate probabilities. In the present example, one ought to say that the FQM probability of a marble in state $|\psi\rangle$ to be in the box is $|\langle\psi|\psi_m\rangle|^2$, and not $|\langle\psi|\psi_{in}\rangle|$ as standard QM has it.[6]

It is now clear where the problem lies: It is true that the probability of finding the system in state $|\psi_{in}\rangle_1...|\psi_{in}\rangle_n$ is vanishingly small, but from this it does *not* follow that the probability of finding all the marbles *in the box* is equally small, because these are not the same probabilities. The argument infers from the fact that the probability of finding the system in state $|\psi_{in}\rangle_1...|\psi_{in}\rangle_n$ is small that the probability of finding it in the box is equally small and thus implicitly associates 'being in the box' with the state $|\psi_{in}\rangle_1...|\psi_{in}\rangle_n$. In doing so, it carries over to FQM a way of thinking about probabilities that contradicts FQM's basic assumptions.

The second way to push back against my conclusion is to appeal to everyday-language practices. Lewis (2003, pp. 140–1; 2016, pp. 93–95) argues that an everyday-language claim like 'all marbles are in the box' involves both a claim about individual marbles *and* the ensemble of marbles because everyday language

---

[6] There is a question which state exactly one uses to calculate probabilities, because according to the Fuzzy Link, all $|\psi_m\rangle$ with $|a|^2 \geq 1 - \epsilon$ have the property at stake. A possible response is to take the state with the smallest admissible a (namely $|a|^2 = 1 - \epsilon$) and stipulate that p equals $|\langle\psi|\psi_m\rangle|^2$ for all states whose coefficient of $|\psi_{in}\rangle$ is smaller than a and 1 for all states with this coefficient greater than a. The last clause is needed to prevent that a state which is closer to $|\psi_{in}\rangle$ than $|\psi_m\rangle$ is assigned a probability smaller than one of being in the box.

does not distinguish between the two. Therefore one cannot drive a wedge between claims about individuals and claims about ensembles in the way that I suggest (Monton [2004] also endorses this argument). Lewis immediately adds that 'this is an empirical claim about everyday language, and requires justification' (2003, p. 140) but expresses confidence that such an investigation would reveal that 'we do not make the distinction in ordinary language between the claim that the ensemble of marbles is in the box and the claim that each of the marbles individually is in the box' (2016, p. 93).

Whether Lewis's claim about ordinary language is true is a factual question that ultimately can only be settled through empirical research, and the jury on this is still out. But let's assume, for the sake of argument, that Lewis is right and ordinary language does not make a distinction between claims about individual marbles and ensembles of marbles. How relevant is this for the question of understanding FQM? Views about the importance of ordinary language vary widely, and the judgement isn't always favourable. In fact, significant strands of analytical philosophy have been concerned with clearing up the ambiguities and inconsistencies of ordinary language in numerous domains of inquiry. So it is no anathema to replace ordinary language by a suitably regimented artificial language if this solves problems, and my claim is that arithmetic is one of those places where such a revisionary attitude is justified. The most common formal system of arithmetic, Peano arithmetic, is a first-order theory (see, for instance, Machover, 1996). The axioms of Peano arithmetic capture all standard truths of arithmetic and, being first order, only involve quantification over individuals. So no appeal to ensembles and ensemble properties is needed to capture the truths of arithmetic. If ordinary language creates an arithmetic anomaly by importing elements into the theory that aren't needed, then the right reaction seems to be to replace ordinary language with a suitably regimented formal language.[7] The counting anomaly shows us that position is a more complex property than we had previously assumed, and we have to update our views about it accordingly. If an air of paradox remains, it has to be dispelled in the same way in which many other fallacies and imprecisions of ordinary language have to be dispelled.

7.4 Conclusion

I have argued that it is sufficient for the marbles to be in the box that $\tilde{B}E$ ($\equiv P_{\epsilon,R_{in}}(e_1)\&\ldots\&P_{\epsilon,R_n}(e_{in})$) holds, and that nothing else is needed. Since counting

---

[7] The claim is only that first-order arithmetic is able to capture standard truths of arithmetic. The claim is not that no other formulations of arithmetic could be given. There are of course second-order formulations of arithmetic which involve quantification over predicate variables. But these are introduced to solve problems that have nothing to do with the counting anomaly, and the existence of these theories does not force us to use them in the current context.

is a process that is concerned with individual objects rather than with ensembles, all that is needed for counting is that the conjunction of all $P_{\epsilon, R_{in}}(e_j)$ is true. The general lesson we learn from this discussion is that the property structure of FQM is more complex than that of standard QM, a price we have to pay for the admission of non-eigenstates as property-bearing states.

Appendix

Note that we recover the E-E rule if we assume $\epsilon = 0$ and replace '$\geq$' is replaced by '$=$.' A system in state $|\psi\rangle$ has the property U iff $|\langle \psi | e_u \rangle|^2 = 1$, where $|e_u\rangle$ is the state in the Hilbert space associated with the property U. This is equivalent to the condition $\langle \psi | \hat{P}_{e_u} | \psi \rangle = 1$, where $\hat{P}_{e_u}$ is the projection operator on $|e_u\rangle$. If there is not just one single vector but an entire subspace S_u of the Hilbert space associated with U, the condition reads $\langle \psi | \hat{P}_{S_u} | \psi \rangle = 1$, where $\hat{P}_{S_u}$ is the projection operator on the subspace S_u. Now choose U to be 'being located within interval $R_1 \times \ldots \times R_n$.' Then this condition reads $\langle \psi | \hat{P}_{R_1 \times \ldots \times R_n} | \psi \rangle = 1$. Now expand both $|\psi\rangle$ and $\hat{P}_{R_1 \times \ldots \times R_n}$ in the position basis: $|\psi\rangle = \int_{-\infty}^{\infty} \ldots \int_{-\infty}^{\infty} d^n r \, \psi(r_1, \ldots, r_n) | r_1 \ldots r_n \rangle$ and $\hat{P}_{R_1 \times \ldots \times R_n} = \int_{-\infty}^{\infty} \ldots \int_R d^n r | r_1 \ldots r_n \rangle \langle r_1 \ldots r_n |$. Plugging this into the above condition yields: $\langle \psi | \hat{P}_{R_1 \times \ldots \times R_n} | \psi \rangle = \int_{R_1 \times \ldots \times R_n} |\psi(r_1, \ldots, r_n)|^2 d^n r = 1$, which obviously is Equ. 7.1 with the aforementioned changes.

We are now in a position to prove that CP holds for properties thus defined.

$\Rightarrow$: Assume $P_{R_1}(e_1)$ and $P_{R_2}(e_2)$ and $\ldots$ and $P_{R_n}(e_n)$ holds, that is, $\int_R |\psi_i(r_i)|^2 dr_i = 1$; $i = 1, \ldots, n$. Since we built up our collective 'n-marble entity' from n non-interacting marbles, the state will not be entangled and can be written as the product of the states of the individual marbles: $\psi(r_1, \ldots, r_n) = \psi_1(r_1) \ldots \psi_n(r_n)$; and since in standard quantum mathematics (SQM) the wave functions of a well-behaved quantum state is integrable, we can factorise the integral in Def. 2: $\int_{R_1 \times \ldots \times R_n} |\psi_1(r_1) \ldots \psi_n(r_n)|^2 d^n r = \int_{R_1} |\psi_1(r_1)|^2 dr_1 \ldots \int_{R_n} |\psi_n(r_n)|^2 dr_n$. But by assumption all terms of this product equal one, hence $\int_{R_1 \times \ldots \times R_n} |\psi_1(r_1) \ldots \psi_n(r_n)|^2 d^n r = 1$.

$\Leftarrow$: Assume $P_{R_1 \times \ldots \times R_n}(e_1, \ldots, e_n)$ holds, that is, $\int_{R_1 \times \ldots \times R_n} |\psi_1(r_1) \ldots \psi_n(r_n)|^2 d^n r = 1$. Factorise the integral as before: $\int_{R_1 \times \ldots \times R_n} |\psi_1(r_1) \ldots \psi_n(r_n)|^2 d^n r = \int_{R_1} |\psi_1(r_1)|^2 dr_1 \ldots \int_{R_n} |\psi_n(r_n)|^2 dr_n = 1$. It is an axiom of SQM that $\int_{R_i} |\psi_i(r_i)|^2 dr_i \leq 1$ for all $i = 1, \ldots, n$. For this reason, the earlier product can equal 1 only if $\int_{R_i} |\psi_i(r_i)|^2 dr_i = 1$ for all $i = 1, \ldots, n$. qed. This completes the proof of CP for SQM.

It is straightforward to see that the first half of the proof no longer goes through if $\epsilon > 0$ and (7.1) is a proper inequality. The second part, however, is not affected by this change. For this reason, CPP fails in FQM, but RCPP holds.

References

Albert, D. Z. and Loewer, B. (1995). Tails of Schrodinger's Cat. In *Perspectives on Quantum Reality: Non-Relativistic, Relativistic, and Field-Theoretic*, ed. Rob Clifton. Dordrecht: Kluwer Academic Publishers, pp. 81–92.

Bassi, A. and Ghirardi, G. (1999). More about dynamical reduction and the enumeration principle. *British Journal for the Philosophy of Science*, **50**, 719–734.

Bassi, A. and Ghirardi, G. (2001). Counting marbles: reply to Clifton and Monton. *British Journal for the Philosophy of Science*, **52**, 125–130.

Bassi, A., and Ghirardi, G. C. (2003). Dynamical reduction models. *Physics Reports*, **379**, 257–426.

Bell, J. S. (1987). Are there quantum jumps? In *Speakable and Unspeakable in Quantum Mechanics*, ed. J. S. Bell. Cambridge: Cambridge University Press, pp. 201–212.

Clifton, R. (1996). The properties of modal interpretations of quantum mechanics. *British Journal for the Philosophy of Science*, **47**, 371–398.

Clifton, R. and Monton, B. (1999). Losing your marbles in wave function collapse theories. *British Journal for the Philosophy of Science*, **50**, 697–717.

Clifton, R. and Monton, B. (2000). Counting marbles with "accessible" mass density: a reply to Bassi and Ghirardi. *British Journal for the Philosophy of Science*, **51**, 155–164.

Dirac, P. A. M. (1930). *The Principles of Quantum Mechanics*. Oxford: Oxford University Press.

Frigg, R. (2003). On the property structure of realist collapse interpretations of quantum mechanics and the so-called counting anomaly. *International Studies in the Philosophy of Science*, **17**, 43–57.

Frigg, R. and Hoefer, C. (2007). Probability in GRW theory. *Studies in History and Philosophy of Modern Physics*, **38**, 2007, 371–389.

Gao, S. (2017). *The Meaning of the Wave Function: In Search of the Ontology of Quantum Mechanics*. Cambridge: Cambridge University Press.

Ghirardi, G. and Bassi, A. (1999). Do dynamical reduction models imply that arithmetic does not apply to ordinary macroscopic objects? *British Journal for the Philosophy of Science*, **50**, 49–64.

Ghirardi, G., Grassi, R., and Benatti, F. (1995). Describing the macroscopic world: closing the circle within the dynamic reduction program. *Foundations of Physics*, **25**, 5–38.

Ghirardi, G., Rimini, A., and Weber, T. (1986). Unified dynamics for microscopic and macroscopic systems. *Physical Review*, **34D**, 470–491.

Lewis, P. J. (1997). Quantum mechanics, orthogonality, and counting. *British Journal for the Philosophy of Science*, **48**, 313–328.

Lewis, P. J. (2003a). Counting marbles: reply to critics. *British Journal for the Philosophy of Science*, **54**, 165–170.

Lewis, P. J. (2003b). Four strategies for dealing with the counting anomaly in spontaneous collapse theories of quantum mechanics. *International Studies in the Philosophy of Science*, **17**, 137–142.

Lewis, P. J. (2016). *Quantum Ontology: A Guide to the Metaphysics of Quantum Mechanics*. Oxford: Oxford University Press.

Machover, M. (1996). *Set Theory, Logic and Their Limitations*. Cambridge: Cambridge University Press.

Monton, B. (2004). The problem of ontology for spontaneous collapse theories. *Studies in History and Philosophy of Modern Physics*, **35**, 407–421.

Neumann, J. v. (1955). *Mathematical Foundations of Quantum Mechanics*. Princeton: Princeton University Press.

Sakurai, J. J. (1994). *Modern Quantum Mechanics*. Reading, MA: Addison Wesley.

8

Paradoxes and Primitive Ontology in Collapse Theories of Quantum Mechanics

RODERICH TUMULKA

Collapse theories are versions of quantum mechanics according to which the collapse of the wave function is a real physical process. They propose precise mathematical laws to govern this process and to replace the vague conventional prescription that a collapse occurs whenever an "observer" makes a "measurement." The "primitive ontology" of a theory (more or less what Bell called the "local beables") are the variables in the theory that represent matter in space–time. There is no consensus about whether collapse theories need to introduce a primitive ontology as part of their definition. I make some remarks on this question and point out that certain paradoxes about collapse theories are absent if a primitive ontology is introduced.

8.1 Introduction

Although collapse theories [1] have been invented to overcome the paradoxes of orthodox quantum mechanics, several authors have set up similar paradoxes in collapse theories. I argue here, following Monton [2], that these paradoxes evaporate as soon as a clear choice of the primitive ontology is introduced, such as the flash ontology or the matter density ontology. In addition, I give a broader discussion of the concept of primitive ontology, what it means, and what it is good for.

According to collapse theories of quantum mechanics, such as the Ghirardi–Rimini–Weber (GRW) theory [3, 4] or similar ones [5, 6, 7], the time evolution of the wave function ψ in our world is not unitary but instead stochastic and non-linear; and the Schrödinger equation is merely an approximation, valid for systems of few particles but not for macroscopic systems, i.e., systems with (say) 10^{23} or more particles. The time evolution law for ψ provided by the GRW theory is formulated mathematically as a stochastic process; see, e.g., [4, 7, 8]; and can be summarized by saying that the wave function ψ of all the N particles in the universe evolves as if somebody outside the universe made, at random times with rate $N\lambda$, an unsharp quantum measurement of the position observable of a randomly chosen particle. "Rate $N\lambda$" means that the probability of an event in time dt is equal to $N\lambda\,dt$; λ is a constant of order $10^{-15}\,\text{sec}^{-1}$. It turns out that the empirical predictions of the GRW theory agree with the rules of standard quantum mechanics up to deviations that are so small that they cannot be detected with current technology [7, 9, 10, 11, 12].

The merit of collapse theories, also known as dynamical state reduction theories, is that they are "quantum theories without observers" [13], as they can be formulated in a precise way without reference to "observers" or "measurements," although any such theory had been declared impossible by Bohr, Heisenberg, and others. Collapse theories are not afflicted with the vagueness, imprecision, and lack of clarity of ordinary, orthodox quantum mechanics (OQM). Apart from the seminal contributions by Ghirardi et al. [3], Bell [4], Pearle [5], Diósi [6, 14], and a precursor by Gisin [15], collapse theories have also been considered by Gisin and Percival [16], Leggett [17], Penrose [18], Adler [9], Weinberg [19], among others. A feature that makes collapse models particularly interesting is that they possess extensions to relativistic space–time that (unlike Bohmian mechanics) do not require a preferred foliation of space–time into spacelike hypersurfaces [20, 21, 22]; see Maudlin [23] for a discussion of this aspect.

Collapse theories have been understood in two very different ways: some authors (e.g., Bell [4], Ghirardi et al. [24], Goldstein [13], Maudlin [25], Allori et al. [8], Esfeld [26]) think that a complete specification of a collapse theory requires, besides the evolution law for ψ, a specification of variables describing the distribution of matter in space and time (called the *primitive ontology* or PO), while other authors (e.g., Albert and Loewer [27], Shimony [28], Lewis [29], Penrose [18], Adler [9], Pearle [30], Albert [31]) think that a further postulate about the PO is unnecessary for collapse theories. The goals of this chapter are to discuss some aspects of these two views, to illustrate the concept of PO, and to convey something about its meaning and relevance. I begin by explaining some more what is meant by ontology (Section 8.2) and primitive ontology (Section 8.3). Then (Section 8.4) I discuss three paradoxes about GRW from the point of view of PO. In Section 8.5, I turn to a broader discussion of PO. Finally in Section 8.6, I describe specifically its relation to the mind-body problem.

8.2 Ontology and Its Relevance to Physics

The "ontology" of a theory means what exists, according to that theory. John S. Bell [35] coined the word "beables" (as opposed to observables) for the variables representing something real, according to that theory; that is, for the ontology. Many researchers, physicists as well as philosophers, find it hard to think in terms of an ontology, partly because we all have practiced for many years to think "in the quantum mechanical way." Physicist Jeremy Bernstein [67] wrote: "Many of the papers I tried to read were written by philosophers and had words like "ontology" sprinkled over them like paprika." The quote conveys hesitation and reservations towards this word. I recommend that physicists overcome this hesitation, as ontology is, in fact, a highly relevant concept.

And a very simple one. Despite the fancy name, the concept of ontology is already familiar from classical physics. For example, Newtonian mechanics talks about particles, and macroscopic objects such as rocks and trees consist, according to Newtonian mechanics, of particles—so one says that the ontology of Newtonian mechanics is particles. For another example, the electric field may at first appear as a mere calculational device (if defined as the force that a unit charge would feel if placed at a certain location), but later developments (specifically the realization that the electric and magnetic fields can carry energy and momentum) convince us that the electric and magnetic fields are something real. That is, the electric and magnetic fields are part of the ontology of classical electrodynamics.

It was long thought that the key to clarity in QM was to avoid talking about ontology and stick to operational statements. In my opinion, that thought has not paid off. Sometimes, we want to talk about events that occurred before humans existed, and then it seems particularly absurd to assume that facts only become definite when observed. But even when we talk about laboratory experiments, it is hard to stick to operational statements and hard to avoid talking about what actually happened in reality. And about that, OQM has only contradictory and unclear statements to offer. The true key to clarity is to set up a hypothesis about what happens in reality, to analyze what the consequences will be if that hypothesis is taken seriously, and to reject the hypothesis if it leads to empirically incorrect consequences. It was long thought that such hypotheses do not exist for QM, but Bohmian mechanics and collapse theories are counterexamples to that thought. Bohmian mechanics nicely illustrates some advantages of a clear ontology:

- The theory can be stated in full on a single page (whereas operational statements are cumbersome to formulate with high precision).
- The usual quantum rules about the probabilities of outcomes of experiments can be derived as theorems instead of being postulated as axioms [32].
- Paradoxes can be resolved simply by taking seriously the fundamental laws of the theory. For example, Bell [33] analyzed how Wheeler's delayed-choice paradox gets resolved in Bohmian mechanics.
- Symmetries (such as Lorentz or Galilean invariance or time reversal invariance or gauge invariance) have a clear meaning and can be derived as theorems [8]. Likewise for superselection rules [34].

8.3 Primitive Ontology in Collapse Theories

Primitive ontology (PO) is a name for that part of the ontology that represents matter in 3 + 1-dimensional space–time. For example, the PO in classical mechanics and Bohmian mechanics are the particles (material points), mathematically represented by their world lines and thus by their coordinate functions $Q_k(t)$ (the position of particle k at time t). The wave function in Bohmian mechanics is part of the ontology (it is something real) but not part of the primitive ontology (as it lives in configuration space, not in physical space).

Bell [35, 4] used the expression "local beables" for the variables representing something real ("beables") associated with a space–time point ("local"). For example, in Bohmian mechanics, ψ is non-local because it refers to several space–time points at once, whereas $Q_k(t)$ (or, equivalently, the number of actual particles at the space–time point (x, y, z, t)) is a local beable. In all theories that I will consider here, the local beables are exactly the primitive ontology (even though it may not necessarily be so in all conceivable theories).

In collapse theories, I will focus on two particular choices of PO for the GRW theory, the flash ontology and the matter density ontology. According to the *flash ontology* [4], matter is fundamentally described by a discrete set of space–time points called flashes. Flashes are material space–time points. In the GRW theory with the flash ontology (GRWf for short), there is one flash for each collapse of ψ; its time is the time of the collapse, and its position is the collapse center; see, e.g., [4, 8] for mathematical descriptions.

According to the *matter density ontology* [24, 13], matter is fundamentally continuously distributed and described by a density function $m(x, y, z, t)$ on space–time. In the GRW theory with the matter density ontology (GRWm for short), the m function at time t is obtained from $|\psi_t|^2$ (which is a density function on $3N$-dimensional configuration space) by integrating out the coordinates of $N - 1$ particles to get a 3-dimensional density function; in order not to prefer any particle, one averages over all sets of $N - 1$ particles (perhaps using a weighted average with the particles' masses as weights). See, e.g., [8] for a mathematical description.

The view with which to contrast GRWf and GRWm is that there is no primitive ontology, in fact no ontology at all in physical 3-space, and that the ontology comprises only ψ (GRWØ for short). To illustrate the difference between GRWf/GRWm and GRWØ, let me make up a creation myth (as a metaphorical way of speaking): Suppose God wants to create a universe governed by GRW theory. He creates a wave function ψ of the universe that starts out as a particular ψ_0 that he chose and evolves stochastically according to a particular version of the GRW time evolution law. According to GRWØ, God is now done. According to GRWf and GRWm, however, a second act of creation is necessary, in which he creates the matter, i.e., either the flashes or continously distributed matter with density m, in both cases coupled to ψ by the appropriate laws. (I will elaborate on this point in Section 8.5.2.)

There are several motivations for considering GRW∅. First, it seems more parsimonious than GRWm or GRWf. Second, it was part of the motivation behind GRW theory to avoid introducing an ontology in additon to ψ; otherwise, we may think, we could have used Bohmian mechanics. In fact, much of the motivation came from the measurement problem of OQM. The measurement problem arises from treating the measurement apparatus as a quantum system (because it consists of electrons and quarks): If the time evolution law for every wave function is linear, then the joint wave function of object and apparatus after a quantum measurement is a non-trivial superposition of states corresponding to different outcomes (except if the object was initially in an eigenstate of the observable measured). If there are no variables in addition to the wave function, then there is no fact about which outcome was the actual one. Thus, if there is an actual outcome, then we must either abandon the linearity of the Schrödinger evolution or introduce further ontology in addition to ψ; this is the measurement problem. The GRW theory was intended to choose the first option, not the second. And yet, I think that GRW∅ is not an acceptable theory, but GRWm and GRWf are.

In the next section, I will exemplify some issues by means of some paradoxes that have been raised as objections to the GRW theory; I will describe what these paradoxes look like from the point of view of a theory with PO and will point out some advantages of the PO view by showing that it resolves these paradoxes. In the section after that, I will discuss the need for a primitive ontology more broadly.

8.4 Three Paradoxes About GRW Theories

In each case, I will first describe a paradox and then explain why it evaporates in GRWf and GRWm. For GRWm, some of the relevant points have already been made by Monton [2].

8.4.1 Paradox 1: Does the Measurement Problem Persist?

Paradox: Here is a reason one might think that the GRW theory fails to solve the measurement problem [27]. Consider a quantum state like Schrödinger's cat, namely a superposition

$$\psi = c_1\psi_1 + c_2\psi_2 \qquad (8.1)$$

of two macroscopically distinct states ψ_i with $\|\psi_1\| = 1 = \|\psi_2\|$, such that both contributions have non-zero coefficients c_i. Given that there is a problem—the measurement problem—in the case in which the coefficients are equal, one should also think that there is a problem in the case in which the coefficients are not exactly equal but roughly of the same size. One might say that the reason why there is a problem is that, according to quantum mechanics, there is a superposition, whereas according to our intuition there should be a definite state. But then it is hard to see how this problem should go away just because c_2 is much smaller than c_1. How small would c_2 have to be for the problem to disappear? No matter if $c_2 = c_1$ or $c_2 = c_1/100$ or $c_2 = 10^{-100}c_1$,' in each case both contributions are there. But the only relevant effect of the GRW process replacing the unitary evolution, as far as Schrödinger's cat is concerned, is to randomly make one of the coefficients much smaller than the other (although it also affects the shape of the suppressed contribution [36]).

Answer: The argument indeed raises questions about GRW0. From the point of view of GRWm or GRWf, however, the argument is flawed, as it pays no attention to the PO. To take the PO seriously means that whether Schrödinger's cat is really dead must be read off from the PO.

In GRWf, if $|c_2| \ll |c_1|$ then the next flash is overwhelmingly likely to be associated with ψ_1 (say, $\psi_1 = |\text{dead}\rangle$), and so on for further flashes, with the result that the flashes form the shape of a dead cat. They form this shape even if there are a few extra flashes associated with $|\text{alive}\rangle$, but in fact, since each flash associated with a collapse favoring $|\text{dead}\rangle$ further reduces the probability of a flash associated with $|\text{alive}\rangle$, it has probability near 1, given that $|c_2| \ll |c_1|$, that the number of later flashes associated with $\psi_2 = |\text{alive}\rangle$ is zero. So yes, the wave function is still a superposition, but the definite facts that our intuition wants can be found in the PO. The flashes represent reality in 4-dimensional space–time; the cat is in the flashes, not in ψ. To be sure, we have no precise definition of which patterns of flashes should be regarded as dead cats; that fact, however, is not worrisome; it arises simply because the words "dead cat" are not precisely defined in ordinary language.

In GRWm, if ψ is close to $|\text{dead}\rangle$ then m equals $m_{|\text{dead}\rangle}$ up to a small perturbation, and that can reasonably be accepted as the PO of a dead cat. Note the following: Whereas the wave function is a superposition of two packets ψ_1, ψ_2 that correspond to *two very different* kinds of (particle) configurations in ordinary QM or Bohmian mechanics, there is only *one* configuration of the matter density m—the definite fact that our intuition wants. A subtlety here arises from the fact that the small contribution to m that comes from ψ_2 often still has internal structure (say, looks like a live cat); I will address this point in Section 8.5.4.

8.4.2 Paradox 2: How Can You Call a Cat Dead if There is a Small Probability of Finding it Alive?

Paradox: As a variant of the first paradox, one might say that even after the GRW collapses have pushed $|c_1|^2$ near 1 and $|c_2|^2$ near 0 in the state vector (8.1), there is still a positive probability $|c_2|^2$ that if we make a "quantum measurement" of the macro-state—of whether the cat is dead or alive—we will find the state ψ_2, even though the GRW state vector has collapsed to a state vector near ψ_1, a state vector that might be taken to indicate that the cat is really dead (assuming $\psi_1 = |\text{dead}\rangle$). Thus, it seems not justified to say that, when ψ is close to $|\text{dead}\rangle$, the cat is really dead. This paradox is another version of the "tail problem" [28, 27, 29, 37, 38].

Answer: It is important here to appreciate the following difference between orthodox quantum mechanics and GRWm/GRWf concerning the meaning of the wave function and how the wave function makes contact with our world: In orthodox quantum mechanics, a system's wave function governs the probabilities for the outcomes of experiments on the system, whereas in GRWm/GRWf, the wave function governs the PO. This difference creates a difference concerning what one means when saying that the cat is dead: In orthodox quantum mechanics, one means that if we made a "quantum measurement" of the cat's macro-state, we would with probability 1 find it dead, whereas in GRWm/GRWf, one means that the PO forms a dead cat. If the cat is dead in this sense, in GRWm/GRWf, and ψ is close but not exactly equal to $|\text{dead}\rangle$, then there is still a tiny but non-zero probability that within the next millisecond the collapses occur in such a way that the cat is suddenly alive! But that does not contradict the claim that a millisecond before the cat was dead; it only means that GRWm/GRWf allows resurrections to occur—with tiny probability! In particular, if we observe the cat after that millisecond, there is a positive probability that we find it alive (simply because it *is* alive) even though before the millisecond it actually was dead.[1]

8.4.3 Paradox 3: Consider Many Systems

Paradox: A variant of the previous paradox was formulated by Lewis [39] in terms of counting marbles; the discussion continued in [38, 40, 41, 42, 43, 44, 45, 46, 47, 48, 2, 49, 30, 50]. Let ψ_1 be the state "the marble is inside the box" and ψ_2 the state "the marble is outside the box"; these wave functions have disjoint supports S_1, S_2 in configuration space (i.e., wherever one is non-zero, the other is zero). Let ψ be given by (8.1) with $0 < |c_2|^2 \ll |c_1|^2 < 1$; finally, consider a system of n (non-interacting) marbles at time t_0, each with wave function ψ, so that the wave function of the system is $\psi^{\otimes n}$. Then for each of the marbles, we would feel entitled to say that it is inside the box, but on the other hand, the probability that all marbles be found inside the box is $|c_1|^{2n}$, which can be made arbitrarily small by making n sufficiently large.

---

[1] For GRWf, I assume here that the reason c_2 is small lies in GRW collapses that have occurred in the past. When that is not the case, for example when ψ is prepared to be (8.1) with small c_2, the situation in GRWf is yet different, as then there is no fact at the initial time from the PO as to whether the cat is dead or is alive. This will play a role in the subsequent paradox.

Answer: For GRWm it follows from the PO, as in the answer to the previous paradox, that each of the marbles is inside the box at the initial time t_0. However, it is known that a superposition like (8.1) of macroscopically distinct states ψ_i will approach under the GRW evolution either a wave function $\psi_1(\infty)$ concentrated in S_1 or another $\psi_2(\infty)$ in S_2 with probabilities $|c_1|^2$ and $|c_2|^2$, respectively. (Here I am assuming $H = 0$ for simplicity. Although both coefficients will still be nonzero after any finite number of collapses, one of them will tend to zero in the limit $t \to \infty$.) Thus, for large n the wave function will approach one consisting of approximately $n|c_1|^2$ factors $\psi_1(\infty)$ and $n|c_2|^2$ factors $\psi_2(\infty)$, so that ultimately about $n|c_1|^2$ of the marbles will be inside and about $n|c_2|^2$ outside the box— independently of whether anybody observes them or not. The occurrence of some factors $\psi_2(\infty)$ at a later time provides another example of the resurrection-type events mentioned above; they are unlikely but do occur, of course, if we make n large enough.

The act of observation plays no role in the argument and can be taken to merely record pre-existing macroscopic facts. To be sure, the physical interaction involved in the act of observation may have an effect on the system, such as speeding up the evolution from ψ towards either $\psi_1(\infty)$ or $\psi_2(\infty)$; but GRWm provides unambiguous facts about the marbles also in the absence of observers.

In GRWf, the story is a little more involved. The fact that the answer depends on the choice of PO illustrates again the relevance of the PO, as well as the necessity to make the PO and its laws explicit. The story is more involved because the PO cannot be considered at only one point in time t_0 but needs to be considered over some time interval (say a millisecond), and because it depends on randomness. First, if we assume that the smallness of c_2 is due to previous collapses centered inside the box, then the flashes during the millisecond before t_0 form n marbles inside the box. Thus, as in GRWm, initially we have n marbles inside the box, of which $n|c_2|^2$ will be outside the box after a while.

Now consider the other case: that the smallness of c_2 is not due to previous collapses, but due to some other method of preparing ψ. Then we may have to consider only flashes after t_0. Consider first a single marble. Something improbable may already happen at this stage; for example, all the flashes might occur outside the box. In that case we would say that the marble *is* outside the box. As well, it might happen that half of the flashes occur outside and half of them inside the box; in that case we would say that half of the marble's matter is located inside the box. The overwhelmingly probable case, of course, is that more than 99% of the marble's flashes occur inside the box, a case in which it is reasonable to say that the marble is inside the box. Thus, if $|c_2|^2 \ll |c_1|^2$ at time t_0 then in GRWf (unlike in GRWm) the marble is not necessarily (only very probably) inside the box, provided the time interval we consider is the millisecond after t_0. Now consider n marbles, with n so big that $|c_1|^{2n} \ll 1$; then it is not probable any more that for *all* marbles, 99% of the flashes occur inside the box. Rather, the overwhelmingly probable case is that for the majority of marbles, 99% of the flashes occur inside the box (so that one should say these marbles are inside), while for a few marbles, a significant fraction of flashes occurs outside, and for extremely few marbles, even all flashes occur outside. In the limit $t \to \infty$, as a consequence of the convergence to either $\psi_1(\infty)$ or $\psi_2(\infty)$, for each marble either almost all flashes occur inside the box or almost all flashes occur outside.

8.4.4 What Is Real and What Is Accessible

Another remark concerns the reply by Bassi and Ghirardi [42, 43, 44] (also 2003, Sec. 11) to the tail problem and the marble problem; see also [2]. Bassi and Ghirardi use GRWm but immediately focus on (what they call) the *accessibility* of the matter density, i.e., on the fact that the matter density cannot be measured by inhabitants of the GRWm world to arbitrary accuracy. In particular, they point out that for the marble wave function, the matter density outside the box is not detectable. I think that Bassi and Ghirardi took two steps at once, thereby making the argument harder to understand for their readers, and that perhaps they did not take the PO seriously enough. To the extent that the worry expressed in the earlier paradoxes (and by Shimony [28], Albert and Loewer [27], Lewis [39], Clifton and Monton [40]) is whether GRW theories do give rise to unambiguous facts about the aliveness of Schrödinger's cat or the location of the marble, it concerns whether GRW theories provide a picture of reality that conforms with our everyday intuition. Such a worry cannot be answered by pointing out what an observer can or cannot measure. Instead, I think, the answer can only lie in what the ontology *is like*, not in what observers see of it. Moreover, I think, it can only lie in what the *primitive ontology* is like, as that is the part of the ontology representing "the cat" and "the marble." It can easily cause confusion to blur the distinction between what is real and what is accessible, as not everything that is real and well defined according to GRW theories is accessible; for example, the number of collapses in a given time interval $[t_1, t_2]$ is well-defined but cannot be measured reliably [51, 52].

Now for the marble state considered earlier, it is a fact for the matter density ontology that the fraction $|c_2|^2 \ll 1$ of the marble's matter lies outside the box—a fact that does not contradict the claim that the marble is inside the box, as can be illustrated by noting that anyway, for thermodynamic reasons, the marble creates a vapor out of some of its atoms (with low partial pressure), an effect typically outweighing $|c_2|^2$. The state of the PO in which the overwhelming majority of matter is inside the box justifies saying that the marble is inside the box. Thus, the PO does provide a picture of reality that conforms with our everyday intuition. All this is independent of whether the PO is observable (accessible) or not.

Bassi and Ghirardi sometimes sound as if they did not take the matter density seriously when it is not accessible; I submit that the PO should always be taken seriously.

The reason Bassi and Ghirardi attribute such importance to whether the matter density is accessible is presumably the following: If the matter density outside the box could be measured to be non-zero, then this would seem to threaten the claim that the marble is inside. But the threat is actually not serious: For example, the vapor created by the marble can in fact be measured to have non-zero density, but this fact does not at all suggest that "the marble" should be regarded as being outside; it only suggests (indeed, it entails) that a small part of the marble's matter is outside the box.

8.5 The Need for a Primitive Ontology

In this section, I offer further considerations about the status and role of the PO.

8.5.1 Is There a Cat?

We all once learned OQM and got used to "quantum think"—certain ways of reasoning employed in OQM. One of these ways of reasoning postulates that, whenever ψ is the wave function of a live cat,[2] there is a live cat. Since collapse theories are not OQM, these ways of reasoning need not apply, and this postulate is questionable. Bohmian mechanics provides an example of a situation in which the PO could be in a configuration of a live cat and move with time in the manner of a live cat, while ψ is a non-trivial superposition of a live cat and a dead cat. This example suggests that

$$\text{it is appropriate to say "there is a live cat" when and only when part of the PO behaves like a live cat.} \tag{8.2}$$

In particular, whether there is a live cat depends on ψ only through the dependence of the PO on ψ by virtue of the laws governing the PO. It then follows that there is no live cat in GRW0, because there is no PO, regardless of ψ. This is perhaps the most fundamental problem about theories without a PO: It is hard to avoid the conclusion that they imply the absence of matter in space–time.

Furthermore, (8.2) implies that there is a logical gap between saying

$$\text{"}\psi\text{ is the wave function of a live cat"} \tag{8.3}$$

and saying

$$\text{"there is a live cat."} \tag{8.4}$$

---

[2] I.e., a wave function concentrated in a certain region of configuration space and with energies of certain subsystems in suitable ranges, etc.

After all, in Bohmian mechanics, (8.4) follows from (8.3) by virtue of a law of the theory, which asserts that the configuration $Q(t)$ of the PO is $|\psi_t|^2$ distributed at every time t. This law entails that if ψ_t is the wave function of a live cat, then $Q(t)$ will be the configuration of a live cat and will continue to behave like the configuration of a live cat for $t' > t$ as long as $\psi_{t'}$ stays the wave function of a live cat. Thus, Bohmian mechanics suggests that (8.4) would not follow from (8.3) if there was not a law connecting the two by means of the PO.

An attractive feature of the PO view is that the word "cat" actually refers to a *thing*: a set of particles in Bohmian mechanics or a set of flashes in GRWf or a part of the continuous matter in GRWm that behaves like a cat under the influence of ψ. Correspondingly, the statement "the cat is alive" simply means that that *thing* behaves in a certain way—like a live cat. In GRW∅, in contrast, we need to assume a re-interpretation of English phrases: We need to assume that the word "cat" does not refer to any *thing* and that the phrase "the cat is alive" does not summarize the behavior of a thing called "the cat," but rather that this phrase really means (8.3). I find that hard to swallow.

Despite all this, there is a sense in which GRW∅ works: The GRW wave function ψ_t is, at almost all times, concentrated, except for tiny tails, on a set of configurations that are macroscopically equivalent to each other. Thus, we can read off from ψ_t what the macro-state is: For example, we could pretend there were particles whose configuration Q is $|\psi_t|^2$ distributed and consider the macroscopic appearance of Q. Alternatively, we could compute $m(x, y, z, t)$ and consider its macroscopic appearance. Some thought reveals that the macroscopic appearance of Q agrees (with high probability) with that of m (and in fact with that of the flashes in GRWf [53]; but see also [54]); for example, they agree about which way the pointer of a measurement instrument points. So ψ_t contains all the information about what the macro-configuration would look like if there were any matter that could have this macro-configuration. This fact explains why GRW∅ provides enough information to read off empirical predictions, although they are, in fact, the empirical predictions of GRWf and GRWm, and not those of GRW∅.

Actually, while GRWf and GRWm are empirically equivalent in the sense that there is no experiment that could test one against the other, their macro-configurations of the PO can differ if the parameters (i.e., constants of nature) of the GRW time evolution (i.e., the collapse width σ and the collapse rate λ) are chosen in extreme ways. This became visible in [10], where we drew a parameter diagram with axes σ and λ for the GRW theories and therein the empirically refuted region and the philosophically unsatisfactory region: it turned out that the latter region is different for GRWf than for GRWm if based on the following definition: "We regard a parameter choice (σ, λ) as philosophically satisfactory if and only if the PO agrees on the macroscopic scale with what humans normally think macroscopic reality is like" [10]. (But see also [54] for a further subtlety about the two regions.)

I remark that an issue parallel to GRWØ versus GRWm/GRWf comes up in many-worlds theories. The most popular understanding of many-worlds is that there is only the wave function ψ which evolves according to the linear Schrödinger equation. The theory actually becomes clearer if one introduces a suitable PO; an example is provided by Schrödinger's many-worlds theory [55].

8.5.2 All Observables?

Pearle [30] suggested, instead of GRWm, the view that *every* observable $\hat{A}$ is attributed as a "true value" the value that in OQM would be the average of $\hat{A}$ in the state ψ_t,

$$A(t) = \langle \psi_t | \hat{A} | \psi_t \rangle, \tag{8.5}$$

and that the $m(x, y, z, t)$ function of GRWm is just the special case for $\hat{A} = \hat{M}(x, y, z)$ the mass density operator, which in the position representation of non-relativistic QM of N particles is multiplication by the function

$$\sum_{i=1}^{N} m_i \, \delta(x_i - x) \, \delta(y_i - y) \, \delta(z_i - z). \tag{8.6}$$

To contrast this view with GRWm, let me use the creation myth again that I mentioned in Section 8.3: For creating a GRWm world, God would need two acts of creation, one for creating ψ and one for creating the matter with density $m(x, y, z, t)$. For creating a world according to Pearle's view, would God just have to create ψ, or would he have to carry out a separate act of creation for every observable $\hat{A}$? Let us compare the situation to the status of energy in a Newtonian universe. If God wants to create a Newtonian universe, he needs to create point particles that move according to the Newtonian equation of motion,

$$m_i \frac{d^2 \boldsymbol{Q}_i(t)}{dt^2} = -\nabla_i V\Big(\boldsymbol{Q}_1(t), \ldots, \boldsymbol{Q}_N(t)\Big) \tag{8.7}$$

for a suitable potential V. Then energy is defined as

$$E(t) = \sum_i \frac{m_i}{2} \left| \frac{d\boldsymbol{Q}_i}{dt} \right|^2 + V\Big(\boldsymbol{Q}_1(t), \ldots, \boldsymbol{Q}_N(t)\Big), \tag{8.8}$$

and no separate act of creation is needed in which God would have to create energy according to this equation. This situation suggests that likewise, no separate act of creation is needed to introduce $A(t)$. So Pearle's suggestion is really a form of GRWØ, notwithstanding the appearance that GRWm be contained in it.

Let me connect Pearle's suggestion to empirical predictions. It is one of the key roles of the PO that when deriving empirical predictions ("with probability p, the pointer will be in position x"), we actually derive statements about the PO, as the pointer is where its flashes/matter density/particles are. It is illuminating to see how this works out in various theories with PO [56]. In Pearle's approach, in contrast, if we use the m function to determine the pointer position, then this is just one among many possibilities of defining what is meant by "pointer position." Other possibilities could be based on the flash ontology, or on taking $\hat{A}$ to be the center-of-mass position operator of the particles forming the pointer. It is not clear, then, which choice would be correct in case they disagree (and, as mentioned earlier, GRWf and GRWm sometimes do disagree about the macro-configuration of the PO for extreme values of σ and λ [10]). But even when they agree, I find it unclear what justifies these choices and prefers them over all other operators $\hat{A}$. It seems within our power to define which pieces of matter we call pointers but not to define what is meant by their position.

8.5.3 Paradoxes

The paradoxes of Section 8.4 also pose a problem for GRW∅. For example, consider Paradox 1 again, the most basic one: After Schrödinger's cat collapses to |alive⟩, the amplitude of |dead⟩ is tiny but not exactly zero. What right then do we have to say that the cat *is alive*? To solve this problem, Albert and Loewer [37] have proposed a rule under the name "fuzzy link" asserting that any observable $\hat{A}$ has value A whenever ψ lies approximately in the eigenspace of $\hat{A}$ with eigenvalue A. Applying this rule to $\hat{A}$ the projection to the space of wave functions of live cats yields that whenever the amplitude of |dead⟩ in a superposition of |alive⟩ and |dead⟩ is tiny, the cat is alive. The question arises, however, why we would have the freedom to introduce a rule such as the fuzzy link. When we formulate a fundamental physical theory, we have the freedom to posit which things exist according to the theory (the ontology) and the laws governing them. For GRW∅, we would posit that ψ and only ψ exists and that it evolves according to the GRW process. It seems that there is no further room, then, for additional postulates such as the fuzzy link. Of course, if we introduce further elements of the ontology, such as a PO, then we can also postulate further laws governing the further ontology.

8.5.4 Worlds in the Tails

Another puzzle about the tails [38, 49, 57, 50, 58] arises from the fact that in GRWm, there is some matter that is governed by the tails of ψ: The m function has small ripples that follow the evolution of suppressed parts of ψ, for example of a dead cat after $2^{-1/2}(|\text{dead}\rangle + |\text{alive}\rangle)$ has collapsed to near $|\text{alive}\rangle$. Arguably, as in Schrödinger's many-worlds theory with a PO [55], this means that GRWm has a many-worlds character: Apart from the live cat, there also exists a dead cat. In GRWf, in contrast, when the tails are too small, there are no flashes associated with them, so that such further worlds do not exist. In GRWm, one of the worlds is the "fat world" (associated with the bulk of the matter), the others are the "faint worlds" (associated with the tails). I think we are forced to admit that the faint worlds are real. As pointed out by Wallace [36], the faint worlds behave very differently than the fat world, in fact catastrophically, due to the gradient in the Gaussian that gets multiplied onto ψ at every collapse. It seems that the upshot is that GRWm predicts the existence of several worlds, the fat and the faint ones, with very different behavior, and that we live in the fat world (as most observers do, since observers do not survive for long in the faint worlds).

8.5.5 Lorentz Invariance

Another consideration concerns Lorentz invariance. Relativistic versions of GRWf and GRWm have been developed [20, 21, 22] for non-interacting particles; relativistic collapse processes for ψ (albeit ultraviolet divergent) had been developed even earlier [59, 14]. The flashes transform like space–time points and the m function like a Lorentz scalar, while there is no simple relation between ψ_Σ and $\psi_{\Lambda(\Sigma)}$ for spacelike hypersurfaces Σ and Lorentz transformations Λ. For GRW0, in contrast, it seems unsatisfactory to merely replace the GRW process by a Lorentz invariant collapse process, for the following reason [22]. For a spacelike 3-surface $\Sigma = A \cup B$ with $A \cap B = \emptyset$, the reduced density matrix of A is

$$\hat{\rho}_A = \text{tr}_B |\psi_\Sigma\rangle\langle\psi_\Sigma|. \tag{8.9}$$

But this depends on B: For $\Sigma' = A \cup B'$, $\psi_{\Sigma'}$ may be very different from ψ_Σ due collapses between Σ and Σ', and $\hat{\rho}'_A = \text{tr}_{B'}|\psi_{\Sigma'}\rangle\langle\psi_{\Sigma'}|$ may be very different from $\hat{\rho}_A$. For example, if $\psi_\Sigma = 2^{-1/2}(|\uparrow\downarrow\rangle - |\downarrow\uparrow\rangle)$ is the spin-singlet state of an EPR pair and Bob's particle passes through a Stern–Gerlach magnet and collapses to either $|\uparrow\downarrow\rangle$ or $|\downarrow\uparrow\rangle$ before reaching Σ', then the reduced state on Alice's side is $\hat{\rho}_A = \frac{1}{2}(|\uparrow\rangle\langle\uparrow| + |\downarrow\rangle\langle\downarrow|)$ before the collapse and either $\hat{\rho}'_A = |\uparrow\rangle\langle\uparrow|$ or $\hat{\rho}'_A = |\downarrow\rangle\langle\downarrow|$ afterward. Now, in GRW0 we would have liked to read off the local state of affairs in region A from ψ, but the answer depends on the choice of B or B'. However, local facts in region A should not depend on whether we consider B or B'! This problem is absent in GRWf and GRWm because, even though reduced density matrices are still subject to the same ambiguity, the PO consists of local variables (the flashes or matter density) and thus provides plenty of local facts, enough for a meaningful physical theory. Albert (personal communication, 2014) has suggested that for GRWØ, one could simply use a preferred foliation of space–time into spacelike hypersurfaces. This would have the disadvantage of making GRWØ less relativistic than GRWf or GRWm, but the bigger problem with this suggestion is that, as with the fuzzy link rule, there does not seem to be any room for further postulates (such as to use a particular foliation for extracting local facts) after it has been stipulated that ψ evolves according to a certain relativistic collapse law and that the ontology contains only ψ.

8.6 Connection to the Mind–Body Problem

The persistent disagreement between two camps, one maintaining that GRWØ is an acceptable theory (e.g., Adler [9], Albert [31], Pearle [30]; Rimini), and the other that a PO is needed (e.g., Allori [60, 61, 62, 63], Ghirardi et al. [24], Bedingham et al. [22], Goldstein et al. [53], Maudlin [57, 64]), is perhaps ultimately related to different views about the mind–body problem. After a brief outline of this problem (for a detailed description, see, e.g., the first few chapters of [65]), I will explain how it comes up in connection with requirements on fundamental physical theories (such as, to have a PO).

8.6.1 What Is the Mind–Body Problem?

The color red looks a particular way to me that I cannot express in words. I would guess it looks the same way to other people, but I cannot really check (after all, it is logically possible that things that look red to me, such as tomatoes, look blue to somebody else and that we would never notice because that other person will call things red when they look blue to her). "The way red looks" is the conscious experience of red, which is different from the knowledge that the light that reached my eye had a wavelength of 800 nanometers and is called "red" in English.

If we try to come up with an explanation of this experience, we encounter a problem: A physical theory may claim that the particles in my brain follow certain trajectories, but this claim would not explain the experience of red, regardless of what the trajectories are. This is the essence of the mind–body problem, also sometimes called the "hard problem of consciousness." In contrast, all the information processing (from light of 800 nm to the English word "red") poses no obstacle in principle for an explanation in terms of particle trajectories. (It is the "easy problem of consciousness.") The mind–body problem is not specific to particles; it would be the same with fields, wave functions, or any other kind of physical object. Here, consciousness does not mean being awake or being aware or knowing oneself but seeing colors.

To appreciate the problem, it may help to try to write a computer program that makes the computer see red. That seems impossible. It is clearly possible that the computer identifies data from a digital camera as corresponding to a wavelength of 800 nanometers and to the English word "red." However, for this information processing, the computer need not see red, and by virtue of this information processing alone, the computer does not see red.

However, there is no consensus about the mind–body problem. Functionalists think that there is no such problem and that there is nothing more to the mind than information processing. I think there is a mind–body problem, and I do not see how it could ever be solved, i.e., how conscious experiences could be reduced to physical processes. I conclude that there are facts in the world beyond the physical facts.

8.6.2 How the Mind–Body Problem Affects Physics

A goal of a fundamental physical theory is to explain our experiences. In order to even connect with our experiences, it may seem that the theory has to solve the mind–body problem, which seems hopeless.

Luckily, there is a simple way out (as emphasized by Maudlin). Suppose the theory implies the existence of macroscopic objects in 3-space in particular macroscopic configurations. And suppose that we are not completely deluded about the world around us, that instrument pointers actually point more or less the way we think, and that we can usually read correctly. Then we can compare the predicted macro-configurations to the perceived macro-configurations, and claim, in case of agreement, that the theory is empirically adequate. It actually works like that in classical physics, and in any theory with a primitive ontology.

Here is a passage from Maudlin [64] in this direction:

> A theory's specification of the fermion density [as an example of a PO] in every region of the universe entails the distribution of matter at macroscopic scale. And if what it predicts at macroscopic scale matches everything we think we know about the world at macroscopic scale (including where the pointers ended up pointing, where the ink is on the paper, the shape of the earth, the dimensions of the Empire State building, etc., etc., etc.) then the theory is empirically adequate in any reasonable sense. There may be objections to such a theory, but they cannot rightfully be called empirical objections.

That is, physical theories with a PO do not need to solve the mind–body problem, whereas theories without a PO seem to be stuck with this unsolvable problem. Thus, physical theories are better off with a PO.

Acknowledgments

I thank David Albert, Eddy Chen, Shelly Goldstein, Ned Hall, Travis Norsen, Philip Pearle, Antoine Tilloy, and Nino Zanghì for helpful discussions.

References

[1] G.C. Ghirardi: Collapse Theories. In E. N. Zalta (ed.), *Stanford Encyclopedia of Philosophy*, published online by Stanford University. http://plato.stanford.edu/entries/qm-collapse/ (2007)

[2] B. Monton: The Problem of Ontology for Spontaneous Collapse Theories. *Studies in History and Philosophy of Modern Physics* **35**: 407–421 (2004) http://philsci-archive.pitt.edu/1410/

[3] G.C. Ghirardi, A. Rimini, and T. Weber: Unified Dynamics for Microscopic and Macroscopic Systems. *Physical Review D* **34**: 470–491 (1986)

[4] J. S. Bell: Are There Quantum Jumps? Pages 41–52 in C. W. Kilmister (ed.), *Schrödinger. Centenary Celebration of a Polymath.* Cambridge University Press (1987a). Reprinted as chapter 22 of [66].

[5] P. Pearle: Combining stochastic dynamical state-vector reduction with spontaneous localization. *Physical Review A* **39**: 2277–2289 (1989)

[6] L. Diósi: Models for universal reduction of macroscopic quantum fluctuations. *Physical Review A* **40**: 1165–1174 (1989)

[7] A. Bassi and G.C. Ghirardi: Dynamical Reduction Models. *Physics Reports* **379**: 257–426 (2003) http://arxiv.org/abs/quant-ph/0302164

[8] V. Allori, S. Goldstein, R. Tumulka, and N. Zanghì: On the Common Structure of Bohmian Mechanics and the Ghirardi–Rimini–Weber Theory. *British Journal for the Philosophy of Science* **59**: 353–389 (2008) http://arxiv.org/abs/quant-ph/0603027

[9] S. L. Adler: Lower and Upper Bounds on CSL Parameters from Latent Image Formation and IGM Heating. *Journal of Physics A: Mathematical and Theoretical* **40**: 2935–2957 (2007) http://arxiv.org/abs/quant-ph/0605072

[10] W. Feldmann and R. Tumulka: Parameter Diagrams of the GRW and CSL Theories of Wave Function Collapse. *Journal of Physics A: Mathematical and Theoretical* **45**: 065304 (2012) http://arxiv.org/abs/1109.6579

[11] A. Bassi and H. Ulbricht: Collapse models: from theoretical foundations to experimental verifications. *Journal of Physics: Conference Series* **504**: 012023 (2014) http://arxiv.org/abs/1401.6314

[12] M. Carlesso, A. Bassi, P. Falferi, and A. Vinante: Experimental bounds on collapse models from gravitational wave detectors. *Physical Review D* **94**: 124036 (2016) http://arxiv.org/abs/1606.04581

[13] S. Goldstein: Quantum Theory Without Observers. *Physics Today*, Part One: March 1998, 42–46. Part Two: April 1998, 38–42.

[14] L. Diósi: Relativistic theory for continuous measurement of quantum fields. *Physical Review A* **42**: 5086–5092 (1990)

[15] N. Gisin: Quantum Measurements and Stochastic Processes. *Physical Review Letters* **52**: 1657 (1984)

[16] N. Gisin and I. C. Percival: The quantum state diffusion picture of physical processes. *Journal of Physics A: Mathematical and General* **26**: 2245–2260 (1993)

[17] A. J. Leggett: Testing the limits of quantum mechanics: motivation, state of play, prospects. *Journal of Physics: Condensed Matter* **14**: R415–R451 (2002)

[18] R. Penrose: Wavefunction Collapse As a Real Gravitational Effect. Pages 266–282 in A. Fokas, T. W. B. Kibble, A. Grigoriou, B. Zegarlinski (ed.s), *Mathematical Physics 2000*. London: Imperial College Press (2000)

[19] S. Weinberg: Collapse of the State Vector. *Physical Review A* **85**: 062116 (2012) http://arxiv.org/abs/1109.6462

[20] R. Tumulka: A relativistic version of the Ghirardi–Rimini–Weber model. *Journal of Statistical Physics* **125**: 821–840 (2006a) http://arxiv.org/abs/quant-ph/0406094

[21] R. Tumulka: Collapse and Relativity. Pages 340–352 in A. Bassi, D. Dürr, T. Weber and N. Zanghì (eds.), *Quantum Mechanics: Are there Quantum Jumps? and On the Present Status of Quantum Mechanics*, AIP Conference Proceedings **844**. American Institute of Physics (2006b) http://arxiv.org/abs/quant-ph/0602208

[22] D. Bedingham, D. Dürr, G.C. Ghirardi, S. Goldstein, R. Tumulka, and N. Zanghì: Matter Density and Relativistic Models of Wave Function Collapse. *Journal of Statistical Physics* **154**: 623–631 (2014) http://arxiv.org/abs/1111.1425

[23] T. Maudlin: *Quantum Non-Locality and Relativity: Metaphysical Intimations of Modern Physics*. 3rd edition. Oxford: Blackwell (2011)

[24] F. Benatti, G.C. Ghirardi, and R. Grassi: Describing the macroscopic world: closing the circle within the dynamical reduction program. *Foundations of Physics* **25**: 5–38 (1995)

[25] T. Maudlin: Completeness, supervenience and ontology. *Journal of Physics A: Mathematical and Theoretical* **40**: 3151–3171 (2007)

[26] M. Esfeld: The primitive ontology of quantum physics: guidelines for an assessment of the proposals. *Studies in History and Philosophy of Modern Physics* **47**: 99–106 (2014) http://philsci-archive.pitt.edu/10711/

[27] D. Albert and B. Loewer: Wanted Dead or Alive: Two Attempts to Solve Schrödinger's Paradox. Pages 277–285 in A. Fine, M. Forbes and L. Wessels (eds.), *Proceedings of the 1990 Biennial Meeting of the Philosophy of Science Association. Vol. 1*. East Leasing: Philosophy of Science Association (1990)

[28] A. Shimony: Desiderata for a Modified Quantum Dynamics. Pages 49–59 in A. Fine, M. Forbes and L. Wessels (ed.s), *Proceedings of the 1990 Biennial Meeting of the Philosophy of Science Association Vol. 2*. East Leasing: Philosophy of Science Association (1990)

[29] P. Lewis: GRW and the tails problem. *Topoi* **14**: 23–33 (1995)

[30] P. Pearle: How Stands Collapse II. Pages 257–292 in W. Myrvold and J. Christian (ed.s): *Quantum Reality, Relativistic Causality, and Closing the Epistemic Circle*, The Western Ontario Series in Philosophy of Science Vol. 73. New York: Springer (2009) http://arxiv.org/abs/quant-ph/0611212

[31] D. Albert: *After Physics*. Harvard University Press (2015)

[32] D. Dürr, S. Goldstein, and N. Zanghì: Quantum Equilibrium and the Role of Operators as Observables in Quantum Theory. *Journal of Statistical Physics* **116**: 959–1055 (2004) http://arxiv.org/abs/quant-ph/0308038

[33] J. S. Bell: De Broglie–Bohm, Delayed-Choice Double-Slit Experiment, and Density Matrix. *International Journal of Quantum Chemistry* **14**: 155–159 (1980). Reprinted as chapter 14 of [66].

[34] S. Colin, T. Durt, and R. Tumulka: On Superselection Rules in Bohm–Bell Theories. *Journal of Physics A: Mathematical and General* **39**: 15403–15419 (2006) http://arxiv.org/abs/quant-ph/0509177

[35] J. S. Bell: The theory of local beables. *Epistemological Letters* **9**: 11 (1976). Reprinted as chapter 7 of [66].

[36] D. Wallace: Life and death in the tails of the GRW wave function. (2014) http://arxiv.org/abs/1407.4746

[37] D. Albert and B. Loewer: Tails of Schrödinger's Cat. Pages 81–91 in R. Clifton (ed.), *Perspectives on Quantum Reality*. Dordrecht: Kluwer (1996)

[38] A. Cordero: Are GRW Tails as Bad as They Say? *Philosophy of Science* **66**: S59–S71 (1999)

[39] P. Lewis: Quantum Mechanics, Orthogonality, and Counting. *British Journal for the Philosophy of Science* **48**: 313–328 (1997)

[40] R. Clifton and B. Monton: Losing your marbles in wavefunction collapse theories. *British Journal for the Philosophy of Science* **50**: 697–717 (1999) http://arxiv.org/abs/quant-ph/9905065

[41] R. Clifton and B. Monton: Counting marbles with 'accessible' mass density: a reply to Bassi and Ghirardi. *British Journal for the Philosophy of Science* **51**: 155–164 (2000) http://arxiv.org/abs/quant-ph/9909071

[42] A. Bassi and G.C. Ghirardi: Do dynamical reduction models imply that arithmetic does not apply to ordinary macroscopic objects? *British Journal for the Philosophy of Science* **50**: 49–64 (1999a) http://arxiv.org/abs/quant-ph/9810041

[43] A. Bassi and G.C. Ghirardi: More about dynamical reduction and the enumeration principle. *British Journal for the Philosophy of Science* **50**: 719 (1999b) http://arxiv.org/abs/quant-ph/9907050

[44] A. Bassi and G.C. Ghirardi: Counting Marbles: Reply to Clifton and Monton. *British Journal for the Philosophy of Science* **52**: 125–130 (2001)

[45] R. Frigg: On the Property Structure of Realist Collapse Interpretations of Quantum Mechanics and the So-Called "Counting Anomaly." *International Studies in the Philosophy of Science* **17**: 43–57 (2003) http://sas-space.sas.ac.uk/1041/

[46] P. Lewis: Counting Marbles: A Reply to Critics. *British Journal for the Philosophy of Science* **54**: 165–170 (2003)

[47] P. Lewis: Interpreting Spontaneous Collapse Theories. *Studies in History and Philosophy of Modern Physics* **36**: 165–180 (2005)

[48] P. Lewis: GRW: A Case Study in Quantum Ontology. *Philosophy Compass* **1**: 224–244 (2006)

[49] D. Wallace: Philosophy of quantum mechanics. Pages 16–98 in D. Rickles (ed.), *The Ashgate companion to contemporary philosophy of physics*. Aldershot: Ashgate (2008) http://arxiv.org/abs/0712.0149

[50] K. J. McQueen: Four Tails Problems for Dynamical Collapse Theories. *Studies in the History and Philosophy of Modern Physics* **49**: 10–18 (2015) http://arxiv.org/abs/1501.05778

[51] C. W. Cowan and R. Tumulka: Can One Detect Whether a Wave Function Has Collapsed? *Journal of Physics A: Mathematical and Theoretical* **47**: 195303 (2014) http://arxiv.org/abs/1307.0810

[52] C. W. Cowan and R. Tumulka: Epistemology of Wave Function Collapse in Quantum Physics. *British Journal for the Philosophy of Science* **67(2)**: 405–434 (2016) http://arxiv.org/abs/1307.0827

[53] S. Goldstein, R. Tumulka, and N. Zanghì: The Quantum Formalism and the GRW Formalism. *Journal of Statistical Physics* **149**: 142–201 (2012) http://arxiv.org/abs/0710.0885

[54] C. Sebens: Killer Collapse: Empirically Probing the Philosophically Unsatisfactory Region of GRW. *Synthese* **192**: 2599–2615 (2015) http://philsci-archive.pitt.edu/11350/

[55] V. Allori, S. Goldstein, R. Tumulka, and N. Zanghì: Many-Worlds and Schrödinger's First Quantum Theory. *British Journal for the Philosophy of Science* **62(1)**: 1–27 (2011) http://arxiv.org/abs/0903.2211

[56] V. Allori, S. Goldstein, R. Tumulka, and N. Zanghì: Predictions and Primitive Ontology in Quantum Foundations: A Study of Examples. *British Journal for the Philosophy of Science* **65**: 323–352 (2014) http://arxiv.org/abs/1206.0019

[57] T. Maudlin: Can the world be only wavefunction? Pages 121–143 in S. Saunders, J. Barrett, A. Kent, and D. Wallace (ed.s): *Many worlds? Everett, quantum theory, and reality.* Oxford University Press (2010)

[58] M. Egg and M. Esfeld: Primitive ontology and quantum state in the GRW matter density theory. *Synthese* **192**: 3229–3245 (2015) http://philsci-archive.pitt.edu/11102/

[59] P. Pearle: Toward a relativistic theory of statevector reduction. Pages 193–214 in A. I. Miller (ed.), *Sixty-Two Years of Uncertainty*, volume 226 of *NATO ASI Series B*. New York: Plenum (1990)

[60] V. Allori: On the Metaphysics of Quantum Mechanics. Pages 116–140 in S. LeBihan (ed.): *Précis de philosophie de la physique*. Paris: Vuibert (2013a) http://philsci-archive.pitt.edu/9343/

[61] V. Allori: Primitive Ontology and the Structure of Fundamental Physical Theories. Pages 58–75 in D. Albert and A. Ney (ed.s), *The Wave Function*. Oxford University Press (2013b) http://philsci-archive.pitt.edu/9342/

[62] V. Allori: Primitive Ontology in a Nutshell. *International Journal of Quantum Foundations* **1**: 107–122 (2015a) http://philsci-archive.pitt.edu/11651/

[63] V. Allori: How to Make Sense of Quantum Mechanics (and More): Fundamental Physical Theories and Primitive Ontology. Preprint (2015b) http://philsci-archive.pitt.edu/11652/

[64] T. Maudlin: Local Beables and the Foundations of Physics. Pages 317–330 in M. Bell and S. Gao (ed.s): *Quantum Nonlocality and Reality – 50 Years of Bell's Theorem* (2016) http://www.ijqf.org/wps/wp-content/uploads/2014/12/Maudlin-Local-Beables.pdf

[65] D. Chalmers: *The Conscious Mind.* Oxford University Press (1996)

[66] J. S. Bell: *Speakable and Unspeakable in Quantum Mechanics*. Cambridge University Press (1987b)

[67] J. Bernstein: More About Bohm's Quantum. *American Journal of Physics* **79**: 601 (2011)

9

On the Status of Primitive Ontology

PETER J. LEWIS

Spontaneous collapse theories provide a promising solution to the measurement problem. But they also introduce a number of problems of their own concerning dimensionality, vagueness, and locality. In response to these problems, advocates of collapse theories have proposed various accounts of the primitive ontology of collapse theories—postulated underlying entities governed by the collapse theory and underwriting our observations. The most prominent of these are a mass density distribution over three-dimensional space, and a set of discrete "flash" events at space–time points. My argument here is that these primitive ontologies are redundant, in the sense that the structures exhibited by the primitive ontologies that allow them to solve the problems facing spontaneous collapse theories are also present in the wave function. But redundancy is not nonexistence; indeed, the fact that the relevant structures are already there in the wave function shows that the mass density ontology and the flash ontology exist whether they are explicitly postulated or not. By the same token, there is no need to decide between a wave function ontology, a mass density ontology, and a flash ontology.

9.1 Introduction

Spontaneous collapse theories provide a promising solution to the measurement problem. But they also introduce a number of problems of their own. First, the primary explanatory entity of a collapse theory—the wave function—inhabits a high-dimensional space rather than the three-dimensional space of experience. Second, the continuity of the wave function introduces a new and potentially problematic form of vagueness when used to describe discrete physical systems such as particles or marbles. Third, the collapse of the wave function is hard to reconcile with special relativity.

In response to these problems, advocates of collapse theories have proposed various accounts of the *primitive ontology* of collapse theories—postulated underlying entities governed by the collapse theory and underwriting our observations. The most prominent of these are a mass density distribution over three-dimensional space and a set of discrete "flash" events at space–time points. These ontologies provide strategies for solving some or all of the problems listed earlier.

My argument here is that, in the case of spontaneous collapse theories, primitive ontology is *redundant*, in the sense that the structures exhibited by the primitive ontologies that allow them to solve the problems facing spontaneous collapse theories are also present in the wave function. But redundancy is not *nonexistence*; indeed, the fact that the relevant structures are already there in the wave function shows that the mass density ontology and the flash ontology exist whether they are explicitly postulated or not. By the same token, there is no need to decide between a wave function ontology, a mass density ontology, and a flash ontology.

In this regard, spontaneous collapse theories can be fruitfully contrasted with Bohm's theory. Bohmian particles are also sometimes described as primitive ontology, in response to the concern about the dimensionality of the wave function in particular. But in the Bohmian case, the primitive ontology is not redundant in the stated sense: the particle structure is not already present in the wave function but must be explicitly postulated. Hence there is a genuine question about whether Bohmian particles exist.

9.2 Collapse Theories

Collapse theories were introduced to solve the measurement problem in quantum mechanics. The measurement problem arises because quantum states can be superposed, whereas measurement results (apparently) cannot. For example, suppose quantum state ψ_A represents a particle as located in some spatial region A, and quantum state ψ_B represents the particle as located in some disjoint spatial region B. Then according to quantum mechanics, there is a third quantum state ψ_{AB} formed by superposing them: $\psi_{AB} = 2^{-1/2}(\psi_A + \psi_B)$. Such states are ubiquitous in quantum mechanics and essential to its explanations.

Now consider a device that measures the location of the particle. When the state is ψ_A the device displays "A," and when the state is ψ_B the device displays "B." But when the state is ψ_{AB}, the state of the device is not a superposition of displaying "A" and displaying "B"; rather, it displays "A" half the time and "B" half the time. Indeed, it is hard to see what a superposition of displaying "A" and displaying "B" could mean.

Von Neumann (1932, 186) dealt with this problem by proposing distinct dynamical laws for measurements and non-measurements. Between measurements, the quantum state evolves according to the continuous, linear Schrödinger dynamics—von Neumann's "process 2." But during a measurement, the quantum state jumps discontinuously into one of the eigenstates of the measured observable, with probabilities given by the square of the coefficient on that eigenstate in the premeasurement state. This is von Neumann's "process 1," otherwise known as the *collapse postulate*. Applied to state $\psi_{AB} = 2^{-1/2}(\psi_A + \psi_B)$, the collapse postulate says that a measurement of the location of the particle causes the state to jump either to ψ_A or to ψ_B, with probabilities 1/2 each.

The trouble with von Neumann's proposal is that the two dynamical laws are incompatible: neither can be reduced to the other. Furthermore, nothing physically distinctive takes place during a measurement: a measurement isn't a distinct kind of physical *process*, it is just one that has particular pragmatic importance *to us*. So there seems to be no *physical* distinction between measurements and non-measurements such that different fundamental laws could apply to measurements and non-measurements.

One response to this predicament is just to reject the collapse postulate, so that the Schrödinger dynamics applies to all systems at all times. This is the response embodied (in different ways) by Bohmian theories (Bohm 1952) and Everettian theories (Everett 1957). But a different response is also available—namely to precisify the collapse postulate so that it makes no essential appeal to measurement. This is the approach pursued by spontaneous collapse theories.

The original spontaneous collapse theory—the GRW theory of Ghirardi, Rimini and Weber (1986)—incorporates two incompatible dynamical processes, just like von Neumann's account. The first is the Schrödinger dynamics, which applies most of the time. The second is a collapse process, which localizes the quantum state in the coordinates of one particle. That is, for each particle in a system, there is a chance of 10^{-16} per second that it will undergo a collapse, and if it does so, then the quantum state is multiplied by a Gaussian of width $10^{-5} cm$ in the coordinates of that particle. The location of the center of the collapse is chosen at random so as to reproduce the statistical results of von Neumann's collapse postulate.

Unlike von Neumann's account, though, the GRW approach does not tie collapse to measurement: collapses just naturally occur at a certain rate, whether anyone is measuring a system or not. But the collapse process can nevertheless *explain* why measurements have unique outcomes. For microscopic systems, the collapse rate is so low that collapses are never observed. But for a macroscopic solid object containing on the order of 10^{23} particles, the collapse rate is 10^7 per second, and since the positions of the particles in a solid object are highly correlated, a collapse for one particle suffices to localize the whole object.

Hence the GRW solution to the measurement problem: a single-particle state like ψ_{AB} is stable, but when the location of the particle is measured, its position is correlated with something we can see, such as the position of a pointer on a dial. The particle and pointer now form a composite macroscopic object, and the spontaneous collapse process very rapidly localizes the state of this object, either to a state in which the particle is in region A and the pointer is pointing at "A" or to a state in which the particle is in region B and the pointer is pointing at "B." No special new process occurs during measurement; the measurement just amplifies the naturally occurring spontaneous collapse process due to the inevitable involvement of a large number of correlated particles.

9.3 Problems

As it stands, the GRW theory describes a world without particles. The representational machinery of the theory consists of the quantum state alone, and the quantum state, expressed as a wave function, is a spread-out entity rather than a spatially discrete particle. After a collapse, the wave function "bunches up" in three coordinates, and under certain circumstances, this "bunching" is relatively stable, so that we can use the language of particles and trajectories with some success to describe the way these "bunches" move around. But if the wave function describes all there is, then particles are not fundamental constituents of reality.

The view that some entity represented by the wave function is all there is (at the fundamental level) is called *wave function monism* (Ney 2013, 37). Wave function monism is a natural way of understanding the ontology associated with spontaneous collapse theories, but it raises some difficulties. The first is that the wave function is not defined over ordinary three-dimensional space but over a $3N$-dimensional configuration space, where N is the number of particles in the system concerned (Bell 1987, 44). More precisely, since there are no particles at the fundamental level, the way to put it is that what we ordinarily think of as an N-particle system is described at the fundamental level by a $3N$-dimensional wave function. In particular, since we ordinarily think of the universe as a system containing a very large number of particles, then according to wave function monism, it is better conceived as a single entity residing in a high-dimensional space.

The difficulty here concerns how a fundamentally high-dimensional world can nevertheless appear to us as three-dimensional. Albert (1996) maintains that the explanation is dynamical: the physical laws for macroscopic objects take a particularly simple form if the $3N$ dimensions are grouped into threes, and so creatures like us should be expected to have an internal model of the world as three-dimensional. However, others are sceptical of this dynamical explanation. Monton (2006), for example, argues that if there is no three-dimensional world, it is not clear that there *are* any "creatures like us" with internal models of the world.

A second difficulty with wave function monism stems from the fact that even after a collapse, the wave function in the coordinates of the collapsed particle is still non-zero everywhere. If we require that *all* the wave-function amplitude associated with an object be contained in a particular region if the object is to count as being located in that region, then neither particles nor macroscopic objects ever have determinate locations according to spontaneous collapse theories, and we have no solution to the measurement problem after all (Albert and Loewer 1990). This is called the *tails problem*.

The obvious response here is to relax the requirement that all the wave function amplitude be contained in the relevant region: for an object to be located in a given region, it is sufficient that some large proportion of the wave function amplitude is contained in it (Albert and Loewer, 1996). This introduces a novel kind of vagueness, since presumably there is no precise matter of fact concerning the proportion that is required. And loosening the location requirement in this way produces other odd consequences: for example, each of a large collection of objects can individually count as being located in a region according to the requirement even when the collection as a whole does not count as being in the region (Lewis 1997). This is because when each of a collection of N objects has proportion p of its wave function amplitude in a region, where p is slightly less than 1, the proportion of the wave function of the whole system that is in the subspace of configuration space corresponding to all the objects being in the region is p^N, which can be considerably less than 1. This latter problem is called the *counting anomaly*.

A third difficulty with wave function monism concerns the reconciliation of the spread-out nature of the wave function with special relativity. A collapse in the coordinates of a particular particle affects the whole of space simultaneously, but according to special relativity, there is no absolute standard of simultaneity, so the collapse seems ill defined. Furthermore, for entangled particles, a collapse for any one of them instantaneously localizes all of them, no matter how far apart they are, which is in prima facie conflict with relativistic prohibitions on faster-than-light causation (Bell 1987).

None of these difficulties is unique to spontaneous collapse theories. The dimensionality problem and the counting anomaly affect any account of quantum mechanics that takes the wave function to represent a physical entity, including most ways of understanding Bohmian and Everettian theories (Lewis 2016, 97 and 151). The difficulty with relativity also affects Bohmian theories, as the Bohmian particle dynamics apparently requires an absolute standard of simultaneity (Bell 1981). Nevertheless, this particular combination of problems is most frequently addressed in the context of spontaneous collapse theories, and a particular kind of solution—adding ontology—is also most prevalent in the context of spontaneous collapse theories. It is this kind of ontological solution that I am interested in here.

9.4 Ontological Solutions

The original proposal for adding ontology to a spontaneous collapse theory was made by Ghirardi, Grassi, and Benatti, (1995). They suggest adding a mass density distribution to the theory, defined in terms of the quantum state. Precisely, if $M(\mathbf{r})$ is an operator for average mass density, then the mass density distribution $\mathcal{M}(\mathbf{r}, t)$ can be defined in terms of the quantum state $\psi(t)$ as $\mathcal{M}(\mathbf{r}, t) = \langle \psi(t) | M(\mathbf{r}) | \psi(t) \rangle$. It is the mass density distribution that directly accounts for the locations of macroscopic objects and our perceptions of them.

Their motivations for adding a mass density distribution to the theory allude to the problems mentioned in the previous section. The mass density, unlike the quantum state, lives "in ordinary space" (Ghirardi, Grassi, and Benatti 1995, 7). They regard this as an improvement over an ontology in $3N$-dimensional space, at least in the sense that it is "more traditional" (Ghirardi, Grassi, and Benatti 1995, 37). Furthermore, they explain in detail how the mass density distribution avoids the tails problem: in essence, for a macroscopic object, the spontaneous collapse process localizes the wave function to such an extent that the mass density outside the immediate location of the object is undetectably small (Ghirardi, Grassi, and Benatti 1995, 22–25).

A later article explains how the mass density ontology disposes of the counting anomaly (Bassi and Ghirardi 1999). If an object is a region of high mass density (surrounded by lower mass density), then when the whole of a collection of objects has a particular location precisely when each of the objects individually has that location. Since the location of objects according to the mass density ontology directly depends on the mass density in a particular region of three-dimensional space and not directly on the proportion of wave function amplitude in a particular subspace of configuration space, the counting anomaly does not arise.

More recently, the mass density ontology has been taken up as an example of *primitive ontology* in quantum mechanics. Proponents of primitive ontology include Dürr, Goldstein, and Zanghì (1992), Allori (2013), and Maudlin (2013). The central constraints on primitive ontology are that its entities reside in three-dimensional space and that the behaviour of these entities underwrites all the properties of the macroscopic objects we interact with (Allori 2013, 60). Clearly the mass density ontology fulfils these desiderata. The claim of the primitive ontologist is that any adequate quantum mechanical theory must be written in terms of the behavior of some or other primitive ontology.

Allori and Maudlin both stress the importance of the three-dimensionality of the primitive ontology. Allori (2013, 62–63) echoes Ghirardi, Grassi, and Benatti's claim that the mass density ontology is "traditional": "if you can account for everything you need to account for employing already successful and well-tested explanatory techniques, why not do so?" Wave function monism, on this view, is just too radical a break with the past. Maudlin goes further: the primitive ontology (or *primary ontology* in his terminology) *has* to be three-dimensional in order to determine our observations of three-dimensional objects (2013, 148). Every empirically adequate theory has to postulate a three-dimensional primitive ontology.

Primitive ontologies have also been suggested for other accounts of quantum mechanics. A mass density ontology has been developed for the branching worlds of Everettian quantum mechanics (Allori et al. 2011). Dürr, Goldstein, and Zanghì (1992) and Maudlin (2013) explore Bohmian quantum mechanics, and identify Bohmian particles as the primitive ontology in this case.

Conversely, other primitive ontologies have been proposed for spontaneous collapse theories. Most prominent is Tumulka's *flash ontology* (2006), based on some comments by Bell (1987), according to which the primitive ontology consists of discrete events at space–time points corresponding to the centers of GRW collapse events. This ontology too, can straightforwardly address the issue of dimensionality, since the flashes are defined in three-dimensional space. It also bypasses the counting anomaly, since an object according to the flash ontology is just a swarm of flashes. Since the location of an object depends directly on the locations of flashes in three-dimensional space and does not depend directly on the proportion of the wave function in a subspace of configuration space, the counting anomaly does not arise.

Additionally, the flash ontology suggests a way forward concerning the conflict with special relativity: since the flashes are not spread out in space, a collapse event per se does not require instantaneous change everywhere. Indeed, the hope of a reconciliation with special relativity is the primary motivation behind Tumulka's adoption of the flash ontology. Tumulka proposes that, given a flash event at a particular space–time point, the probability distribution for the location of the next flash event is defined over surfaces that are relativistically invariant—that is, over surfaces such that $x^2 + y^2 + z^2 - c^2t^2$ is a constant—rather than over simultaneity surfaces for which t is a constant. Hence for a single particle, a "flashy" version of the GRW theory can be made consistent with special relativity. But for two or more particles, there is still the worry that for entangled particles, a flash for one particle can instantly affect the future flash probabilities of the other, no matter how distant.

9.5 The Primitive Ontologist's Dilemma

The story so far is that wave function monism is problematic as an ontology for spontaneous collapse theories and that either the mass density or the flash ontology can address some or all of these problems. This story can be resisted by arguing (as I have done elsewhere) that the dimensionality of the wave function and the counting anomaly are not really *problems* that need to be addressed (Lewis 2003; Lewis 2013). But perhaps some of them still look like problems—and certainly the problem of reconciling spontaneous collapse theories with special relativity remains. Here I take a different line: to the extent that dimensionality, the counting anomaly and consistency with special relativity are problems for spontaneous collapse theories, primitive ontology cannot provide distinctive solutions, despite initial appearances.

The argument is inspired by Hempel's, (1965) theoretician's dilemma. Hempel points out that theoretical entities, if they do the job they are supposed to do, are redundant. And if they don't do the job they are supposed to do, they are (trivially) redundant. So they are redundant. The job, as Hempel sees it, is to serve in explanations of observable phenomena. And the reason they are redundant if they do their job is that an explanation is a deductive argument from premises describing observable initial conditions to a conclusion describing an observable phenomenon, by way of suitable universal generalizations (laws). So an argument that uses laws mentioning theoretical entities can be replaced without loss by an argument whose laws mention only observable things.

Hempel's theoretician's dilemma is only a challenge to the extent that one buys into his deductive-nomological account of explanation. If explanations in physics are not typically deductive arguments, then the fact that one can replace a deductive argument mentioning theoretical entities with one that does not has no significant consequences. In particular, theoretical entities might well be essential elements in physical explanations.

But now consider the use of primitive ontology to solve the problems facing spontaneous collapse theories. These problems are not (directly) causal in nature, and the role of primitive ontology is not to act as a *causal* intermediary between the wave function and the phenomena. Rather, the relevant dependency is something like determination rather than causation. The wave function at a time determines a mass density distribution at that time over three-dimensional space, solving the dimensionality problem and the counting anomaly. Similarly, the evolution of the wave function over a period of time determines a distribution of flashes over the relevant chunk of four-dimensional space–time, solving the dimensionality problem and the counting anomaly and hopefully getting us closer to a reconciliation of spontaneous collapse theories with special relativity.

Even if causal explanations are not typically deductive arguments, explanations involving determination relations quite plausibly do typically involve deductive arguments. In a classical world, a suitable arrangement of particles *entails* that there is a table here. Similarly in a spontaneous collapse world with primitive ontology: a suitable wave function *entails* a particular mass density distribution, which in turn *entails* that there is a table here. Alternatively, a suitable wave function evolution *entails* a particular distribution of flashes, which in turn *entails* that there is a table here.

In either case, the wave function entails the state of the primitive ontology, and the state of the primitive ontology entails the state of observable things. Hence Hempel's dilemma seems to apply here: if the primitive ontology succeeds in its role as an intermediate step in this deductive chain, then it is redundant. If the table supervenes on the flashes and the flashes on the wave function, then the table supervenes directly on the wave function. If the primitive ontology doesn't succeed in this role, then it is trivially redundant. So it is redundant.

In particular, since the state of the primitive ontology is entailed by the wave function, whatever structure there is in the primitive ontology that allows it to address the dimensionality problem, the counting problem, and the conflict with relativity is structure that is already present in the wave function. So if primitive ontology succeeds at solving these problems, it is redundant. And of course, if it *fails* to solve the problems, then it is redundant. The upshot is that primitive ontology plays no crucial role in solving the problems afflicting spontaneous collapse theories; the wave function by itself can perform all the relevant work. So why did we think we needed it in the first place?

The response may be: The primitive ontology is not supposed to be an additional layer of reality residing between the fundamental wave function and the observable world. Rather, the primitive ontology is an ontological *replacement* for the wave function. In particular, supporters of the primitive ontology approach often try to reconceive the wave function as a *law* governing the evolution of the primitive ontology (Dürr, Goldstein, and Zanghì 1992; Allori 2013).

But this is a difficult road to follow. The wave function in nonrelativistic quantum mechanics changes over time, so in this context one would have to admit that the laws of nature change over time. Perhaps this is not impossible to contemplate (Callender 2015), but it is certainly at odds with the claim often made in defense of primitive ontology that it is traditional, whereas wave function monism is a radical break with the past. If the primitive ontology approach requires time-evolving laws, this is a quite radical break too—arguably more radical than admitting a fundamental entity that is most readily defined over a high-dimensional space.

Alternatively, one can hope that in the ultimate relativistic quantum theory, the quantum state will not change over time (Dürr, Goldstein, and Zanghì 1992). Goldstein and Zanghì (2013) show how, in a Bohmian theory with a time-independent wave function, the particles can still move, and hence the wave function can act as a law directing the primitive ontology. But no such approach is applicable to a spontaneous collapse approach, in which the dynamical role of the wave function in undergoing collapses seems essential to the theory. Certainly the most straightforward approach to understanding spontaneous collapse theories is that the quantum state is indeed a *state* and hence a representation of the ontology of the system.

9.6 Primitive Ontology Made Easy

The argument of the previous section is that primitive ontology is redundant. In causal contexts, redundancy arguments are typically followed by non-existence claims: if a postulated entity is explanatorily redundant, then we shouldn't say it exists. For example, if an absolute rest frame plays no explanatory role in mechanics, then we should say that there is no absolute rest frame. But that wouldn't be appropriate in the case of primitive ontology. The issue here isn't causal redundancy but what we might call *compositional* redundancy: if tables are nothing but distributions of mass density, and mass density distributions are nothing but wave function distributions, then tables are nothing but wave function distributions (and similarly for the flash ontology). Compositional redundancy doesn't lend itself to non-existence claims. I am nothing but an arrangement of cells (let's say), and my cells are nothing but an arrangement of molecules, but it doesn't follow that cells don't exist. Indeed, it trivially follows from the existence and arrangement of the molecules that my cells exist.

This is an instance of the *easy* approach to ontology of Thomasson (2014). Tables exist, because all it takes for a table to exist is that the more fundamental ontology is arranged in a certain way. So, by the same token, if the wave function represents something physical, then the mass density ontology exists—and so does the flash ontology. The wave function distribution instantiates a particular mass density distribution over three-dimensional space. And the evolution of the wave function over time instantiates a set of flashes in space–time, since collapse events exist, and a flash is nothing but the center of a collapse event.

So the redundancy argument of the previous section should not be taken to be an argument against the *existence* of either the mass density ontology or the flash ontology. Rather, it should be taken as an argument that we don't have to explicitly *postulate* them or to decide between them, since they both come along "for free" with wave function monism.

Indeed, both ontologies are *important*. The wave function is a highly structured entity. It contains the structures exemplified by the mass density distribution and by the flash distribution, in the sense that it entails both these structures. These structures do not *exhaust* the structure of the wave function, of course: the wave function entails the mass density distribution and the flash distribution, but not vice versa.

This is reflected in the fact that the quantum state cannot be represented in four-dimensional space–time. (At least, it cannot *easily* be represented in a four-dimensional space–time: see Wallace and Timpson (2010) for a proposal about how to do so.) But nevertheless, the space–time structure is *there*, and the existence of three-dimensional primitive ontology *shows* that it is there. Similarly, the wave function contains structures we can use to solve the counting anomaly, and the use of the mass density language or the flash language to solve the counting anomaly shows this.

What about the conflict with special relativity? This is not so easily laid to rest. But the fact that the flash structure is present in the wave function gives some hope for the success of Tumulka's program. We do not need to say in addition that the wave function itself doesn't exist, even if there is a sense in which it is spread out in three-dimensional space (i.e., the sense expressed by the mass density distribution). Because of this spread-out nature, the wave function has different properties in different frames of reference, but this doesn't impugn it (Myrvold 2002). After all, cars and tunnels have different properties in different frames of reference: in one frame, the car is longer than the tunnel, and in another, the tunnel is longer than the car (Maudlin 1994, 54). And they certainly exist.

It is worth contrasting this with the case of Bohm's theory. Dürr, Goldstein, and Zanghì (1992) and Maudlin (2013) count the particles of Bohm's theory as an example of primitive ontology, but it is important to note that, unlike the primitive ontologies proposed for spontaneous collapse theories, the existence and locations of the Bohmian particles are not entailed by the wave function. Rather, they have to be postulated separately, along with a new dynamical law to govern their motion. If you don't postulate particles in addition to the wave function in a no-collapse context, you end up with the ontology of a many-worlds theory. Bohm's theory and many-worlds theories are very different, with different strengths and weaknesses. So here there is a genuine and difficult ontological question. In the case of spontaneous collapse theories, though, questions about the existence of primitive ontology are easily answerable in the affirmative.

9.7 Conclusion

Often the ontology of spontaneous collapse theories is presented as a choice: either one adopts wave function monism, or one adopts a mass density ontology, or one adopts a flash ontology (e.g., Ney 2013, 40). My argument here is that there is no pressing need to make such a choice. The wave function ontology entails the mass density ontology and the flash ontology, and with it the solutions to the apparent problems facing spontaneous collapse theories. Indeed, using the language of a mass density distribution or a flash distribution may be useful in explaining exactly how the problems can be resolved. But using these languages doesn't undermine the idea that ultimately, for a spontaneous collapse theory, all there is is the wave function.

References

Albert, David Z. (1996), "Elementary quantum metaphysics," in J. Cushing, A. Fine and S. Goldstein (eds.), *Bohmian Mechanics and Quantum Theory: An Appraisal*. Dordrecht: Kluwer, 277–284.

Albert, David Z. and Barry Loewer (1990), "Wanted dead or alive: two attempts to solve Schrödinger's paradox," in A. Fine, M. Forbes and L. Wessels (eds.) *PSA 1990: Proceedings of the 1990 Biennial Meeting of the Philosophy of Science Association*, vol. 1. Chicago: University of Chicago Press, 277–285.

Albert, David Z., and Barry Loewer (1996), "Tails of Schrödinger's cat," in R. Clifton (ed.), *Perspectives on Quantum Reality*. Dordrecht: Kluwer, 81–92.

Allori, Valia (2013), "Primitive ontology and the structure of fundamental physical theories," in Ney and Albert (2013), 58–75.

Allori, Valia, Sheldon Goldstein, Roderich Tumulka, and Nino Zanghì (2011). "Many worlds and Schrödinger's first quantum theory," *British Journal for the Philosophy of Science* 62: 1–28.

Bassi, Angelo, and GianCarlo Ghirardi (1999), "More about dynamical reduction and the enumeration principle," *British Journal for the Philosophy of Science* 50: 719–734.

Bell, John S. (1981), "Quantum mechanics for cosmologists," in C. Isham, R. Penrose and D. Sciama (eds.), *Quantum Gravity 2*. Oxford: Clarendon Press, 611–637.

Bell, John S. (1987), "Are there quantum jumps?" in C. W. Kilmister (ed.), *Schrödinger: Centenary Celebration of a Polymath*. Cambridge: Cambridge University Press, 41–52.

Bohm, David (1952), "A suggested interpretation of the quantum theory in terms of "hidden" variables, parts I and II," *Physical Review* 85: 166–193.

Callender, Craig (2015), "One world, one beable," *Synthese* 192: 3153–3177.

Dürr, Detlef, Sheldon Goldstein, and Nino Zanghì (1992), "Quantum equilibrium and the origin of absolute uncertainty," *Journal of Statistical Physics* 67: 843–907.

Everett III, Hugh (1957), "'Relative state' formulation of quantum mechanics," *Reviews of Modern Physics* 29: 454–462.

Ghirardi, GianCarlo, Renata Grassi, and Fabio Benatti (1995), "Describing the macroscopic world: closing the circle within the dynamical reduction program," *Foundations of Physics* 25: 5–38.

Ghirardi, GianCarlo, Alberto Rimini, and Tullio Weber (1986), "Unified dynamics for microscopic and macroscopic systems," *Physical Review D* 34: 470–491.

Goldstein, Sheldon, and Nino Zanghì (2013), "Reality and the role of the wave function in quantum theory," in Ney and Albert (2013), 91–109.

Hempel, Carl (1965), "The theoretician's dilemma: a study in the logic of theory construction," in *Aspects of Scientific Explanation*. New York: Free Press, 173–226.

Lewis, Peter J. (1997), "Quantum mechanics, orthogonality, and counting," *British Journal for the Philosophy of Science* 48: 313–328.

Lewis, Peter J. (2003), "Four strategies for dealing with the counting anomaly in spontaneous collapse theories of quantum mechanics," *International Studies in the Philosophy of Science* 17: 137–142.

Lewis, Peter J. (2013), "Dimension and illusion," in Ney and Albert (2013), 110–125.

Lewis, Peter J. (2016), *Quantum Ontology: A Guide to the Metaphysics of Quantum Mechanics*. Oxford: Oxford University Press.

Maudlin, Tim (1994), *Quantum Non-Locality and Relativity*. Oxford: Blackwell.

Maudlin, Tim (2013), "The nature of the quantum state," in Ney and Albert (2013), 126–153.

Monton, Bradley (2006), "Quantum mechanics and 3*N*-dimensional space," *Philosophy of Science* 73: 778–789.
Myrvold, Wayne C. (2002), "On peaceful coexistence: Is the collapse postulate incompatible with relativity?" *Studies in History and Philosophy of Modern Physics* 33: 435–466.
Ney, Alyssa (2013), "Introduction," in Ney and Albert (2013), 1–51.
Ney, Alyssa and David Z. Albert (eds.) (2013), *The Wave Function*. Oxford: Oxford University Press.
Thomasson, Amie L. (2014), *Ontology Made Easy*. Oxford: Oxford University Press.
Tumulka, Roderich (2006), "Collapse and relativity," in A. Bassi, D. Dürr, T. Weber and N. Zanghì (eds.), *Quantum Mechanics: Are There Quantum Jumps?* and *On the Present Status of Quantum Mechanics*. AIP Conference Proceedings 844: 340–352.
von Neumann, Johann (1932), *Mathematische Grundlagen der Quantenmechanik*. Berlin: Springer.
Wallace, David, and Christopher G. Timpson (2010), "Quantum mechanics on spacetime I: Spacetime state realism," *British Journal for the Philosophy of Science* 61: 697–727.

10

Collapse or No Collapse? What Is the Best Ontology of Quantum Mechanics in the Primitive Ontology Framework?

MICHAEL ESFELD

Recalling the state of the art in the interpretation of quantum physics, this chapter emphasizes that one cannot simply add a collapse parameter to the Schrödinger equation in order to solve the measurement problem. If one does so, one is also committed to a primitive ontology of a configuration of matter in physical space in order to have something in the ontology that constitutes the determinate measurement outcomes. The chapter then argues that in the light of this consequence, the collapse postulate loses its attractiveness in comparison to an ontology of persisting particles moving on continuous trajectories according to a deterministic law.

10.1 The State of the Art: The Measurement Problem and Beyond

The measurement problem is the central issue in the formulation and understanding of any quantum theory, since it concerns the link between the theory and the data. The standard setup of this problem for non-relativistic quantum mechanics (QM) is provided by ([1], p. 7):

(1) The wave-function of a system is *complete*, i.e., the wave-function specifies (directly or indirectly) all of the physical properties of a system.

(2) The wave-function always evolves in accord with a linear dynamical equation (e.g. the Schrödinger equation).

(3) Measurements of, e.g., the spin of an electron always (or at least usually) have determinate outcomes, i.e., at the end of the measurement the measuring device is either in a state which indicates spin up (and not down) or spin down (and not up).

Any two of these propositions are consistent with one another, but the conjunction of all three of them is inconsistent. This can be easily illustrated by means of Schrödinger's cat paradox ([2], p. 812): if the cat is completely described by the wave function and if the wave function always evolves according to the Schrödinger equation, then, due to the linearity of this wave equation, superpositions and entangled states will in general be preserved. Consequently, a measurement of the cat will in general not have a determinate outcome: at the end of the measurement, the cat will not be in the state of either being alive or being dead.

Hence, the measurement problem is not just a—philosophical—problem of the interpretation of a given formalism. It concerns also the very formulation of a consistent quantum theory. Even if one abandons (3), one has to put forward a formulation of quantum physics that establishes a link with at least the appearance of determinate measurement outcomes. If one retains (3), one has to develop a formulation of a quantum theory that goes beyond a theory in which only a wave function and a linear dynamical equation for the evolution of the wave function figure. Accordingly, the formulation of a consistent quantum theory can be divided into many-worlds theories, rejecting (3), collapse theories, rejecting (2), and additional variable theories, rejecting (1).

However, research in the last decade has made clear that we do not face three equally distinct possibilities to solve the measurement problem but just two: the main dividing line is between endorsing (3) and rejecting it. If one endorses (3), the consequence is not that one has to abandon *either* (1) *or* (2) but that one has to amend *both* (1) *and* (2). Determinate measurement outcomes as described in (3) are outcomes occurring in ordinary physical space, that is, in three-dimensional space or four-dimensional space–time. Hence, endorsing (3) entails being committed to the existence of a determinate configuration of matter in physical space that constitutes measurement outcomes (such as a live cat or an apparatus configuration that indicates spin up, etc.). If one does so, one cannot stop at amending (2). The central issue then is not whether a collapse term for the wave function has to be added to the Schrödinger equation, because even with the addition of such a term, this equation still is an equation for the evolution of the wave function, by contrast to an equation for the evolution of a configuration of matter in physical space. Consequently, over and above the Schrödinger equation—however amended—a law or rule is called for that establishes an explicit link between the wave function and the configuration of matter in physical space. By the same token, (1) has to be changed in such a way that reference is made to the configuration of matter in physical space and not just the quantum state as encoded in the wave function.

Here and in the following, the configuration of matter in physical space is intended to be the *entire* configuration of matter of the universe at a given time (recall that we are concerned with non-relativistic QM); accordingly, the wave function is intended to be the *universal* wave function, that is, the wave function of the entire universe. In other words, the quantum theory we are after is a universal theory—that is, a theory whose laws apply to the entire universe— like classical mechanics. It is of course not the final theory, again like classical mechanics.

In the literature, the configuration of matter in physical space is known as the *primitive ontology* of quantum physics.[1] There are three proposals to spell out what that configuration is:

[1] *Point particles* that consequently always have a determinate position in physical space (there cannot be point particles in physical space without these particles being located somewhere in that space, independently of whether or not we are able to determine that location). The corresponding quantum theory is *Bohmian mechanics (BM)*, which adds particle positions to the quantum state as given by the wave function and a law for the evolution of these positions to the Schrödinger equation. That law is known as the guiding equation. The wave function figures in that law, its job being to determine the velocity of the particles at a time t, given their positions at t. This theory goes back to [6] and [7]. Its dominant contemporary version is the one of [3] (and see the textbook [8]).

[2] *A matter density field* that stretches all over physical space, having varying degrees of density at different points or regions of space. The corresponding quantum theory is the *GRW matter density theory (GRWm)*, which uses the collapse formalism of Ghirardi, Rimini, and Weber ([9]) in the version of a Continuous Spontaneous Localization (CSL) of the wave function in configuration space ([10]). It establishes a link between the wave function in configuration space and the matter in physical space in such a way that the wave function and its evolution according to the GRW-CSL equation describe a wave or field filling all of physical space and the evolution of that field in physical space ([11]).

[3] *Isolated point events in physical space, called "flashes."* These flashes are ephemeral. They do not stretch beyond the space–time point at which they occur. The corresponding quantum theory is the *GRW flash theory (GRWf)*, which uses the original collapse formalism of Ghirardi, Rimini, and Weber ([9]) with the wave function making occasional jumps so that it undergoes a spontaneous localization in configuration space. A rule then is added to the GRW equation stating that whenever a spontaneous localization of the wave function occurs in configuration space, a flash shows up at a point in physical space. The flash theory has been proposed as an ontology of the GRW formalism by Bell [5, ch. 22, originally published 1987]. The term "flash" was coined by Tumulka [12, p. 826].

---

[1] This term goes back to Dürr et al. [3, ch. 2], originally published 1992. A forerunner of this notion can be found in Mundy [4, p. 46]. Cf. also Bell's notion of "local beables" in Bell [5, ch. 7], originally published 1975.

Despite their differences, these three quantum theories have a number of important features in common ([13] were the first to work out their common structure). In the first place, (a) they are all committed to the matter in physical space being primitive objects. That is to say, these objects do not have any physical properties over and above their being localized in physical space. Hence, by contrast to what one can maintain about classical particles, these objects cannot be considered as having an intrinsic mass or an intrinsic charge (or an intrinsic spin)[2]. They do not have any intrinsic properties. As regards BM, experimental considerations involving interference phenomena—for instance in the context of the Aharonov-Bohm effect and of certain interferometry experiments—show, in brief, that mass and charge are effective at all the possible particle locations that the wave function admits; they can hence not be considered as properties that are localized at the Bohmian particle positions (see, e.g., [15, 16] and references therein; cf. also most recently [17]). In brief, any property that one may contemplate attributing to the particles over and above their position is in fact situated at the level of their quantum state as represented by the wave function instead of being a candidate for an intrinsic property of the particles. The same goes for the flash-events on the GRWf theory. When it comes to the GRWm theory, the matter density field is a primitive stuff filling all of space, which admits different degrees of density as a primitive matter of fact, but which has no further physical properties, as [18] point out:

> Moreover, the matter that we postulate in GRWm and whose density is given by the m function does not ipso facto have any such properties as mass or charge; it can only assume various levels of density. ([18], pp. 331–332)

Furthermore, (b) the fact that all the primitive ontology theories of quantum physics infringe upon proposition (1) of the measurement problem comes out clearly when one enquires into their consequences for our knowledge: as [19] establish, not only the Bohmian particle positions—and hence the particle trajectories—are not always accessible to an observer, but also the GRWm matter density field and the GRWf flash distribution are not entirely accessible to an observer, although the latter are specified by the wave function, whereas the Bohmian particle positions are not specified by the wave function. Hence, one does not avoid what may seem to be a drawback of subscribing to so-called hidden variables, namely a limited epistemic accessibility, by amending the Schrödinger equation and taking the wave function as it figures in such an amended Schrödinger equation to represent the configuration of matter in physical space. In a nutshell, if one is committed to a configuration of matter in physical space in quantum physics, one also has to

---

[2] See Bell [5, ch. 4], originally published 1971, and [14] for the Bohmian treatment of spin. Similar remarks apply to the GRWm and GRWf theories.

endorse a commitment to a limited epistemic accessibility of that configuration, in whatever way one spells out the theory of that configuration and its evolution.

By way of consequence, (c) probabilities enter into any primitive ontology theory of QM through our ignorance of the exact configuration of matter in physical space. This ignorance implies that we immediately have to resort to probabilistic descriptions in QM, independently of whether the dynamical law for the evolution of the configuration of matter in physical space is deterministic (as in Bohmian mechanics) or stochastic (as in the GRW theory) (see, e.g., [20] on that introduction of probabilities).

Finally, (d) endorsing a primitive ontology theory of quantum physics does not commit us to subscribing to an ontological dualism of matter in physical space (the primitive ontology, consisting in primitive objects) and the quantum state (the wave function) in configuration space. Such a dualism would be highly implausible; for instance, it would be unintelligible how a wave function, being a field on configuration space, could interact with matter in physical space by influencing its motion. There are several proposals in the literature how to conceive the wave function in the primitive ontology framework that all avoid such an implausible dualism: (i) The most straightforward proposal is to ban the wave function as an additional entity from the ontology altogether. The claim thus is that the primitive ontology is the entire ontology. Given the spatial distribution of the elements of the primitive ontology throughout the whole history of the universe, the universal wave function and its evolution are part of the best descriptive system, that is, the system that achieves the best balance between being simple and being informative about that distribution. The wave function hence has only a descriptive role by contrast to an ontological one. This stance is known as quantum Humeanism. It has been worked out in the philosophy literature mainly by taking BM as an example ([21], [22], [23], [24]), but there also exists a concrete model in the physics literature that uses the GRWf theory ([25]). (ii) When one admits the wave function to the ontology, one can still regard it as referring to an entity in physical space. The most prominent proposal in that respect in the philosophical literature is to take the wave function to refer to dispositions of the elements of the primitive ontology for a certain evolution under certain circumstances, namely either one holistic disposition of the entire configuration ([26]) or a multitude of dispositions of each individual element whose manifestations depend on the other elements ([27]). (iii) The corresponding proposal in the physics literature, conceiving the wave function as an entity in physical space, is the one of regarding it either as a multi-field in physical space—that is, a field that attributes a value not to single space–time points, but only to an entire bunch of them ([28], ch. 6.2)—or as referring to a multitude of fields associated with each element of the primitive ontology (such as each Bohmian particle; [29]).

Given this state of the art, the next task then is to evaluate the theories that solve the measurement problem by being committed to a primitive ontology of matter distributed in physical space. That is, one seeks an answer to the following two questions: What is the best proposal for the physical objects? What is the best proposal for the dynamics? This chapter sets out to answer these two questions (see also [30]).

10.2 What Is the Best Proposal for the Physical Objects?

Atomism is the oldest and most influential tradition in natural philosophy, going back to the pre-Socratic philosophers Leucippus and Democritus and having been turned into a precise physical theory by Newton. Atomism offers a clear and simple explanation of the realm of our experience. Macroscopic objects are composed of point particles. All the differences between the macroscopic objects—at a time as well as in time—are accounted for in terms of the spatial configuration of these particles and its change, which is subject to certain laws. That is why Feynman famously writes at the beginning of the *Feynman lectures on physics*:

If, in some cataclysm, all of scientific knowledge were to be destroyed, and only one sentence passed on to the next generations of creatures, what statement would contain the most information in the fewest words? I believe it is the *atomic hypothesis* (or the atomic *fact*, or whatever you wish to call it) that *all things are made of atoms—little particles that move around in perpetual motion, attracting each other when they are a little distance apart, but repelling upon being squeezed into one another.* In that one sentence, you will see, there is an enormous amount of information about the world, if just a little imagination and thinking are applied. ([31], ch. 1–2)

When it comes to quantum physics, atomism—that is, an ontology of permanent particles moving on continuous trajectories—is implemented in BM. What changes in BM with respect to classical mechanics is the law of motion for these particles, namely a non-local law that takes quantum entanglement into account.

There are three main reasons to take an ontology of particles to be the best proposal also in the domain of quantum physics: (i) In the first place, also in this domain, all experimental evidence is evidence of discrete objects (i.e., particles)—from dots on a display to traces in a cloud chamber. Entities that are not particles—such as waves or fields—come in as figuring in the explanation of the behavior of the particles, but they are not themselves part of the experimental evidence. For instance, the double slit experiment is made apparent by particles hitting on a screen. (ii) The argument from composition is not touched by the transition from classical to quantum physics: from the chemical elements on to all macroscopic objects, everything is composed of elementary quantum particles. (iii) Consequently, all other proposals are parasitic on the particle proposal: even if they do not admit particles in the ontology—such as the GRWm theory—their formalism for QM works in terms of a fixed number of N permanent particles, which defines the configuration space; its dimension is $3N$, with each point of that space corresponding to a possible particle configuration in physical space. Hence, these other proposals have to interpret a particle configuration formalism as not representing a particle configuration.

However, why should one retain particles and thus trajectories in the light of the evidence from quantum physics? The mentioned arguments speak in favor of an ontology of discrete objects, but one may wonder whether this has to be an ontology of particles. The GRWf ontology of single, discrete events can be considered as a particle ontology without the trajectories so that what remains of the particles are isolated events in space–time. However, the three aforementioned arguments can be adapted in such a way that they single out particles over flashes: (iii) applies not only to the GRWm ontology, but also to the GRWf ontology: the flashes are discontinuous, and there is no intertemporal identity of them, since the flashes do not persist in time. But the formalism is based on a fixed number of persistent particles. (ii) This observation applies also to the argument from composition: there is nothing in the GRWf ontology that could ground the intertemporal identity of macroscopic objects.

As regards (i), it is true that the particle evidence in quantum physics is evidence of discrete objects, which could be point events like point particles. But the evidence of discrete objects in physical space is not evidence of a physical space into which discrete objects are inserted; it is evidence only of relative positions of discrete objects, that is, distances among discrete objects. However, by renouncing on persisting objects, the GRWf ontology is committed to the existence of an absolute background space into which the flashes are inserted and an absolute background time in which the flashes show up (there can even be times at which there are no flashes at all in the universe). By contrast, although BM is usually also formulated in terms of the particles moving in a background space and a background time, the ontological commitment to a background space and time can be dispensed with in BM. The ontology of BM can be conceived only in terms of distance relations among particles that change, with the representation of that change in terms of particle trajectories in a background space and time being only a means of representation of the primitive ontology instead of implying the ontological commitment to an absolute space and time (see [32] and [33]). By the same token, one can regard the universal wave function in BM as a means of representation only instead of as an element of the ontology over and above the particles, as mentioned at the end of the preceding section.

The ontology of matter being one continuous stuff, known as gunk, that fills all of space is as old as the ontology of atomism, going back to the pre-Socratic natural philosophers as well. This view does not have to commit itself to points, neither to material points nor to points of space (see [34]). It is hence not tied to endorsing an absolute background space: it can be construed as being committed to a continuous stuff that is extended, but not to an absolute space that is distinct from that stuff and into which that stuff is inserted.

In order to accommodate variation, gunk cannot be conceived as being homogeneous throughout space. To take variation into account, one has to maintain that there is more stuff in some parts of space and less stuff in others. Atomism conceptualizes variation in terms of different distances among the discrete point particles so that some particles are situated close to one another, whereas others are further apart: there are clusters of point particles with distances among them that are smaller than the distances that these particles bear to particles outside such a cluster. By contrast, the gunk ontology cannot accommodate variation throughout space in terms of the concentration of primitive point particles. It therefore faces this question: What constitutes the fact of there being more matter in some regions of space and less matter in others?

The view of matter being gunk has to acknowledge as a further primitive a variation of the density of gunk throughout space with gunk being more dense in some parts of space and less dense in other parts. That is to say, gunk admits of degrees, as expressed by the m function in the GRWm formalism: there is more stuff in some parts of space than in others, with the density of matter in the parts of space changing in time; otherwise, the theory would not be able to accommodate variation. Formally, one can represent the degrees of density in terms of attributing a value of matter density to the points of space (the m function as evaluated at the points of space), although the matter density stuff, being gunk, is infinitely divisible, and this ontology is not committed to the existence of points of space. The main problem is that it remains unclear what could constitute the difference in degrees of stuff at points of space, if matter just is primitive stuff. The gunk theory thus is committed to the view of matter being a bare substratum with its being a primitive fact that this substratum has various degrees of density in different parts of space. In a nutshell, there is a primitive stuff-essence of matter that furthermore admits of different degrees of density. In comparison to the gunk view of matter, atomism is the simpler and clearer proposal for an ontology of fundamental physics, because it avoids the dubious commitment to a bare substratum or primitive stuff-essence of matter with different degrees of density.

It seems that QM favors a particle ontology, given that the formalism is conceived in terms of a fixed number of permanent particles, but that quantum field theory (QFT) clearly favors the view of matter being a field and thus one continuous stuff filling all of space instead of discrete objects. However, a quantum field never is a field in the sense of a continuous entity filling space that has definite values at the points of space. The fields figuring in the textbook formalism of QFT are operator valued fields, that is, mathematical objects employed to calculate probabilities for obtaining certain measurement outcomes at certain points of space if certain procedures are applied. They hence do not represent properties that occur at points of space or space–time. No ontology can be built on operator valued fields. As regards the experimental evidence, the foregoing remarks apply, namely that also in the domain of QFT, all the experimental evidence is one of discrete objects, such as particle traces in a cloud chamber. When it comes to the wave function, again, like in QM, it is a field on configuration space and not a continuous stuff in physical space.

Consequently, the measurement problem hits QFT in the same way as QM (see [35]). In particular, there is no relativistic theory of measurement. Consequently, problems that primitive ontology theories of quantum physics may face with respect to relativistic physics cannot be counted as an argument against these theories: they solve the measurement problem, and no one has produced a solution to this problem that (a) acknowledges determinate measurement outcomes and (b) is a relativistic theory of interactions, including in particular measurement interactions (see again [35]). BM has to rely on a privileged foliation of space–time into spatial hypersurfaces in its ontology, which can be introduced through the universal wave function (see [36]). However, this foliation does not show up in the deduction of the statistics of measurement outcomes from the dynamical laws. As regards GRWf and GRWm, there are relativistic versions as long as one considers only a distribution of the elements of the primitive ontology throughout the whole of space–time (see [12] for GRWf and [37] for GRWm), but no relativistic versions of interactions (see [38]).

The Bohmian solution of the quantum measurement problem works for QFT in the same way as for QM: as it is a *non sequitur* to take particle trajectories to be ruled out in QM due to the Heisenberg uncertainty relations, so it is a *non sequitur* to take permanent particles moving on definite trajectories according to a deterministic law to be ruled out in QFT due to the statistics of particle creation and annihilation phenomena. In both cases, such an underlying particle ontology is in the position to explain the statistics of measurement outcomes (see [39] and [40] as to how this is achieved in a Bohmian QFT in the framework of what is known as the Dirac sea model). In brief, QFT does not change anything with respect to the evaluation of the proposals for a primitive ontology of quantum physics: the arguments that favor a particle ontology remain valid in the domain of QFT.

Nonetheless, atoms *qua* point particles are theoretical entities. They are not seen by the naked eye when one sees, for instance, dots on a screen as outcomes of the double slit experiment. They are admitted because they provide the best explanation of the observable facts. The simplicity and parsimony of this proposal are part of the case for its being the best explanation. To put it in a nutshell, *particle evidence is best explained in terms of particle ontology*. However, this explanation is not given by the ontology of point particles alone but by this ontology together with the dynamics that is put forward to describe the motion of the particles: it is the dynamics that provides for the stability of the macroscopic objects with which we are familiar. That is why assessing the proposals for a primitive ontology of quantum physics depends not only on the ontology put forward for the physical objects but also on the dynamics that is conceived for these objects.

10.3 What Is Best Proposal for the Dynamics?

BM is usually presented by formulating the laws of motion on the configuration space $\mathbb{R}^{3N}$, where N is the number of particles and $x_1, \ldots, x_k, \ldots, x_N \in \mathbb{R}^{3N}$ represents their positions at time t. The configuration then evolves according to the guiding equation

$$\frac{dx_k}{dt} = \operatorname{Im} \frac{\hbar}{m_k} \frac{\psi^* \nabla_k \psi}{\psi^* \psi}(x_1, \ldots, x_N), \tag{10.1}$$

where $\psi(x_1, \ldots, x_n)$ is the wave function representing the quantum state of the system and Im denotes the imaginary part. The time-evolution of this wave function, in turn, is given by the Schrödinger equation

$$i\hbar \frac{\partial \psi}{\partial t} = \left(-\sum_{j=1}^{N} \frac{\hbar^2}{2m_j} \Delta_j + V(x_1, \ldots, x_n) \right) \psi, \tag{10.2}$$

familiar from standard QM. The non-local character of the guiding equation is manifested in the fact that the velocity of any particle at time t depends on the position of every other particle at time t; the law of motion, in other words, describes the evolution of the particle configuration *as a whole*. This is necessary in order to take quantum non-locality—as illustrated for instance by Bell's theorem—into account.

In GRW, the evolution of the wave function Ψ_t is given by a modified Schrödinger equation. The latter can be defined as follows: the wave function undergoes spontaneous jumps in configuration space at random times distributed according to the Poisson distribution with rate $N\lambda$. Between two successive jumps, the wave function Ψ_t evolves according to the usual Schrödinger equation. At the time of a jump, the kth component of the wave function Ψ_t undergoes an instantaneous collapse according to

$$\Psi_t(x_1, \ldots, x_k, \ldots, x_N) \mapsto \frac{(L_{x_k}^x)^{1/2} \Psi_t(x_1, \ldots, x_k, \ldots, x_N)}{\|(L_{x_k}^x)^{1/2} \Psi_t\|}, \tag{10.3}$$

where the localization operator $L^x_{x_k}$ is given as a multiplication operator of the form

$$L^x_{x_k} := \frac{1}{(2\pi\sigma^2)^{3/2}} e^{-\frac{1}{2\sigma^2}(x_k-x)^2}, \tag{10.4}$$

and x, the centre of the collapse, is a random position distributed according to the probability density $p(x) = \|(L^x_{x_k})^{1/2}\Psi_t\|^2$. This modified Schrödinger evolution captures in a mathematically precise way what the collapse postulate in textbook QM introduces by a *fiat*, namely the collapse of the wave function so that it can represent localized objects in physical space, including in particular measurement outcomes. GRW thereby introduce two additional parameters, the mean rate λ as well as the width σ of the localization operator, which can be regarded as new constants of nature whose values can be inferred from (or are at least bounded by) experiments (such as chemical reactions on a photo plate, double slit experiments, etc.). An accepted value of the mean rate λ is of the order of $10^{15} s^{-1}$. This value implies that the spontaneous localization process for a single particle occurs only at astronomical timescales of the order of $10^{15} s$, while for a macroscopic system of $N \sim 10^{23}$ particles, the collapse happens so fast that possible superpositions are resolved long before they would be experimentally observable. Moreover, the value of σ can be regarded as localization width; an accepted value is of the order of $10^{-7} m$. The latter is constrained by the overall energy increase of the wave function of the universe that is induced by the localization processes.

However, as explained in the first section, modifying the Schrödinger equation is, by itself, not sufficient to solve the measurement problem: to do so, one has to answer the question of what the wave function and its evolution represent. One therefore has to add to the GRW equation a link between the evolution of the mathematical object Ψ_t in configuration space and the distribution of matter in physical space in order to account for the outcomes of experiments and, in general, the observable phenomena. [11] accomplish this task by taking the evolution of the wave function in configuration space to represent the evolution of a matter density field in physical space. This then constitutes the GRWm theory. It amounts to introduce in addition to Ψ_t and its time evolution a field $m_t(x)$ on physical space $\mathbb{R}^3$ as follows:

$$m_t(x) = \sum_{k=1}^{N} m_k \int d^3x_1 \ldots d^3x_N \, \delta^3(x - x_k)|\Psi_t(x_1, \ldots, x_N)|^2. \tag{10.5}$$

This field $m_t(x)$ is to be understood as the density of matter in physical space $\mathbb{R}^3$ at time t (see [13], section 3.1).

By introducing two new dynamical parameters—lambda and sigma—whose values have to be put in by hand, the GRW theory abandons the simplicity and elegance of the Schrödinger equation and the Bohmian guiding equation, without amounting to a physical benefit (there is of course a benefit in comparison to stipulating the collapse postulate by a simple *fiat*, but doing so is no serious theory). Indeed, there is an ongoing controversy whether the GRWm ontology of a continuous matter density field that develops according to the GRW equation is sufficient to solve the measurement problem. The reason is the so-called problem of the tails of the wave function. This problem arises from the fact that the GRW theory mathematically implements spontaneous localization by multiplying the wave function with a Gaussian, such that the collapsed wave function, although being sharply peaked in a small region of configuration space, does not actually vanish outside that region; it has tails spreading to infinity. In the literature starting with [41] and [42], it is therefore objected that the GRW theory does not achieve its aim, namely to describe measurement outcomes in the form of macrophysical objects having a determinate position. However, there is nothing indefinite about the positions of objects according to GRWm. It is just that an (extremely small) part of each objects matter is spread out through all of space. But since the overwhelming part of any ordinary objects matter is confined to a reasonably small spatial region, we can perfectly well express this in our (inevitably vague) everyday language by saying that the object is in fact located in that region (see [43], pp. 418–419, and [44]). Thus, the GRWm ontology offers a straightforward solution to what Wallace [45, p. 56] calls the problem of bare tails.

However, there is another aspect, which is known as the problem of structured tails (see [45], p. 56). Consider a situation in which the pure Schrödinger evolution would lead to a superposition with equal weight of two macroscopically distinct states (such as a live and a dead cat). The GRW dynamics ensures that the two weights do not stay equal, but that one of them (e.g., the one pertaining to the dead cat) approaches unity while the other one becomes extremely small (but not zero). In terms of matter density, we then have a high-density dead cat and a low-density live cat. The problem is that it seems that the low-density cat is just as cat-like (in terms of shape, behavior, etc.) as the high-density cat, so that in fact there are two cat-shapes in the matter density field, one with a high and another one with a low density. There is an ongoing controversy about this problem: Maudlin [46, pp. 135–138] takes it to be knock-down objection against the GRW matter density ontology, whereas others put forward reasons that aim at justifying to dismiss the commitment to there being a low-density cat that is as cat-like as the high-density cat in the matter density field (see notably [47], [48], pp. 150–154, and [49], section 3).

Be that as it may, there arguably is another, more important drawback of the GRW dynamics that concerns the meaning of the spontaneous localization of the wave function in configuration space for the evolution of the matter density field in physical space. To illustrate this issue, consider a simple example, namely the thought experiment of one particle in a box that Einstein presented at the Solvay conference in 1927 (the following presentation is based on de Broglie's version of the thought experiment in [50], pp. 28–29, and on [51]): the box is split in two halves which are sent in opposite directions, say from Brussels to Paris and Tokyo. When the half box arriving in Tokyo is opened and found to be empty, there is on all accounts of QM that acknowledge that measurements have outcomes a fact that the particle is in the half box in Paris.

On GRWm, the particle is a matter density field that stretches over the whole box and that is split in two halves of equal density when the box is split, these matter densities travelling in opposite directions. Upon interaction with a measurement device, one of these matter densities (the one in Tokyo in the example given) vanishes, while the matter density in the other half box (the one in Paris) increases so that the whole matter is concentrated in one of the half boxes. One might be tempted to say that some matter travels from Tokyo to Paris; however, since it is impossible to assign any finite velocity to this travel, the use of the term "travel" is inappropriate. For lack of a better expression let us say that some matter is delocated from Tokyo to Paris (this term has been proposed by Matthias Egg, see [52], p. 193); for even if the spontaneous localization of the wave function in configuration space is conceived as a continuous process as in [10], the time it takes for the matter density to disappear in one place and to reappear in another place does not depend on the distance between the two places. This delocation of matter, which is not a travel with any finite velocity, is quite a mysterious process that the GRWm ontology asks us to countenance.

On BM, by contrast, in this example, there always is one particle moving on a continuous trajectory in one of the two half boxes, and opening one of them only reveals where the particle was all the time. In other words, BM provides a local account of the case of the particle in a box. However, when moving from Einstein's thought experiment with one particle in a box (1927) to the EPR experiment ([53]), even BM can no longer give a local account, as proven by Bell's theorem ([5], ch. 2; see also notably chs. 7 and 24). On the GRWm theory, again, the measurement in one wing of the experiment triggers a delocation of the matter density, more precisely a change in its shape in both wings of the experiment, so that, in the version of the experiment by Bohm [54, pp. 611–622], the shape of the matter density constitutes two spin measurement outcomes. On BM, fixing the parameter in one wing of the EPR experiment influences the trajectory of the particles in both wings via the wave function of the whole system, which consists of the measured particles as well as of the particles that make up the measuring devices.

Hence, in this case, it clearly comes out that according to the Bohmian velocity equation (10.1), the velocity of any particle depends, strictly speaking, on the position of all the other particles. However, each particle always moves with a determinate, finite velocity that is not greater than the velocity of light so that its motion traces out a continuous trajectory, without anything jumping—or being delocated—in physical space. The best conjecture for a velocity field that captures this motion that we can make, namely (10.1), requires acknowledging that the motions of these particles are correlated with each other, but this does not imply a commitment to there being some spooky agent or force in nature that instantaneously coordinates the motions of all the particles in the universe. Quantum physics just teaches us that it is a fact about the universe that when we seek to write down a simple and general law that accounts for the empirical evidence, we have to conceive a law that represents the motions of the particles to be correlated with one another.

However, [55] is certainly right in pointing out that a complete suspension of the principles of separability and local action would make it impossible to do physics: a theory that says that the motion of any object is effectively influenced by the position of every other object in the configuration of matter of the universe would be empirically inadequate and rule out any experimental investigation of nature. In order to meet Einstein's requirement, it is not necessary to rely on a collapse dynamics, as does GRW. BM fulfills this condition because decoherence will in general destroy the entanglement between large and/or distant systems, allowing us to treat them, for all practical purposes, as evolving in an independent manner. Moreover, BM is able to recover classical behavior in the relevant regimes (see [3], ch. 5). Since BM is a theory about the motion of particles, this classical limit does not involve or require any change in the ontological commitment but consists in the proposition that typical Bohmian trajectories look approximately Newtonian on macroscopic scales (if the characteristic wavelength associated to ψ is small compared to the scale on which the interaction potential varies). Altogether, the Bohmian theory, against the background of an ontology of point particles that are characterized only by their relative positions and a dynamics for the change of these positions, illustrates that there is nothing suspicious about a non-local dynamics.

Let us turn now to GRWf. As mentioned in the preceding section, the flashes can be conceived as Bohmian particles deprived of their trajectories, so that all that is left are isolated point events in space–time. The main problem for this ontology is that the flashes are too sparsely distributed. Consider what this means for the dynamics: the account that the original GRW theory envisages for measurement interactions does not work on the flash ontology—in other words, this ontology covers only the spontaneous appearance and disappearance of flashes but offers no account of interactions. On the original GRW proposal, a measurement apparatus is supposed to interact with a quantum object; since the apparatus consists of a great number of quantum objects, the entanglement of the wave function between the apparatus and the measured quantum object will be immediately reduced due to the spontaneous localization of the wave function of the apparatus. However, even if one supposes that a measurement apparatus can be conceived as a galaxy of flashes (but see the reservations of [56], pp. 257–258), there is on GRWf nothing with which the apparatus could interact: there is no particle that enters it, no mass density and in general no field that gets in touch with it either (even if one conceives the wave function as a field, it is a field in configuration space and not a field in physical space where the flashes are). There only is one flash (standing for what is usually supposed to be a quantum particle) in its past light cone, but there is nothing left of that flash with which the apparatus could interact.

10.4 Conclusion

This chapter has made evident that one cannot simply add a collapse parameter to the Schrödinger equation in order to solve the measurement problem. If one does so, one also has to specify a primitive ontology of a configuration of matter in physical space in order to have something in the ontology that constitutes the determinate measurement outcomes and that evolves as described by the modified Schrödinger equation. This then is the important ingredient for the ontology and not the wave function, independently of whether it undergoes collapses. As explained at the end of section 1, one can endorse the wave function as a parameter that figures centrally in the law of motion for the primitive ontology but nevertheless regard it only as a bookkeeping device of that motion.

In the light of this consequence, the collapse principle loses its attractiveness. If one needs a primitive ontology over and above the collapse postulate anyway, one can retain a particle ontology and a deterministic law of motion for the particles with a universal wave function that never collapses and deduce the QM probability calculus from that law. Recall that, as mentioned in Section 8.1, any primitive ontology implies that we do not enjoy full epistemic accessibility to the configuration of matter, so that probabilities come in anyway through our ignorance of the exact initial conditions, independently of whether the law for the evolution of the wave function is stochastic. Of course, this assessment would change if experimental tests of collapse theories like GRW against theories that exactly produce the predictions of textbook QM—such as BM—were carried out successfully and confirmed the collapse theories where they deviate from the standard predictions (see [57] for such experiments).

As things stand, the arguments for the particle ontology notably are that all the experimental evidence is particle evidence, that all composed objects are made of particles, and that any QM formalism is conceived in terms of a definite number of persisting particles. As regards the dynamics, a law for the particle motion on continuous trajectories such as the Bohmian guiding equation gives an account of the non-local correlations as brought out by Bell's theorem and the EPR experiment in terms of correlated particle motion without anything ever jumping or being delocated in physical space.

References

[1] Maudlin, T. (1995). Three measurement problems. *Topoi*, 14:7–15.
[2] Schrödinger, E. (1935). Die gegenwärtige Situation in der Quantenmechanik. *Naturwissenschaften*, 23:807–812.
[3] Dürr, D., Goldstein, S., and Zanghì, N. (2013b). *Quantum physics without quantum philosophy*. Berlin: Springer.
[4] Mundy, B. (1989). Distant action in classical electromagnetic theory. *British Journal for the Philosophy of Science*, 40(1):39–68.
[5] Bell, J. S. (2004). *Speakable and unspeakable in quantum mechanics*. Cambridge: Cambridge University Press, second edition.
[6] de Broglie, L. (1928). La nouvelle dynamique des quanta. *Electrons et photons. Rapports et discussions du cinquième Conseil de physique tenu à Bruxelles du 24 au 29 octobre 1927 sous les auspices de l'Institut international de physique Solvay*, pages 105–132. Paris: Gauthier-Villars. English translation in Bacciagaluppi, G. and Valentini, A., editors (2009). *Quantum theory at the crossroads. Reconsidering the 1927 Solvay conference*, pages 341–371. Cambridge: Cambridge University Press.
[7] Bohm, D. (1952). A suggested interpretation of the quantum theory in terms of "hidden" variables. 1. *Physical Review*, 85(2):166–179.
[8] Dürr, D. and Teufel, S. (2009). *Bohmian mechanics: the physics and mathematics of quantum theory*. Berlin: Springer.
[9] Ghirardi, G. C., Rimini, A., and Weber, T. (1986). Unified dynamics for microscopic and macroscopic systems. *Physical Review D*, 34(2):470–491.
[10] Ghirardi, G. C., Pearle, P., and Rimini, A. (1990). Markov processes in Hilbert space and continuous spontaneous localization of systems of identical particles. *Physical Review A*, 42:78–89.
[11] Ghirardi, G. C., Grassi, R., and Benatti, F. (1995). Describing the macroscopic world: closing the circle within the dynamical reduction program. *Foundations of Physics*, 25(1):5–38.
[12] Tumulka, R. (2006). A relativistic version of the Ghirardi–Rimini–Weber model. *Journal of Statistical Physics*, 125(4):821–840.
[13] Allori, V., Goldstein, S., Tumulka, R., and Zanghì, N. (2008). On the common structure of Bohmian mechanics and the Ghirardi-Rimini-Weber theory. *British Journal for the Philosophy of Science*, 59(3):353–389.
[14] Norsen, T. (2014). The pilot-wave perspective on spin. *American Journal of Physics*, 82(4):337–348.
[15] Brown, H. R., Dewdney, C., and Horton, G. (1995). Bohm particles and their detection in the light of neutron interferometry. *Foundations of Physics*, 25(2):329–347.
[16] Brown, H. R., Elby, A., and Weingard, R. (1996). Cause and effect in the pilot-wave interpretation of quantum mechanics. In Cushing, J. T., Fine, A., and Goldstein, S., editors, *Bohmian mechanics and quantum theory: an appraisal*, volume 184 of *Boston Studies in the Philosophy of Science*, pages 309–319. Dordrecht: Springer.
[17] Pylkkänen, P., Hiley, B. J., and Pättiniemi, I. (2015). Bohm's approach and individuality. In Guay, A. and Pradeu, T., editors, *Individuals across the sciences*, chapter 12, pages 226–246. Oxford: Oxford University Press.

[18] Allori, V., Goldstein, S., Tumulka, R., and Zanghì, N. (2014). Predictions and primitive ontology in quantum foundations: a study of examples. *British Journal for the Philosophy of Science*, 65(2):323–352.

[19] Cowan, C. W. and Tumulka, R. (2016). Epistemology of wave function collapse in quantum physics. *British Journal for the Philosophy of Science*, 67:405–434.

[20] Oldofredi, A., Lazarovici, D., Deckert, D.-A., and Esfeld, M. (2016). From the universe to subsystems: Why quantum mechanics appears more stochastic than classical mechanics. *Fluctuations and Noise Letters*, 15(3):164002: 1–16.

[21] Miller, E. (2014). Quantum entanglement, Bohmian mechanics, and Humean supervenience. *Australasian Journal of Philosophy*, 92:567–583.

[22] Esfeld, M. (2014b). Quantum Humeanism, or: physicalism without properties. *The Philosophical Quarterly*, 64(256):453–470.

[23] Callender, C. (2015). One world, one beable. *Synthese*, 192(10):3153–3177.

[24] Bhogal, H. and Perry, Z. R. (2016). What the Humean should say about entanglement. *Noûs*, page DOI 10.1111/nous.12095.

[25] Dowker, F. and Herbauts, I. (2005). The status of the wave function in dynamical collapse models. *Foundations of Physics Letters*, 18:499–518.

[26] Esfeld, M., Lazarovici, D., Hubert, M., and Dürr, D. (2014). The ontology of Bohmian mechanics. *British Journal for the Philosophy of Science*, 65(4):773–796.

[27] Suárez, M. (2015). Bohmian dispositions. *Synthese*, 192:3203–3228.

[28] Forrest, P. (1988). *Quantum metaphysics*. Oxford: Blackwell.

[29] Norsen, T., Marian, D., and Oriols, X. (2015). Can the wave function in configuration space be replaced by single-particle wave functions in physical space? *Synthese*, 192:3125–3151.

[30] Esfeld, M. (2014a). The primitive ontology of quantum physics: guidelines for an assessment of the proposals. *Studies in History and Philosophy of Modern Physics*, 47:99–106.

[31] Feynman, R. P., Leighton, R. B., and Sands, M. (1963). *The Feynman lectures on physics. Volume 1*. Reading, MA: Addison-Wesley.

[32] Vassallo, A., Deckert, D.-A., and Esfeld, M. (2016). Relationalism about mechanics based on a minimalist ontology of matter. *European Journal for Philosophy of Science*, pages DOI 10.1007/s13194–016–0160–2, preprint http://philsci–archive.pitt.edu/12398/, arXiv:1609.00277[physics.hist–ph].

[33] Vassallo, A. and Ip, P. H. (2016). On the conceptual issues surrounding the notion of relational Bohmian dynamics. *Foundations of Physics*, 46(8):943–972.

[34] Arntzenius, F. and Hawthorne, J. (2005). Gunk and continuous variation. *The Monist*, 88:441–465.

[35] Barrett, J. A. (2014). Entanglement and disentanglement in relativistic quantum mechanics. *Studies in History and Philosophy of Modern Physics*, 48:168–174.

[36] Dürr, D., Goldstein, S., Norsen, T., Struyve, W., and Zanghì, N. (2013a). Can Bohmian mechanics be made relativistic? *Proceedings of the Royal Society A*, 470:2162.

[37] Bedingham, D., Dürr, D., Ghirardi, G. C., Goldstein, S., Tumulka, R., and Zanghì, N. (2014). Matter density and relativistic models of wave function collapse. *Journal of Statistical Physics*, 154:623–631.

[38] Esfeld, M. and Gisin, N. (2014). The GRW flash theory: a relativistic quantum ontology of matter in space–time? *Philosophy of Science*, 81:248–264.

[39] Colin, S. and Struyve, W. (2007). A Dirac sea pilot-wave model for quantum field theory. *Journal of Physics A*, 40(26):7309–7341.

[40] Deckert, D.-A., Esfeld, M., and Oldofredi, A. (2016). A persistent particle ontology for QFT in terms of the Dirac sea. *British Journal for the Philosophy of Science*, pages Preprint http://philsci–archive.pitt.edu/12375/, arXiv:1608.06141[physics.hist–ph].

[41] Albert, D. Z. and Loewer, B. (1996). Tails of Schrödinger's cat. In Clifton, R. K., editor, *Perspectives on quantum reality*, pages 81–91. Dordrecht: Kluwer.

[42] Lewis, P. J. (1997). Quantum mechanics, orthogonality, and counting. *British Journal for the Philosophy of Science*, 48:313–328.

[43] Monton, B. (2004). The problem of ontology for spontaneous collapse theories. *Studies in History and Philosophy of Modern Physics*, 35(3):407–421.

[44] Tumulka, R. (2011). Paradoxes and primitive ontology in collapse theories of quantum mechanics. *http://arxiv.org/quant-ph/1102.5767v1*.

[45] Wallace, D. (2008). Philosophy of quantum mechanics. In Rickles, D., editor, *The Ashgate companion to contemporary philosophy of physics*, pages 16–98. Aldershot: Ashgate.

[46] Maudlin, T. (2010). Can the world be only wave-function? In Saunders, S., Barrett, J., Kent, A., and Wallace, D., editors, *Many worlds? Everett, quantum theory, and reality*, pages 121–143. Oxford: Oxford University Press.

[47] Wallace, D. (2014). Life and death in the tails of the GRW wave function. *arXiv:1407.4746 [quant-ph]*.

[48] Albert, D. Z. (2015). *After physics*. Cambridge, MA: Harvard University Press.

[49] Egg, M. and Esfeld, M. (2015). Primitive ontology and quantum state in the GRW matter density theory. *Synthese*, 192(10):3229–3245.

[50] de Broglie, L. (1964). *The current interpretation of wave mechanics. A critical study*. Amsterdam: Elsevier.

[51] Norsen, T. (2005). Einstein's boxes. *American Journal of Physics*, 73:164–176.

[52] Egg, M. and Esfeld, M. (2014). Non-local common cause explanations for EPR. *European Journal for Philosophy of Science*, 4:181–196.

[53] Einstein, A., Podolsky, B., and Rosen, N. (1935). Can quantum-mechanical description of physical reality be considered complete? *Philosophical Review*, 47:777–780.

[54] Bohm, D. (1951). *Quantum theory*. Englewood Cliffs, NJ: Prentice-Hall.

[55] Einstein, A. (1948). Quanten-Mechanik und Wirklichkeit. *Dialectica*, 2:320–324.

[56] Maudlin, T. (2011). *Quantum non-locality and relativity. Third edition*. Chichester: Wiley-Blackwell.

[57] Curceanu, C. and alteri (2016). Spontaneously emitted x-rays: an experimental signature of the dynamical reduction models. *Foundations of Physics*, 46(3):263–268.

Part III
Origin

11

Quantum State Reduction via Gravity, and Possible Tests Using Bose–Einstein Condensates

IVETTE FUENTES AND ROGER PENROSE

It has been proposed that because of a fundamental conflict between basic principles of quantum mechanics and general relativity, a superposition between stationary quantum states that differ in mass distribution would reduce spontaneously to one or the other in a timescale inversely proportional to the gravitational self-energy of this mass-distribution difference. Here, we present arguments in favour of this proposal and suggest a possible test using Bose–Einstein condensates.

11.1 Quantum State Reduction and Gravity

The idea that the phenomenon of *quantum state reduction*—or "collapse of the wave function"—might be a unitarity-violating process *objectively* occurring in nature due to gravitational influences has a fairly long history. Among the earliest proposals of this type are those by Karolyhazy and his collaborators, dating back to 1966 [1], [2], [3]. For other early ideas on this issue see Kibble [4] and Diósi [5]. In 1987, Diósi [6] (see also [7]) put forward a gravitational state-reduction scheme which followed, in certain respects, the well-known previous *non*-gravitational state-reduction proposal of Ghirardi, Rimini, and Weber (in 1986 [8]), but it had an advantage over theirs in that their two free parameters were replaced, in effect, by simply the gravitational constant. Objections were raised by Ghirardi, Grassi, and Rimini [9] that this scheme would be unrealistic as it stood, but Diósi was able to address these matters through the introduction of additional parameters; see his articles [10] and [11] for further discussion. The work of Pearle and Squires [12], [13] is of relevance here, and so also is that of Percival [14], with different takes on the role of gravity in quantum state reduction.

There are many other proposals for theories in which the collapse of the wave function is taken to be an objective physical process, and only a minority of these regard gravitation as being the deciding factor. Even among the gravitational proposals, there are many different motivational arguments, of varying degrees of persuasiveness, and some of these could be considered to be decidedly *ad hoc*, without very compelling physical reasons given to support the particular type of scheme being proposed, beyond a desire to modify the standard quantum formalism so that quantum state reduction might be the result of some kind of (not necessarily gravitationally induced) objective physical process, going beyond the current quantum formalism. While we are strongly sympathetic with this general aim, we are of the opinion that some clear motivation, preferably from well-established areas of physics, needs to be playing a central role. Indeed, the desire for providing a physically justified extension of the current quantum formalism, in order that quantum state reduction becomes replaced by some objective physical process, is, in our opinion, extremely desirable.

Moreover, we would agree with John Bell's negative assessment of the common viewpoint that one can somehow simply *dismiss* quantum state reduction as being some kind of "illusion" due to incomplete consideration of external environmental factors. That dismissal would normally be a resort to "environmental decoherence," which does not, however, really resolve the ontological confusions of the interpretation of the quantum formalism. That seemingly pragmatic standpoint was what Bell referred to as "FAPP" ("for all practical purposes") [15]. In fact, we would go further than this and maintain that there should be experiments (such as will be described in Section 11.4), not too far from the constraints of current technology, that could be expected to yield results that *differ* from those for which such a FAPP standpoint might imply. But in order to have much confidence that any proposed objective state-reduction dynamics could have a good chance of experimental confirmation, we would need to have some strong physical arguments from other accepted areas of physics.

The one area of physical understanding that, in our opinion, can provide us with clear clues as to how one might move beyond the framework of current quantum theory is Einstein's *general theory of relativity*. That theory, which is now extremely well confirmed in many experiments (see, for example, [16], [17], and [18]), not only provides a unique role for gravity among physical processes, but also, as we shall see in Section 11.2, suggests that the current framework of quantum mechanics (and quantum field theory) is indeed in need of modification. We ought surely to expect that the nature of any such change in quantum theory must be such as to be geared to a greater harmony with the basic principles of general relativity. If the central procedures of quantum mechanics are to be *actually* changed—and we must bear in mind that quantum theory is not infrequently proclaimed to be the most successful theory in the history of physics—then we do need very clear physical underpinnings for any proposal to effect such a change. We argue, here, that the underlying principles of general relativity do indeed provide us with some clear clues as to the directions in which the principles of quantum theory indeed require modification.

Yet we must also bear in mind that it is often argued that gravitational forces are normally absolutely insignificant when compared with the chemical and elastic forces that dominate the normal structural and dynamical behaviourof ordinary material bodies, so that the mere incorporation of gravitational *forces* can, at best, simply provide some minute correction to what would take place when these gravitational forces are ignored. Thus, it needs to be clear that the influence that Einstein's general relativity has on the quantum behaviourof physical systems has to be of a different character from the mere incorporation of gravitational forces. We are not looking at circumstances in which gravitational forces might in any way be in competition with other forces. Indeed, it is *not* the case, in the situations under consideration here, that gravitational forces are likely to be at all comparable with other forces, nor is this relevant to the essential issue. The key point is that general relativity is based on certain underlying principles which not only demand that gravitational effects are of a very different character from those of all other physical interactions but that these general principles also bear upon all other physical actions. We shall be seeing that there is a certain profound *tension* between these principles and those of conventional unitary quantum theory. The conclusion that we arrive at seems to demand a time limitation on strict unitary evolution, which is reciprocally related to the magnitude of the gravitational self-energy of the difference between mass distributions of pairs of states in quantum superposition (Penrose [20], [21], [22]; compare also Diósi [5], [6]).

11.2 Principles of General Covariance and Equivalence

Let us consider a small rock, lying on a flat horizontal surface, but where the rock has been put into a quantum superposition of two different locations on that surface. We may imagine that this "Schrödinger's cat" type of state came about through the action of a device which is capable of sliding the rock away, a certain distance from an initial location, and which does so only when it is activated by a high-energy photon. In this case, the photon has come initially from a laser, geared to emit individual high-energy photons, but in order to reach the device, the photon, after leaving the laser, has to be reflected by a 50% beam splitter, the transmitted part of the photon's split state going somewhere else that does not influence the experiment. Of course, unless many detailed and difficult precautions have been taken, the resulting quantum state will involve many entanglements, so that it would not simply consist of our rock alone in a superposition of two separate locations, but each component of the superposition would involve many other things than just the rock, such as the rock-moving device itself, the atoms in the material of the horizontal surface, the molecules in the surrounding air, and so forth. These extra factors do not really affect the argument in any substantial way (despite the fact that their actual elimination could be crucial for a realistic experiment designed to *test* the conclusions that we shall be coming to in this section), since the superposition under consideration need not consist simply of the superposed rock locations alone but, more properly, of two states, each of which consists of the rock in a particular location together with all the other atoms in the air, etc., etc., all in their respective locations, as entangled with that particular rock location. For clarity of thought, let us suppose that the mass of the rock completely dominates the combined masses of all the other relevant materials. This will enable us to discuss things as though the superposition is basically of just the rock locations, despite the fact that each component of the superposition would actually involve many other atoms of the surrounding material.

We shall also suppose, for the sake of convenience, that each of the components of the superposition can be treated as though it is a completely *stationary* state, let us say the left-hand one $|L\rangle$ and the right-hand one $|R\rangle$. We assume that the energy E is the same for each, so we have, according to the procedures of standard quantum mechanics,

$$i\hbar \frac{\partial}{\partial t}|L\rangle = E|L\rangle \text{ and } i\hbar \frac{\partial}{\partial t}|R\rangle = E|R\rangle \tag{11.1}$$

whence,

$$i\hbar \frac{\partial}{\partial t}\{\alpha|L\rangle + \beta|P\rangle\} = E\{\alpha|L\rangle + \beta|R\rangle\} \tag{11.2}$$

where the complex amplitudes α and β are constant (and not both zero). Thus, any superposition of the two rock locations would also be stationary, with the same energy E.

However, this standard conclusion does not take into account the gravitational field of the rock and the fact that, according to general relativity, the space–time for each rock location would be slightly deformed in accordance with the rock's gravitational field. The superposition under consideration ought therefore to involve, also, a superposition of these slightly differing space–times. But how is this to be described? In this article, we shall explore some of the implications of this awkward issue.

In studies of quantum gravity in which the *principle of general covariance* is to be taken seriously (i.e., the principle of the irrelevance of any particular choice of coordinate system), it is normally considered that space–times (or sometimes the initial 3-spaces of evolving space–times) are to be considered as *geometrical objects* in their own right, so that any two of them which are metrically equivalent (i.e., such that there exists a coordinate atlas for each, in which their metric expressions are the same for both of them) should be considered to be identical to each other. However, this would be problematic in the present situation if we consider that the "horizontal plane" of this situation were completely featureless, perhaps being considered as the surface of a completely featureless, exactly spherical Earth. For then, each of the states $|L\rangle$ and $|R\rangle$ would have to be treated as "the same"—because having the rock on one place on this sphere would provide an identical geometry to having it at another point on the sphere.

Clearly this is an inappropriate standpoint to take with regard to standard quantum mechanics, where a quantum state must be described in relation to a fixed background space (or space–time). Otherwise, for example, the state of a single particle located at any one particular place in Euclidean 3-space would be taken as the "same" as it being located at any other place in that space, whence the forming of quantum linear superpositions of such things to make general quantum states of that particle would not make any sense. To get around this seeming conflict between the principle of general covariance and the superposition principle in quantum mechanics, we may provisionally assume that the background space(–time) has some defining irregularities about it (such as, in our current situation, that given by the rock-moving device or the laser). This, however, seems to be a very artificial resolution of the problem, and it seems more natural to remove the degeneracy by fixing things at infinity, so that in some sense the location of any object in the space can be regarded as being somehow located by its relation to some preassigned "frame at infinity." In fact, in general relativity, this is not so easy, because for an asymptotically flat space–time in which there is a positive total mass, the structure at spatial infinity is, in a definite sense, *singular* (see [19], [23]). It is not clear to us whether this issue is important for the current considerations, but we feel that it is not fruitful for us to pursue this matter more thoroughly at this juncture, in the belief that more local considerations are likely to have greater importance.

Accordingly, let us regard the Earth itself as supplying the needed background framework, against which the rock movements are to be considered. There is, however, one further issue that should not be completely ignored, as it would have relevance to situations that are a little different from the one under consideration here. When the rock-moving device moves the rock, there would be a very tiny back reaction on the Earth itself (so that the motion of the mass center of the entire system remains unaffected, being the same for both components of the superposition). Thus, the Earth itself would be in a quantum superposition of two extremely slightly displaced positions. Although for situations different from the one under consideration here, this kind of issue can have theoretical implications, this utterly minute compensatory motion of the Earth can be ignored for our current considerations.

In order to simplify our continuing discussion, it will be helpful for us to assume, provisionally, that (apart from the beam-split photon itself) all motions are extremely slow by comparison with the speed of light—which is natural, after all, since we are basically concerned with states that are actually stationary rather than with the motions leading to the mass separation of these states. Then, to some extent, we are able to discuss the issues within the framework of Newtonian gravitational theory, but treated by paying due respect to the principles underlying Einstein's approach to gravitational theory: specifically, the principles of equivalence and general covariance. Technically, this would mean that we should be treating gravitational theory according to the description put forward by Élie Cartan in 1923 [24] and 1924 [25] and independently by Friedrichs a little later [26]. These ideas were studied, more recently, by Trautman [27] and by Ehlers [28], extending the idea of Newtonian gravitational theory to situations in which the *spatial* geometry can be curved while retaining Newton's universal time. There will be no need to go into the details of the kind of 4-dimensional geometry that this involves, but we may bear in mind that in these schemes, there is indeed a universal time t. (Such a universal time would, at least, enable us to evade conceptual issues that arise when a state reduction takes place with an entangled "EPR" state, where puzzling non-local features arise. Nevertheless, this is merely putting off the central mystery of non-local entangled states.)

It should be pointed out, however, that having a universal "t" does not absolve us of addressing the issue of making sense of the operator "$\partial/\partial t$" that we find in equations (11.1) and (11.2). The meaning of this operator, after all, has rather more to do with the *spatial* variables (say, x, y, z) than with t, as it needs the notion of what it means to "hold the spatial variables fixed" as t increases. If the spatial geometries in the two states in superposition differ, then there will be no natural way, in general, to identify the spatial coordinates of one of the states in superposition with those of the other—where we assume that at infinity, these spatial coordinates agree, so that we are not allowed simply to "translate" these coordinates along with the rock. When the coordinate values begin to enter the vicinity of the rock location, they would have to accommodate the spatial curvature, but this happens for different coordinate values for each of the two different rock locations. It is here that the principle of *general covariance*, as applied to the two different spatial geometries, begins to come into play. The coordinate values themselves should not be playing any particular role. The general covariance principle is really telling us that there is no physical significance to be attached to the values of the spatial coordinates in |L⟩ or in |R⟩ and, accordingly, there is no physical meaningfulness to a pointwise identification between the spatial geometry in these two states. How are we to handle this difficulty, where it is a foundational principle of quantum mechanics that linear superpositions between different states should always be possible (provided that some superselection principle is not violated—of no real relevance here), and that, for given amplitude coefficients (the α and β of equation [11.2]), the resulting state should be uniquely defined. Here we seen to see a profound *tension*, if not an actual contradiction, between fundamental principles of general relativity and of quantum theory. How do we deal with this difficulty?

What might be regarded as a standard "quantum gravity" point of view would be to regard the superposition between these two curved spaces as being purely *formal* (in other words, we simply write a "+" sign between the two states without asking what this means). However, this point of view is not very helpful for resolving the issue that started off this whole discussion, namely, the issue of whether the rock superposition $\alpha|L\rangle + \beta|R\rangle$ could be regarded as being stationary if both $|L\rangle$ and $|R\rangle$ are stationary with the same energy E. To address that issue, we must ask what the term "stationary" actually means in general relativity.

In graphic terms, what it means is that one can "slide" the space–time over itself, in a temporal direction, without changing its geometry. The direction of this sliding is defined by what is technically known as a *timelike Killing vector* (indicated in Figure 11.1). In fact, the Killing vector **k** of the figure can really be identified with the operator $\partial/\partial t$, so the problem of asserting stationarity for the superposition amounts to the problem of identifying this operator for the state $|L\rangle$ with that for $|R\rangle$, in order for one to be able to assert that the superposition $\alpha|L\rangle + \beta|R\rangle$ is actually stationary (with the same energy E). Our point of view here will be that Einstein's principle of general covariance raises a serious problem in identifying the Killing vector (field) **k** in $|L\rangle$ with that in $|R\rangle$. This basic principle of Einstein's theory tells us that the Killing vectors cannot be meaningfully identified with each other simply because there is no meaningful (precise) identification between the spatial geometries of $|L\rangle$ and $|R\rangle$, and this forbids a meaningful pointwise identification of the spatial geometries of $|L\rangle$ and $|R\rangle$. We regard this difficulty as telling us that there is a *fundamental uncertainty* involved in trying to perform such a pointwise

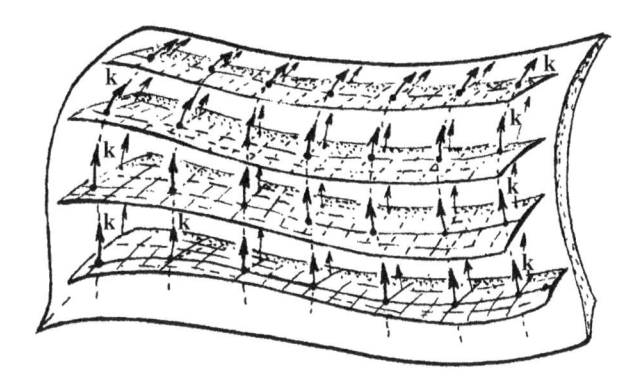

Figure 11.1 A Killing vector **k** is a vector field on a space whose metric, locally, is unaltered by the motion generated by **k**. A stationary space-time is one possessing a timelike Killing vector.

identification between these two geometries. Thus the Killing vector in the space–time of our stationary $|L\rangle$ cannot be meaningfully identified, pointwise, with the Killing vector that expresses the stationarity of the space–time of $|R\rangle$.

How do we estimate the degree of this uncertainty? The proposal here is that we first "cheat" by ignoring the spatial curvature and simply make a pointwise identification using the flat-space coordinates for the states $|L\rangle$ and $|R\rangle$; then we try to estimate the "error" involved in doing so. How do we go about estimating this error? This is where Einstein's *principle of equivalence* comes in. Consider any space–time coordinate point for each of the states $|L\rangle$ and $|R\rangle$, and we try identifying the 4-geometries pointwise simply by identifying points with the same coordinate values in each space–time. Einstein's equivalence principle would tell us that we should really only identify two space–times physically in the vicinity of some point if their local free-fall rest frames are the same at that point. This is because Einstein's notion if a (local) "inertial frame" is a reference frame that falls freely under gravity.

Figure 11.2, illustrates, in space–time terms, how the "free-fall" geodesics differ for each rock location. The black arcs illustrate free fall for the location of the rock shown in black, and the grey arcs free fall for the location of the rock shown in grey. The discrepancies between the free-fall motions are here seen in the space–times for the two states $|L\rangle$ and $|R\rangle$. The "error" in making the pointwise identification between the two frames at each coordinate point is simply a measure of the incompatibility between these two sets of local inertial motions.

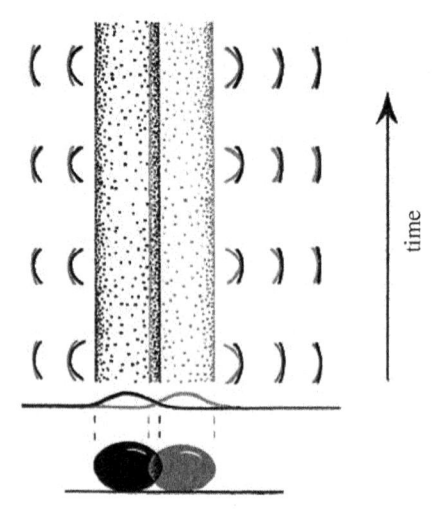

Figure 11.2 The space-time for a rock, in a quantum superposition of two horizontally displaced locations (marked in black and grey). The incompatibility between the superposed free-fall motions is indicated in black and grey, in accordance with the two rock locations.

At such a (coordinate) point, let **a** and **a**′ be the acceleration 3-vectors for the states |L⟩ and |R⟩, respectively. Then we take this local measure of error involved in this identification to be

$$(\mathbf{a} - \mathbf{a}')^2 = (\mathbf{a} - \mathbf{a}') \cdot (\mathbf{a} - \mathbf{a}')$$

at each point in space (the "dot" denoting 3-dimensional coordinate scalar product). We note that this error measure is unaltered if a local acceleration is applied that is the same for both states. To get our measure of *total* error, or "uncertainty" Δ, we integrate this expression over the whole of (coordinate) 3-space:

$$\Delta = \int (\mathbf{a} - \mathbf{a}')^2 \, d^3\mathbf{x}$$
$$= \int (\nabla\Phi - \nabla\Phi')^2 \, d^3\mathbf{x}$$
$$= \int (\nabla\Phi - \nabla\Phi') \cdot (\nabla\Phi - \nabla\Phi') \, d^3\mathbf{x}$$
$$= -\int (\Phi - \Phi')(\nabla^2\Phi - \nabla^2\Phi') \, d^3\mathbf{x}$$

(assuming appropriate falloff at spatial infinity) where $\mathbf{a} = \nabla\Phi$ and $\mathbf{a}' = \nabla\Phi'$, Φ and Φ' being the respective gravitational potentials for the respective states |L⟩ and |R⟩, where we are now adopting a Newtonian approximation for estimating the required error. By Poisson's formula (G being Newton's gravitational constant)

$$\nabla^2\Phi = -4\pi G\rho,$$

so we get

$$\Delta = 4\pi G \int (\Phi - \Phi')(\rho - \rho') \, d^3\mathbf{x},$$

where ρ and ρ' are the respective mass densities of the two states—and we shall take these mass densities in the sense of expectation values for the respective quantum states. Using the formula

$$\Phi(\mathbf{x}) = -\int \frac{\rho(\mathbf{y})}{|\mathbf{x} - \mathbf{y}|} \, d^3\mathbf{y},$$

we get

$$\Delta = E_G = 4\pi G \iint \frac{(\rho(\mathbf{x}) - \rho'(\mathbf{x}))(\rho(\mathbf{y}) - \rho'(\mathbf{y}))}{|\mathbf{x} - \mathbf{y}|} \, d^3\mathbf{x} \, d^3\mathbf{y}, \qquad (11.3)$$

where E_G is just the gravitational self-energy of the difference between the mass distributions of each of the two states.

Recall that this quantity Δ was introduced originally, at the beginning of this discussion, as some kind of a measure of a limitation to regarding the quantum superposition $\alpha|L\rangle + \beta|R\rangle$ as being a stationary state, in accordance with principles of general relativity. Thus, we may take it to be a reasonable inference from general-relativistic principles to regard Δ^{-1} as providing some kind of measure of a limit to the length of time that the superposition might persist; the shorter that this timescale should presumably be, the larger the value Δ is found to have. This conclusion comes about from considerations of classical general relativity, as applied simply to the notion of a quantum superposition of states, no consideration of quantum dynamics being involved except for the quantum notion of *stationarity*. Moreover, no actual measure of a timescale for a "lifetime" of the superposition has yet been provided by these considerations.

However, a significant clue is provided by Heisenberg's time–energy uncertainty principle, where we note that the quantity $\Delta = E_G$ is indeed an *energy*. We take note of the fact that the lifetime of an unstable atomic nucleus is reciprocally related to an energy uncertainty in the nucleus—this being normally regarded as a manifestation of Heisenberg's time–energy uncertainty principle. In a similar way, we may regard E_G as a fundamental uncertainty in the energy of the superposition $\alpha|L\rangle + \beta|R\rangle$. Thus, in a similar way, we can take the view that the "energy uncertainty" E_G is reciprocally related to a *lifetime* of this superposition between the states $|L\rangle$ and $|R\rangle$. Thus, we can regard the macroscopic superposition $\alpha|L\rangle + \beta|R\rangle$ as having an average *lifetime* τ that is roughly given by

$$\tau \approx \frac{\hbar}{E_G},$$

upon which time (on average) the state $\alpha|L\rangle + \beta|R\rangle$ spontaneously "decays" into one or the other of $|L\rangle$ or $|R\rangle$.

There is an issue, in the case of a more general state $|\psi\rangle$, which is not so manifestly a superposition of two distinct stationary states like $|L\rangle$ and $|R\rangle$. One proposal [29] has been to consider that the stationary states into which $|\psi\rangle$ could spontaneously reduce would have to be stationary solutions of the Schrödinger–Newton equation [30]. This is a non-linear version of the Schrödinger equation in which there is a gravitational potential term provided by the expectation value of the mass distribution. We do not discuss this issue further in this chapter, however.

11.3 Superposition of Differing Timescales

So far, we have not considered the effect that mass distributions have on timescales, which is an important feature of Einstein's general relativity. The only estimate of a lifetime for a quantum superposition of states with different mass distributions that we have so far obtained has been via an appeal to Heisenberg's time–energy uncertainty relation. In this section, we present an argument that more directly indicates that such a superposition encounters some kind of time limitation on what would have been its expected persistence as a superposition in accordance with standard Schrödinger evolution.

In recent articles (e.g., [31], [32]), various authors have raised an issue concerning the operation of the Schrödinger equation when different parts of a quantum state are in different gravitational potentials, and therefore appear to have to follow different time rates, so that the *timescale* aspect of the operator "$\partial/\partial t$" (as opposed to its relation to the spatial coordinates, as discussed in Section 11.2) does not have a consistent overall meaning. In our opinion, these arguments do not present a convincing case that there is an obstruction to the continuation of unitary evolution in situations when the quantum-superposed states do not involve a superposition of differing space–time geometries. Here, we present what we believe to be a much stronger case that there is a profound conflict between standard unitary evolution and Einstein's principle of equivalence (see also [33] for a different but related approach).

Let us consider, first, a simple situation of a tabletop quantum experiment, where it is required that the Earth's gravitational field is to be taken into consideration. See Figure 11.3. There are basically two different ways to incorporate the Earth's field (which is to be considered as *constant*, both spatially and temporally, and to be treated non-relativistically). The first—the *Newtonian perspective*—would

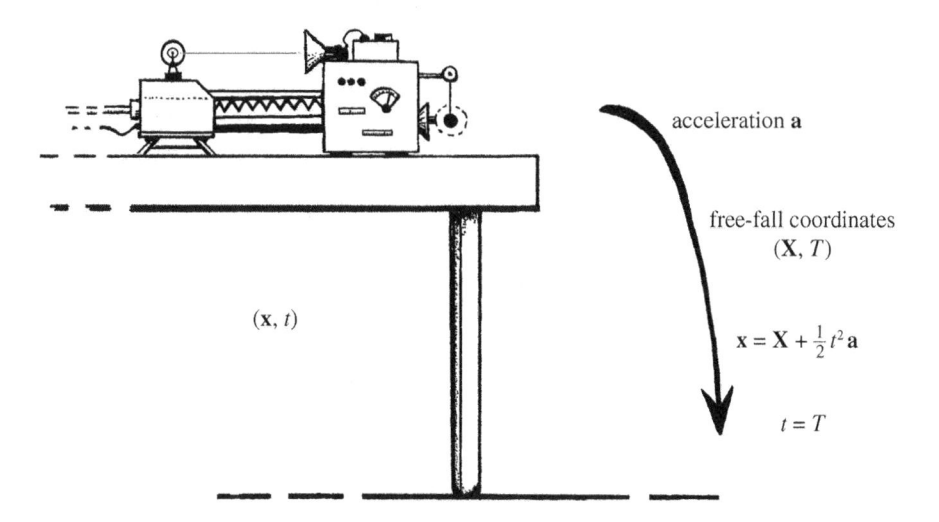

Figure 11.3 An imagined quantum experiment for which the Earth's gravitational field is to be taken into consideration. The Newtonian perspective uses the laboratory coordinates $(\mathbf{x}, t)$, while the Einsteinian perspective use the free-fall coordinates $(\mathbf{X}, T)$.

simply be to incorporate a term into the Hamiltonian representing the Newtonian potential (this being the normal prescription that most physicists would adopt) and use standard Cartesian coordinates (x, y, z, t), or rather $(\mathbf{x}, t)$ in 3-vector form. The second—the *Einsteinian perspective*—would be to adopt a freely falling reference system $(\mathbf{X}, T)$, in accordance with which the Earth's gravitational field *vanishes*. The relation between the two is taken to be

$$\mathbf{X} = \mathbf{x} - \frac{1}{2}t^2 \mathbf{a}, \quad T = t,$$

where the constant 3-vector $\mathbf{a}$ is the acceleration due to the Earth's gravity. The wave function in the $(\mathbf{x}, t)$ system, using the Newtonian perspective, is ψ, whereas for the $(\mathbf{X}, T)$ system, using the Einsteinian perspective, we adopt the wave function Ψ. For a free particle of mass m, we have, according to the Newtonian perspective, the Schrödinger equation

$$i\hbar \frac{\partial \psi}{\partial t} = -\frac{\hbar^2}{2m} \nabla^2 \psi - m\, \mathbf{x} \cdot \mathbf{a}\, \psi$$

whereas, according to the Einsteinian perspective

$$i\hbar \frac{\partial \Psi}{\partial T} = -\frac{\hbar^2}{2m} \nabla^2 \Psi$$

the operator ∇^2 being the same in both coordinate systems. To get consistency between the two perspectives, we need to relate ψ to Ψ by a phase factor

$$\Psi = e^{im\hbar^{-1}(\frac{1}{6}t^3 a^2 - t\mathbf{x} \cdot \mathbf{a})} \psi.$$

For a system involving many particles of total mass M and mass centre $\bar{\mathbf{x}}$ (or $\bar{\mathbf{X}}$ in the Einstein system), this generalizes to

$$\Psi = e^{iM\hbar^{-1}(\frac{1}{6}t^3 a^2 - t\bar{\mathbf{x}} \cdot \mathbf{a})} \psi,$$

and we have the inverse relation (i.e., Newtonian in terms of Einsteinian)

$$\psi = e^{iM\hbar^{-1}(\frac{1}{3}T^3 a^2 + T\bar{\mathbf{X}} \cdot \mathbf{a})} \Psi.$$

(See [34], [35], [36], [22].) Since the difference between the Newtonian and Einsteinian perspectives is merely a phase factor, one might form the opinion that it makes no difference which perspective is adopted. Indeed, the famous experiment by Colella, Overhauser, and Werner [37] (see also [38], [39], [40]) performed originally in 1975 did provide some impressive confirmation of the consistency (up to a certain point) of quantum mechanics with Einstein's principle of equivalence.

However, it is important to notice that the phase factor that is encountered here is not at all harmless, as it involves the time-dependence involved in the term

$$\frac{1}{6}t^3 a^2$$

in the exponent, which affects the splitting of field amplitudes into positive and negative frequencies. In other words, the Einsteinian and Newtonian wave functions belong to different Hilbert spaces, corresponding to *different* quantum field theoretic *vacua*!

In fact, this situation is basically just a limiting case of the familiar relativistic (Fulling–Davies–)*Unruh* effect [33], [34], [35], where in a uniformly accelerating (Rindler) reference frame, we get a non-trivial *thermal vacuum* of temperature

$$\frac{\hbar a}{2\pi kc},$$

where a is the acceleration (k being Boltzmann's constant and c the speed of light). In the current situation, we are considering the Newtonian approximation $c \to \infty$, so the temperature goes to zero. Nevertheless, as a direct calculation shows, the Unruh vacuum actually goes over to the Einsteinian one in the limit $c \to \infty$, in agreement with what has been shown already, and is thus still different from the Newtonian one, even though in the limit the temperature goes to zero.

At this stage, we could still argue that it makes no difference whether the Newtonian or Einsteinian perspective is adopted, so long as one sticks consistently to one perspective or the other overall (since the formalism is maintained within a single Hilbert space). However, the situation becomes radically different when one considers situations in which, as in Section 11.2, we consider quantum superpositions between pairs of states in which the gravitational fields *differ*. If we were to adopt the Newtonian perspective for each member of a pair of such states, we would encounter no problem with the formalism of quantum mechanics, the standard linear framework of unitary evolution applying as well to the Newtonian gravitational field as it does to electromagnetism or to any other standard field theory of physics. But it is another matter altogether if we insist on adopting the Einsteinian perspective. Our standpoint here is that, owing to the enormous support that observations have now provided for Einstein's general theory of relativity in macroscopic physics (referred to in Section 11.1), one must try to respect the Einsteinian perspective as far as possible, in quantum systems, especially in view of the foundational role that the principle of equivalence has for Einstein's general theory (see [22]).

Let us try to imagine the quantum description of the physics taking place in some small region in the neighbourhood of the rock of Section 11.2, where we consider that the rock can persist for some while in a superposition of two different locations. We are not now trying to compare the Einsteinian perspective with a Newtonian one, since our point of view will be that the latter is not relevant to our current purposes, where we regard the Einsteinian perspective to be closer to nature's ways. Instead, we attempt to adopt an Einsteinian perspective for each of the two superposed quantum states. What we now have to contend with is a superposition of two *different* Einsteinian wave functions, each inhabiting a Hilbert space that will turn out to be incompatible with the other!

However, the preceding discussion does not hold exactly, because for each of the two components of the superposition of rock locations, the gravitational field of the rock is not completely uniform. Nevertheless, we shall consider, first, that we are examining the nature of the wave function in some spatial region that is small by comparison with the rock itself, so that we can take it that the gravitational field of each component of the superposition can be taken to be spatially uniform to a good approximation. Adopting the Einsteinian perspective, what we are now confronted with is the fact that the gravitational acceleration fields for the two rock locations will be different from each other, so that the difference between these local acceleration fields **a** and **a**′ will lead to a difference between the Einsteinian vacuum for each rock location. In the neighbourhood of each spatial point, there will be a phase difference between the two states that are supposed to be in superposition that has the form

$$e^{iM\hbar^{-1}(\frac{1}{6}t^3(\mathbf{a}-\mathbf{a}')^2 - t\bar{\mathbf{x}}\cdot(\mathbf{a}-\mathbf{a}'))}.$$

Although the presence of the $\frac{1}{6}t^3(\mathbf{a}-\mathbf{a}')^2$ term tells us, strictly speaking, that when **a** ≠ **a**′ the superposition is *illegal* (the states belonging to different Hilbert spaces), we adopt the view that this incompatibility takes some time to cause trouble (as would eventually become manifest in divergent scalar products, etc.). The idea is that in order to resolve this incompatibility of Hilbert spaces, the superposed state must eventually reduce to one alternative or the other, this incompatibility taking a while to build up. We compare the troublesome term $\frac{1}{6}t^3(\mathbf{a}-\mathbf{a}')^2$ with the harmless one $t\bar{\mathbf{x}}\cdot\mathbf{a}$, the latter ($\times M/\hbar$) being linear in t and therefore not altering the vacuum but, in effect, just corresponding to incorporating the Newtonian gravitational potential term into the Hamiltonian. We take the view that so long as t is small enough, the trouble arising from t^3 remains insignificantly small, where the measure of this smallness comes from comparing $\frac{1}{6}t^3(\mathbf{a}-\mathbf{a}')^2$ with the harmless $t\bar{\mathbf{x}}\cdot\mathbf{a}$. Thus, we take the coefficient $\frac{1}{6}(\mathbf{a}-\mathbf{a}')^2$ as some kind of measure of the propensity for the state to reduce, as a contribution to the overall reduction process. To get the overall state-reduction rate, we must integrate this propensity over the whole of space, which gives the same result (up to a small overall factor) as the E_G given in Section 11.2.

11.4 Bose–Einstein Condensates to Test the State Reduction

What about experimental tests of this state-reduction proposal? The gravitational reduction of quantum states could be demonstrated in the experiment by preparing a superposition state of a system that is large enough to produce a non-negligible gravitational field while being sufficiently small to be suitable for control in the quantum regime. Such a system could be, for example, a tiny mirror consisting of 10^{14} atoms. In an early experimental proposal [41], called the free-orbit experiment with laser interferometry using X-rays (FELIX), the superposition is obtained by placing the mirror in an X-ray laser interferometer tens of thousands of miles long. One photon would be in principle capable of generating a mirror superposition state that could remain coherent long enough to undergo gravitational collapse. The reduction time in this case was estimated to be one second. A more practical experiment has been constructed using optical cavities, which are capable of trapping photons for long times [42], [43], [44]. However, we worry that this system might be too hot to achieve the long coherence times required. The oscillator ground state of the mirror is at micro-Kelvin temperatures, and the spatial separation of the superposition may not be large enough. It can reach at most about one picometer—a tiny fraction of the size of an atom.

In this chapter, we propose the use of a quantum system that can be cooled down to ultra-cold temperatures. Physical systems cannot be cooled down to absolute zero. Heisenberg uncertainties in quantum theory forbid localized systems, such as atoms, to have vanishing momentum dispersion. Bose–Einstein condensates (BECs) reach the coldest temperatures reported so far in experiments, and the clouds can be prepared in spatial superpositions as large as 0.5m [45].

BECs consist of 10^2 to 10^{10} atoms trapped by an electromagnetic potential and cooled down to their ground state (for a review on BECs, see [46]). Recent experiments show that BECs can be cooled below the nanoKelvin regime [47],[48]. At temperatures below the critical temperature,

$$T_c = 3.3125 \frac{\hbar \rho^{2/3}}{m_a k}$$

the system can be described quantum mechanically, since the individual wave functions of each atom overlap, generating collective quantum behavior. We use ρ to denote the particle density and m_a for the mass per atom. Current BEC experiments are capable of producing quantum superpositions of the nature of those required for the observation of the required state reduction. The superpositions are produced through the interaction of the condensate cloud with a laser field. After some time, the atoms are recombined, producing an interference pattern. Such experiments are known as matter–wave interferometers, and the interest in them has been growing since there are important applications in gravimetry and sensing [49]. If one of the superposition states undergoes a different evolution due to an external gravitational gradient or the interaction with a magnetic field, the interference pattern will carry information about the field strength [50]. However, if the superposition undergoes a state reduction due to gravitational self-energy, the interference pattern would not be observed. Unfortunately, environmental noise can also produce decoherence [51], [52]. This can be produced by the interaction of the condensed atoms with a thermal cloud of uncondensed atoms, by the interaction of the system with air molecules, and by interatomic interactions that lead to particle loss in the condensate (such as three-body recombination). It is therefore necessary to show that the effects of environmental decoherence have timescales longer than the state reduction rate.

We consider a BEC consisting of N atoms in a spherical geometry. Typically, atoms are trapped by crossing Gaussian laser beams that produce ellipsoid-shaped clouds. However, for simplicity, we consider a spherical distribution of mass centered in position r_0 by assuming that the beams could be engineered to produce a cloud as spherical as possible. A quantum superposition of two spherical clouds, one with $N_a = a^\dagger a$ atoms centered at r_a and a second one with $N_b = b^\dagger b$ atoms at r_b, can be produced through the interaction Hamiltonian $H_I = \lambda(a^\dagger b + b^\dagger a)$. The creation operators, $a^\dagger$ and $b^\dagger$, and the annihilation operators, a and b, obey canonical commutation relations, and λ is the interaction strength between the atoms and the laser field. The total number of atoms is $N = N_a + N_b$. The Hamiltonian H_I is the matter equivalent of a beam splitter. Such a Hamiltonian describes the tunneling of atoms through a potential barrier of height λ. We denote the eigenstates of the free Hamiltonian $H_o = \hbar\omega_a a^\dagger a + \hbar\omega_b b^\dagger b$ by,

$$|\psi_{ab}\rangle = |N_a N_b\rangle = \left|\frac{N-\mu}{2}, \frac{N+\mu}{2}\right\rangle.$$

Where $-N \leq \mu \leq N$, $\mu \in \mathbb{Z}$. We have neglected particle collisions that can be experimentally suppressed through Feshbach resonances [53]. These should be included in a more detailed analysis. A general superposition of two atomic clouds generated by H_I corresponds to a state,

$$|\phi_{ab}\rangle = \sum_{\mu=-N}^{N} C_\mu \left|\frac{N-\mu}{2}, \frac{N+\mu}{2}\right\rangle.$$

where C_μ are complex coefficients satisfying, $\sum |C_\mu| = 1$. In order to test the state reduction, ideally one would prepare the state,

$$|\psi_{ab}\rangle = \frac{1}{\sqrt{2}}(|N0\rangle + |0N\rangle)$$

known as a NOON state. The state is a superposition of N particles in a cloud centered at r_a with no particles at r_b and vice versa. These states play an important role in quantum sensing since they provide a quadratic improvement over the standard quantum limit in high-precision measurements using atomic interferometry [54]. Unfortunately, such states are very hard to prepare and decohere easily.

Recent experiments report the preparation of state superpositions in a BEC with decoherence times of 2 seconds at a temperature $T = 10$ nanoKelvin [45]. The record separation distance $\Delta r = r_b - r_a$ is 54cm. However, the self-energy increases when the two spherical mass distributions are separated by $\Delta r = r$, where r is the sphere radius. In this case, the gravitational self-energy is given by

$$E_G = \frac{7GNm_a^2}{10r}$$

where m_a is the atomic mass of the condensed atoms. Assuming coherence times $t = 1$ second for the state superposition, the state reduction could be demonstrated if the corresponding rate is smaller. Assuming that $r = 100\mu m$, and an atomic mass of ^{87}Rb is 86.9091835 u, we estimate that the BEC should have $N = 10^{11}$ atoms. Rubidium BECs are reported to have at most $N = 10^7$ atoms. Increasing the number of atoms in the near future to $N = 10^9$ seems challenging but conceivable. BECs can be produced with other atomic species such as helium, sodium, cesium, lithium, and hydrogen, among others. The maximum number of atoms that can be condensed depends on the atomic species. Condensates consisting of 10^{10} hydrogen atoms can be produced [55]. However, the atomic mass of hydrogen is 1.00794 u, increasing the state reduction rate to levels above the state superposition decoherence time. Therefore, it would be convenient to consider heavier atoms such as cesium ($m_a = 132.90545$ u) or erbium ($m_a = 167.259$ u). A $20\mu m$ ceasium condensate consisting of 3×10^{10} atoms would exhibit state reduction after 1 second. We estimate the same reduction time for a $30\mu m$ erbium condensate with a comparable number of atoms.

Although state reduction of quantum superpositions cannot be demonstrated with the experimental conditions described here, we consider that an exhaustive analysis of decoherence rates for different atomic species at ultra-cold temperatures and an analysis of gravitational reduction of superposition states other than NOON states will enable us to find the right conditions to demonstrate the effect in the near future. In this chapter, we considered only spherical condensates, but a study of other geometries might also yield better results.

Acknowledgements

We would like to thank Joseph P. Cotter, Lucia Hackermüller, Thorsten Schumm, and Richard Howl for interesting discussions and guidance.

References

[1] Károlyházy, F. (1966) Gravitation and quantum mechanics of macroscopic bodies, *Nuovo Cim.* A42, 390.

[2] Károlyházy, F. (1974) Gravitation and quantum mechanics of macroscopic bodies, *Magyar Fizikai Polyoirat* 12, 24.

[3] Károlyházy, F., Frenkel, A. and Lukács, B. (1986) On the possible role of gravity on the reduction of the wave function. In *Quantum Concepts in Space and Time*, R. Penrose and C. J. Isham, eds. (Oxford University Press, Oxford), p. 109.

[4] Kibble, T.W.B. (1981) Is a semi-classical theory of gravity viable? In *Quantum Gravity 2: A Second Oxford Symposium*; eds. C. J. Isham, R. Penrose, and D.W. Sciama (Oxford Univ. Press, Oxford), 63–80.

[5] Diósi, L. (1984) Gravitation and quantum-mechanical localization of macro-objects, *Phys. Lett.* 105A, 199–202.

[6] Diósi, L. (1987) A universal master equation for the gravitational violation of quantum mechanics, *Phys. Lett.* 120A, 377–81.

[7] Diósi, L. and Lucács, B. (1987) In favor of a Newtonian quantum gravity, *Ann. Phys.* 44, 488.

[8] Ghirardi, G.C., Rimini, A. and Weber, T. (1986) Unified dynamics for microscopic and macroscopic systems, *Phys. Rev.* D34, 470.

[9] Ghirardi, G.C., Grassi, R., and Rimini, A. (1990) Continuous-spontaneous-reduction model involving gravity, *Phys. Rev.* A42,1057–64.

[10] Diósi, L. (1989) Models for universal reduction of macroscopic quantum fluctuations, *Phys. Rev.* A40, 1165–74.

[11] Diósi, L. (1990) Relativistic theory for continuous measurement of quantum fields, *Phys. Rev.* A42, 5086.

[12] Pearle, P., and Squires, E.J. (1994) Bound-state excitation, nucleon decay experiments and models of wave-function collapse, *Phys. Rev. Letts.* 73(1), 1–5.

[13] Pearle, P., and Squires, E.J. (1996) Gravity, energy conservation and parameter values in collapse models, *Found. Phys.* 26, 291–305.

[14] Percival, I.C. (1995) Quantum spacetime fluctuations and primary state diffusion, *Proc. R. Soc. Lond.* A451, 503–13.

[15] Bell, J.S. (1987) *Speakable and Unspeakable in Quantum Mechanics* (Cambridge University Press, Cambridge).

[16] Will, C.M. (1986) *Was Einstein Right? Putting General Relativity to the Test* (Basic Books, Inc., New York; paperback, 1988, Oxford Univ. Press, Oxford).

[17] Abbott, B.P. et al. (LIGO Scientific collaboration). (2016) Observation of gravitational waves in binary black hole merger, arXiv:1602.03837.

[18] Bahder, T.B. (2009) Clock synchronization and navigation in the vicinity of the earth. *Nova Science Publishers, Inc.*, ISBN: 978-1-60692-114-2.

[19] Penrose, R. (1965) Zero rest-mass fields including gravitation: asymptotic behaviour, *Proc. Roy. Soc. London* A284, 159–203.

[20] Penrose, R. (1993) Gravity and quantum mechanics. In *General Relativity and Gravitation 13. Part 1: Plenary Lectures 1992. Proceedings of the Thirteenth International Conference on General Relativity and Gravitation held at Cordoba, Argentina, 28 June–4 July 1992*. Eds. R.J. Gleiser, C.N. Kozameh, and O.M. Moreschi (Inst. of Phys. Publ. Bristol & Philadelphia), 179–89.

[21] Penrose, R. (1996) On gravity's role in quantum state reduction, *Gen. Rel. Grav.* 28, 581–600.

[22] Penrose, R. (2014) On the gravitization of quantum mechanics 1: quantum state reduction, *Found Phys* 44, 557–575.
[23] Friedrich, H. (1986) On the existence of n-geodesically complete or future complete solutions of Einstein's field equations with smooth asymptotic structure, *Commun. Math. Phys.* 107, 587–609.
[24] Cartan, É. (1923) Sur les variétés à connexion affine et la théorie de la relativité generalisée (premiére partie), *Ann. École Norm. Sup.* 40, 325–412.
[25] Cartan, É. (1924) Sur les variétés à connexion affine et la théorie de la relativité generalisée (suite), *Ann. École Norm. Sup.* 41, 1–45.
[26] Friedrichs, K. (1927) Eine invariante Formulierung des Newtonschen Gravititationsgesetzes und des Grenzüberganges vom Einsteinschen zum Newtonschen Gesetz, *Math. Ann.* 98, 566.
[27] Trautman, A. (1972, 1973) On the Einstein–Cartan equations I–IV, *Bull. Acad, Pol. Sci.*, Ser. Sci. Math. Astron. Phys. 20, 185–90; 503–6; 895–6; 21, 345–6.
[28] Ehlers, J. (1991) The Newtonian limit of general relativity. In *Classical Mechanics and Relativity: Relationship and Consistency (Int. Conf. in memory of Carlo Cataneo, Elba 1989) Monographs and Textbooks in Physical Science, Lecture Notes 20*. Ed. Giorgio Ferrarese (Bibliopolis, Napoli).
[29] Moroz, I.M., Penrose, R., and Tod, K.P. (1998) Spherically-symmetric solutions of the Schrödinger–Newton equations, *Class. and Quantum Grav.* 15, 2733–42.
[30] Tod, K.P., and Moroz, I.M. (1999) An analytic approach to the Schrödinger–Newton equations, *Nonlinearity* 12, 201–16.
[31] Zych, M., Costa, F., Pikovski, I., Ralph, T.C., and Brukner, C. (2012) General relativistic effects in quantum interference of photons, *Class. Quantum Grav.* 29, 224010.
[32] Dimopoulos *et al.* (2007) Testing general relativity with atom interferometry, *Phys. Rev. Letters*, 98, 111102
[33] Oosterkamp, T.H., and Zaanen, J. (2016) A clock containing a massive object in a superposition of states; what makes Penrosian wavefunction collapse tick? arXiv:1401.0176v4 [quant-ph] 3 Oct 2016.
[34] Greenberger, D.M., and Overhauser, A.W. (1979) Coherence effects in neutron diffraction and gravity experiments, *Rev. Mod. Phys.* 51, 43.
[35] Beyer, H., and Nitsch, J. (1986) The non-relativistic cow experiment in the uniformly accelerated reference frame, *Phys. Lett. B* 182, 211–15.
[36] Rosu, H.C. (1999) Classical and quantum inertia: a matter of principle. *Gravitation and Cosmology* 5 No. 2(18) (June 1999), 81–91.
[37] Colella, R., Overhauser, A.W., and Werner, S.A. (1975) Observation of gravitationally induced quantum interference, *Phys. Rev. Lett.* 34, 1472.
[38] Colella, R., and Overhauser, A.W. (1980) Neutrons, gravity and quantum mechanics, *Am. Sci.* 68, 70.
[39] Werner, S.A. (1994) Gravitational, rotational and topological quantum phase shifts in neutron interferometry, *Class. Quant. Grav.* 11, A207.
[40] Rauch, H., and Werner, S.A. (2015) *Neutron Interferometry: Lessons in Experimental Quantum Mechanics, Wave-Particle Duality, and Entanglement* (2nd edition) (Oxford University Press, Oxford).
[41] Penrose, R. (2000) Wavefunction collapse as a real gravitational effect. In *Mathematical Physics 2000,* Eds. A. Fokas, T.W.B. Kibble, A. Grigouriou, and B. Zegarlinski (Imperial College Press, London), 266–82.
[42] Marshall, W., Simon, C., Penrose, R., and Bouwmeester, D. (2003) Towards quantum superpositions of a mirror, *Phys. Rev. Letters* 91, 13–16; 130401.

[43] Weaver, M.J., Pepper, B., Luna, F.M, Eerkens, H.J., Welker, G., Perock, B., Heeck, K., de Man, S., and Bouwmeester, D. (2016) Nested trampoline resonators for optomechanics, *Applied Physics Letters* 108.033501 (doi:1063/1.4939828).

[44] Eerkens, H.J., Buters, F.M., Weaver, M.J., Pepper, B., Welker, G., Heeck, K., Sonin, P., de Man, S., and Bouwmeester, D. (2015) Optical side-band cooling of a low frequency optomechanical system, *Optics Express* 23(6), 8014–20 (doi:10.1364/OE.23.008014).

[45] Asenbaum, P. et al. (2017) Phase shift in an atom interferometer due to spacetime curvature across its wave function, *Phys. Rev. Lett.* 118, 183602 (doi:10.1103/PhysRevLett.118.183602).

[46] Anglin, J., and Ketterle, W. (2002) Bose–Einstein condensation of atomic gases, *Nature* 416, 211–18 (doi:10.1038/416211a).

[47] Steinhauer, J. (2016) Observation of quantum Hawking radiation and its entanglement in an analogue black hole, *Nature Physics* 12, 959 (doi:10.1038/nphys3863).

[48] Schuldt, T. et al. (2015) Design of a dual species atom interferometer for space, *Experimental Astronomy*, 39, 167–206 (doi=10.1007/s10686-014-9433-y).

[49] Arndt, M. et al. (2012) Focus on modern frontiers of matter wave optics and interferometry, *New Journal of Physics* 14, 125006.

[50] Peters, A., Chung, K. Y., and Chu, S. (1999) Atom gravimeters and gravitational redshift, *Nature* 400, 849–52 (doi:10.1038/nature09340).

[51] Pitaevskii, L., and Stringari, S. (2001) Thermal vs quantum decoherence in double well trapped Bose–Einstein condensates, *Phys. Rev. Lett.* 87, 180402 (doi: 10.1103/PhysRevLett.87.180402).

[52] Dalton, B. J. (2011) Decoherence effects in Bose–Einstein condensate interferometry, *Annals of Physics* 326, 668–720 (doi: 10.1016/j.aop.2010.10.006).

[53] Fattori, M. et al. (2008) Atom interferometry with a weakly interacting Bose–Einstein condensate, *Phys. Rev. Lett.* 100, 080405 (doi:10.1103/PhysRevLett.100.080405).

[54] Cable, H. Laloë, F., and Mullin, W. J. (2011) Formation of NOON states from Fock-state Bose–Einstein condensates, *Physical Review A* 83, 053626 (doi:10.1103/PhysRevA.83.053626).

[55] Fried, D.G. et al. (1998) Bose–Einstein condensation of atomic hydrogen, *Phys. Rev. Lett.* 81, 3811 (https://doi.org/10.1103/PhysRevLett.81.3811); Fried, D.G. (1998) PhD thesis.

12
Collapse. What Else?

NICOLAS GISIN

We present the quantum measurement problem as a serious physics problem. Serious because without a resolution, quantum theory is not complete, as it does not tell how one should—in principle—perform measurements. It is physical in the sense that the solution will bring new physics, i.e., new testable predictions; hence it is not merely a matter of interpretation of a frozen formalism. I argue that the two popular ways around the measurement problem, many-worlds and Bohmian-like mechanics, do, de facto, introduce effective collapses when "I" interact with the quantum system. Hence, surprisingly, in many-worlds and Bohmian mechanics, the "I" plays a more active role than in alternative models, like e.g., collapse models. Finally, I argue that either there are several kinds of stuffs out there, i.e., physical dualism, some stuff that respects the superposition principle and some that doesn't, or there are special configurations of atoms and photons for which the superposition principle breaks down. Or, and this I argue is the most promising, the dynamics has to be modified, i.e., in the form of a stochastic Schrödinger equation.

12.1 The Quantum Measurement Problem

Quantum theory is undoubtedly an extraordinarily successful physics theory. It is also incredibly fascinating: somehow, by "brute mental force," one can understand the strange and marvelous world of atoms and photons! Furthermore, it is amazingly consistent, in the sense that it is amazingly difficult to modify the formalism: apparently, any change here or there activates non-locality, i.e., allows one to exploit quantum entanglement for arbitrary fast communication [1, 2, 3]. How could the fathers develop such a consistent theory based on the very sparse experimental evidence they had? In the landscape of theories, whatever that means, quantum theory must be quite isolated, so that if one looks for a theory in the neighborhood, one has to meet it. However, the quantum formalism is not consistent if one demands that a physics theory tells how one should, in principle, make measurements, as we develop in this article; it is also not consistent if one treats the observer as a quantum system [4].

Quantum theory is a physics theory, and all physics theory should tell what is measurable and how to perform measurements. About the first of these two points, quantum theory tells that all self-adjoint operators correspond to a measurable quantity. More precisely and probably more correctly (it depends on the textbooks), quantum theory claims that every physical quantity is represented by a self-adjoint operator, and every physical quantity can be measured (almost by definition of a physical quantity). Note that often these measurable physical quantities are called observables. So far so good. But let's turn to the second point: how to perform measurements. Here quantum theory is surprisingly silent. Often it is said that one should couple the system under investigation to a measurement apparatus, frequently called a pointer, and then measure the latter [5] (for a recent application of this, related to the measurement problem, see [6]). Hence, to measure a physical quantity of interest of your quantum system, you should measure another system. This is the infamous shifty split, as the pointer itself should be measured by coupling it to yet another measurement system, and so on.

If one insists, the theory remains silent. But the defenders of the theory get virulent: "If you don't know how to perform measurements," they claim, "then you are not a good physicist!." Okay, physicists do know how to perform measurements, indeed, especially experimental physicists. But shouldn't all physics theories tell how to perform measurements, at least in principle?

Somehow, quantum theory is incomplete. I belong to the generation that learned that one should never write such a claim in a paper, at least if one wants to publish it in a respectable journal.[1] Admittedly, one has to be careful with such incompleteness claims. The idea is clearly not to go back to classical physics, i.e., to a mechanistic theory in which cogwheels and billiard balls push other cogwheels and billiard balls. The idea is also not to complement quantum theory with local elements of reality, using EPR's terminology [8], nor with local beables in Bell's terminology [9]. The idea is simply to complement quantum theory in such a way that it tells how, in principle, one performs measurements.

To illustrate the kind of complements I am looking for, one could, for example, postulate that the world is made out of two sorts of stuff, one to which the quantum mechanical superposition principle[2] applies and one to which it doesn't apply. Quantum theory would describe only the first kind of stuff and measurements that happen when one couples (somehow) the two sorts of stuffs. The readers will have recognized standard Copenhagen interpretation of quantum theory: the theory applies only to the "small stuff," while the superposition principle doesn't

---

[1] Or on the arxiv [7]?
[2] The superposition principle states, in words, that if some stuff can be either in one state **or** in another, it can also be in the first state **and** in the second one, i.e., in superposition of the two states.

apply to the "large stuff." Measurements happen when one couples a small quantum system to a large measurement apparatus. As sketched here, this dualistic idea, i.e., that there are two kinds of stuffs, is not yet a complete theory. First, because it doesn't tell how to recognize the two sorts of stuffs (besides that the superposition principle applies to one and not to the other). Second, because it doesn't tell how to couple these two kinds of stuffs. Moreover, one may argue that a complete theory should also describe the other—non-quantum—kind of stuff. Nevertheless, I believe that this line of thought deserves to be investigated more in depth, see section 12.6.

What are the alternatives to some sort of dualism? Assume there is only one sort of stuff, but certain arrangements of this stuff make it special. For example, assume everything is made out of elementary particles, but certain arrangements of atoms and photons make them act as measuring apparatuses.[3] Hence the question:

Which configurations of atoms and photons characterize measurement setups? *Looking for such special configurations is an interesting line of research. Thanks to the worldwide development in quantum technologies, we should soon be able to investigate highly complex configurations of (natural or artificial) atoms and (optical or microwave) photons. Will these developments lead to a breakthrough in quantum physics? Possibly, though most physicists bet on the contrary, i.e., I bet that arbitrarily complex quantum processors will be developed, showing no sign of "collapses," no sign of any breakdown of the superposition principle.*

12.2 What Is Physics

Quantum theory explains very well why it is more and more difficult to keep coherence when the complexity increases: because of the so-called decoherence phenomena. So are we facing the end of "clean physics"? Will we have to stay with the fact that, apparently, quantum theory holds at all scales, i.e., the superposition principle is truly universal, but for all practical purposes (FAPP, as Bell would have said [10]) there is a sort of complexity law of Nature—a sort of second law—that states that it is ultimately harder and harder to demonstrate the superposition principle experimentally for larger and more complex systems? Who knows. But for sure, we—the physics community—should not give up the grand enterprise that easily.[4] Recall that

---

[3] Note that this would imply that the property of acting like a measurement apparatus is an emergent property. Moreover, once this emergent property obtains, the proper arrangement of atoms would gain the capacity of top-down causality, as such arrangements of atoms would have the power to collapse superpositions.

[4] In [10] John S. Bell wrote, "In the beginning natural philosophers tried to understand the world around them. Trying to do that they hit upon the great idea of contriving artificially simple situations in which the number of factors involved is reduced to a minimum. Divide and conquer. Experimental science was born. But experiment is a tool. The aim remain: to understand the world. To restrict quantum mechanics to be exclusively about

Physics is all about extracting information about How Nature Does It. *And for physicists, extracting information means performing measurements. Hence the measurement problem has to be taken seriously.*

Let me stress that I consider the quantum measurement problem as a serious and real physics problem. It is serious because without a solution quantum theory is incomplete, as discussed earlier. It is real in the sense that its solution will provide new physics, with new and testable predictions. Hence it is not merely a matter of interpretation of a given formalism: to solve it, one has to go beyond today's physics.

To conclude this section and to be transparent, I should state that I am a naive realist (as most physicists): there is a world out there, and the grand enterprise of physics aims at understanding it, see footnote 4. Additionally, following Schrödinger [11], I consider that "I" am not part of it: my aim is to understand the outside world, but I am not including myself in that outside world. Of course, I am made out of atoms and other stuff that can and should be studied by physics. But physics is not about explaining my presence. As much as possible (and I believe it is entirely possible), physics theories should not postulate that "I" have to exist for the world to function. This may seem too philosophical, but we shall see that it has consequences for possible solutions to the measurement problem. Let me stress that this is not dualism in a physics sense: the world out there could well be made out of a single kind of stuff.

In summary, I believe that the scientific method will never explain why there is something rather than nothing, nor will it explain why "I" am here. Physics must assume both that "I" exist and that there is a world out there, so that "I" can gain better and better understanding of the outside world, i.e., of How Nature Does It.

Let's return to the quantum measurement problem and look for alternatives to what we already discussed, i.e., physical dualism (the assumption that there is more than one sort of stuffs out there), to the existence of "special" configurations of atoms and photons that make them act like measurement devices, and to the end of clean physics.[5]

---

piddling laboratory operations is to betray the great enterprise. A serious formulation will not exclude the big world outside the laboratory."

[5] In order to avoid receiving a km-long e-mail from Chris Fuchs, let me say a few words about QBism [12]. QBism changes the goal of physics. It is no longer about finding out How Nature Does It; QBism restricts physics to what "I" can say about the future. More precisely, about how "I" should bet on future events. For me this is not only a betrayal of the great enterprise, it is almost a sort of solipsism where everything is about "me" and my beliefs. Well, at the end, I am not sure I'll avoid the km-long e-mail.

12.3 Many-Worlds

Why not simply assume that quantum theory is complete and the superposition principle universal? This leads straight to some many-worlds interpretations of quantum theory [13, 14]. Indeed, since quantum theory is amazingly successful and since quantum theory without any addition (i.e., without any vague collapse postulate) leads to the many-worlds, why not merely adopt a many-worlds view?

In the many-worlds view, the measurement problem is circumvented by the claim that everything that has a chance to happen, whatever tiny chance, does actually happen. Hence, it is a sort of huge catalog of everything that could happen. More precisely, it is the catalog of everything that has happened and of everything that is happening and of everything that will ever happen. Simply, we are not aware of the entire catalog, only of that part of the catalog corresponding to the world in which "we" happen to live. But isn't physics precisely about, and only about, that part of the catalog? What is the explanatory power of claiming that everything happens, but "we" are not aware of everything? And what is that "we"?

It is a fact that "I" exist. Actually, it is the fact that I know best. Should "I" be satisfied with a theory that tells that I exist in a hugely enormous number of copies and that all the theory provides is a catalog of everything that "I" or a copy of myself experiences? Not to mention the vast majority of worlds in which the atoms of my body don't make up a human, probably not even a thing. Actually, the theory says a bit more; it also tells about correlations. If "I" see this now, then there are only some events "I" may see in the future. And vice versa, as time doesn't properly exist in the many-worlds. Note that to achieve this, one conditions the catalog on what "I see now," i.e., one uses an effective collapse: one limits the analysis to that part of the catalog in which "I" see this now.

In summary, in many-worlds theories, it is "I" that continuously collapse the state-vector, at least for the purpose of allowing the theory to make predictions about what "I" am observing. In other worlds, in many-worlds the "I" is not merely a passive observer but plays an active role.

Admittedly, many-worlds is a logically consistent interpretation, at least as long as one doesn't insist that "I" exist. Moreover, it is the most natural one if one sticks to standard Hilbert-space quantum theory (i.e., without measurements). But logical consistency is only a necessary condition for a physics theory. Solipsism is another example of a logically consistent theory, somehow on the other extreme to many-worlds: in solipsism, only "I" exist. But, as I stated in the previous section, I am a realist: I just don't see how one can do physics without assuming the "I" and the "world."

Let me address another issue with many-worlds. It is a deterministic theory, even a hyper-deterministic theory, i.e., determinism applies to everything in the entire universe. Indeed, since there can't be any influences coming from outside and since the Schrödinger equation—the only dynamical equation of the theory—is deterministic, everything that happens today, e.g., what I am writing, the way each reader reacts, the details of all solar eruptions, etc, was all encoded in some "quantum fluctuations"[6] of the initial state of the universe.[7] Given the complexity of the (many-) worlds, it had to be encoded in some infinitesimal digits of some quantum state, possibly in the billionths of billionths decimal place. I am always astonished that some people seriously believe in that. Mathematical real numbers are undoubtedly very useful when doing our theory. But are they physically real [15][8]? Do these infinitesimal digits have a real impact on the real world? Is this still proper physics? For sure, such assumptions can't be tested. Hence, for me, hyperdeterminism is a nonsense [15], though it is the dominant trend in today's high-energy physics and cosmology (though see [18, 17]). Apparently, the many followers of today's trend elevate (unconsciously) the linearity of the Schrödinger equation and the superposition principle to some sort of ultimate quasi-religious truth, some truth in which they believe even more than in their own free will. Note that it is not the first time in science history that some equations get elevated above reason: followers of Laplace did also elevate the deterministic Newton equations to some sort of ultimate truth. We know what was the destiny of that belief.

In summary, in order to make predictions in the many-worlds, one introduces some effective collapses that happen when the system is coupled to "I." Hence, the theory is not complete but relies—somehow—on "I," i.e., on some concept foreign to the theory.

12.4 Bohmian Quantum Mechanics

There is yet another way to avoid the quantum measurement problem. Assume that at all times there is one and only one "event" that is singled out.[9] As time passes, the list of singled-out events must be consistent, as in consistent histories [19]. A nice example assumes that, at the end of the day, everything we ever observe is the position of some stuff. Hence, let's assume that the physical quantity "position" is always well determined by some additional variable (additional with respect to standard quantum theory). Interestingly, this can be made consistent [20, 21], though at the cost of some counterintuitive phenomena [22, 23] and assuming it applies to the entire universe (as soon as one cuts out some piece of the universe, one may encounter paradoxes [4]).

---

[6] Don't ask me what that means.
[7] Or, equivalently, in the final state of the Universe.
[8] Recall that the assumption that real numbers are physically real implies that there could be an infinite amounts of information in a finite volume of space [16, 15].
[9] Or one collection of events that are singled out. We may name this collection as one "big" or "composed" event.

Note that one may also apply similar ideas to other physical quantities than position, leading to various modal interpretations of quantum theory [24, 25]. With position as the special physical quantity, the reader has recognized Bohmian quantum mechanics [20, 21]. It is a nice existence proof of non-local hidden variables that deserves to be more widely known [26]. It is non-local despite the fact that the additional variables are points in space, i.e., highly localized. But the dynamics of these point-particles is non-local: by acting here one can instantaneously influence the trajectories of point-particles there, at a distance. At first, this might be considered as quite odd. But quantum physics is non-local, in the sense of violating Bell inequalities. Hence, the non-locality of Bohmian mechanics is quite acceptable. Actually, there is just no choice: in order to recover the predictions of quantum theory and the experimental data, all theories must incorporate the possibility of Bell inequality violations, i.e., some non-locality.

One ugly aspect, in my opinion, of Bohmian mechanics is that the additional variables must remain hidden for ever. If not, if one could somehow collect information about their locations beyond the statistical predictions of quantum theory, then one could activate non-locality, i.e., one could use entanglement not only to violate some Bell inequality but to send classical information at an arbitrarily large speed [27]. But can one add variables to a physics theory while claiming that they are ultimately not accessible? Bohmians answer that the hidden positions determine the results of measurements and hence are not entirely hidden. Indeed, when one observes a result, one can apply an effective collapse as one knows that the hidden positions are now distributed within the reduced wave function corresponding to quantum statistics. But there is no way to know more about the location of that particle.[10] This raises the question when should one apply such an effective collapse? The answer presumably is: "when 'I' register a measurement result," a bit like in many-worlds.

I find it tempting to compare Bohmian mechanics with a toy theory in which one has added as additional variables all the results of all the measurements that will ever be performed in the future, though with the restriction that none of these additional variables can be accessed before the corresponding measurements take place. Note that in such a way, one can turn any theory into a deterministic one.[11] But, for sure, no physicists would take such a toy theory seriously. Admittedly, Bohmian mechanics is much more elegant than the sketched one. But is it fundamentally different?

Let's return to Bohmian quantum mechanics. As said, it is a remarkable existence proof of non-local hidden variables. But does it answer the deep question of the

---

[10] Moreover, there are situations in which the hidden particle leaves a trace where it was not [22, 23].
[11] Note though, that with the sketched construction, time is necessarily built into the toy theory.

quantum measurement problem? I don't think so. As with the many-worlds, it assumes hyperdeterminism and relies on infinitesimal digits for its predictions. Hence, despite the deterministic equations, it is not a deterministic theory [15], as I elaborate in the next section. Moreover and disappointingly, it doesn't make any new prediction.

Finally, Bohmian mechanics with its non-local hidden variable is at great tension with relativity.

12.5 Newtonian Determinism

Some readers may wonder whether I would also have argued against classical Newtonian mechanics had I lived 150 years ago. After all, it also relies on deterministic equations, and Newton's universal gravitation theory is also non-local; and what about the "I"? Let me start with the second aspect, non-locality. I have no problem with quantum non-locality (the possibility to violate Bell inequalities), because quantum randomness precisely prevents the possibility to use quantum non-locality to send classical information [28, 29, 30, 31]. However, Newton's non-locality[12] can be used, in principle, to send information without any physical support carrying this information: move a rock on the moon (with a small rocket) and measure the gravitational field on earth. According to Newton's theory, this allows one to communicate in a non-physical way—i.e., without any physical stuff carrying the information—and at an arbitrarily high speed [32]. This is deeply disturbing and did already disturb Newton himself [33]:

That Gravity should be innate, inherent and essential to Matter, so that one Body may act upon another at a Distance thro a Vacuum, without the mediation of any thing else, by and through which their Action and Force may be conveyed from one to another, is to me so great an Absurdity, that I believe no Man who has in philosophical Matters a competent Faculty of thinking, can ever fall into it.

Admittedly, when I first learned about Newton's universal gravitation theory at high school, I found it beautiful, not noticing how absurd it is. May I suggest that one should always teach Newton together with the comment that it is efficient but absurd? I believe this would be great pedagogy.

Today we know that relativity solved the issue of Newton's non-locality and that experiments have confirmed quantum non-locality beyond any reasonable doubts [34, 35, 36].

Let's now turn to the other similarity between Bohmian mechanics and classical physics, that is the deterministic nature of Newton's equations. For clocks, harmonic oscillators and generally integrable dynamical systems, the stability is

---

[12] Which, by the way, predicts the possibility to violate Bell's inequality.

such that the infinitesimal digits of the initial condition do play no role. For chaotic systems, on the contrary, these infinitesimal digits quickly dominate the dynamics. Hence, since these mathematical infinitesimal digit do not physically exist,[13] chaotic dynamical system are not deterministic.[14] This fact doesn't change anything in practice (FAPP, as Bell would have shouted [10]), but it demonstrates that classical Newtonian mechanics is simply not a deterministic physics theory: despite the use of deterministic equations, it does not describe deterministic physics [15].

There is, however, a huge difference between Newton's determinism (of the equations) and Bohmian or many-worlds. In the former there is no entanglement. Hence, one can separate the world into systems, hence "I" can act on each of them individually. In an enormously entangled world, on the contrary, there is no way to separate subsystems, there is no way to act on just one subsystem. Determinism plus entanglement make things intractable [37]. Accordingly, either "I" can not act, or "I" do induce effective collapses that disentangle the subsystems. But then, why not include these effective collapses in the theory?

12.6 Dualism

In summary, so far we saw four sorts of attempts to circumvent the quantum measurement problem:

[1] dualism as in orthodox Copenhagen quantum mechanics,
[2] some configurations of atoms and photons make them act as measurement setups that break superpositions,
[3] all possible results coexist in some many-worlds,
[4] all results were already encoded in some additional non-local variables, hidden for ever, as in Bohmian mechanics.

---

[13] Several colleagues complained that this "physical existence" is badly defined and/or confuses physical existence with measurability. Let me try to clarify. Obviously real numbers can't be measured, neither today nor in any future. But my claim goes way beyond that. The world out there is pretty well described by today's physics. Actually, it is also pretty well described by the physics of one or two centuries ago and will be even better described by the physics in some centuries. But this doesn't allow us to identify the world out there with its physical description. The world out there is infinitely richer than any physical description and than any human description. Somehow, the world out there is "free," i.e., it doesn't let itself get trapped in our theories, it does not depend on our description. In particular the fact that we use real numbers doesn't imply that real numbers govern the world out there, nor does the fact that some of our descriptions are based on deterministic equations imply that the evolution of the world out there is deterministic. In brief, the world out there can't be confined (locked up) in any finite-time physics theory.

[14] One may argue that there is no quantum chaos. But this is not entirely true, though quantum chaos differs deeply from classical chaos. In the quantum case, it is the high sensitivity to the exact Hamiltonian that should be considered. One may claim that there is one exact fundamental Hamiltonian, thus no indeterminacy. But this is wrong, since, whatever units one chooses, the Hamiltonian contains constants, like the masses of particles, and these constants are described by real numbers. These real numbers are themselves indetermined (random), and hence even the fundamental Hamiltonian leads to chaos.

Let me recall that since I consider the quantum measurement problem as a real physics problem, its solution will necessarily lead to new physics, including new and testable predictions. It is a fact that so far attempts 2, 3, and 4 did not bring up as much good and new physics as attempt 1 did. But I should add that this argument might be a bit unfair, because attempt 1 came first and had thus a significant advantage. Anyway, let's consider attempt number 1, i.e., dualism.

It is probably fair to say that most physicists would reject dualism.[15] But could it be that they go too fast here? Clearly, dividing the world out there into "small" and "large" is not good enough. But couldn't there be stuff to which the superposition principle doesn't apply? Some have argued that the hypothetical non-quantum stuff is space–time and/or gravity [38, 39, 40, 41, 42]. This is certainly a possibility. But I am reluctant to put my bets on this, because everything is connected to space–time and to gravity. Hence, if it is the coupling between the "quantum stuff" and the hypothetical "non-quantum stuff" that determines when a measurement happens, then, continuously, everything always undergoes measurements. In such a case, either the superposition principle is continuously broken and one should never have seen superpositions, or the non-quantum stuff undergoes a bit of superposition.

More formally, denoting $|QS_0\rangle$ the initial state of some quantum stuff that interacts with some non-quantum stuff $|NQS_0\rangle$, then, after an arbitrary short time, the quantum and non-quantum stuff get entangled:

$$|QS_0\rangle|NQS_0\rangle \stackrel{t=\epsilon}{\rightarrow} \sum_j |QS_j\rangle|NQS_j\rangle \qquad (12.1)$$

But if the non-quantum stuff can't at all be in superposition, then state (12.1) can't exist, not even for a split of a second. Hence, there would be instantaneously collapse also for the quantum stuff.

People have speculated that this bit of superposition gets quickly, though not instantaneously, washed out [38, 39, 40, 41, 42]. Why not? But then, why introduce such a non-quantum stuff in the first place? Why not merely assume that all stuff undergoes superpositions, but only in some (precisely) limited way? Readers recognize here spontaneous collapse theories; more on this in section 12.7.

Before closing this section, let's see whether there is not another plausible way to divide the stuff into several sorts, i.e., dualism.[16] There is obviously one that goes back all the way to Descartes: "material stuff" and "non-material stuff." The superposition principle would apply only to the material stuff. This is admittedly

---

[15] By the way, many would even do so virulently, while at the same time claiming to adhere to the Copenhagen interpretation, which is dualist. The same would simultaneously claim with joy how proud they are to work in a field in which rational thinking dominates. Okay, I leave that line of thought to sociologists.

[16] Physics divides "me" and the outside world. I do not consider this as fundamental dualism, but only as the scientific method. Here I am asking whether a real physical dualism is a viable path towards a resolution of the quantum measurement problem.

extremely crude, certainly not yet a theory, not even a valid sketch of a theory, because essentially nothing is said about the "non-material stuff." Moreover, one should not make the situation more confused by thinking that the "non-material stuff" is our "mind," as this would imply that the first measurement that ever happened had to wait for us. However, I like to argue that one should also not reject dualism too quickly. After all, it might well be that there is stuff out there to which the superposition principle does not apply.

Let's return to the quantum measurement problem. Although I am sort of a dualist from a philosophical point of view,[17] I don't think that dualism is the right solution for the measurement problem. It might be that in some decades, if the measurement problem remains without significant progress, one may have to revisit a dualistic solution, but at present we better stick to the assumption that there is one and only one sort of stuff out there in the real world and that the superposition principle applies to it.

12.7 Modified Schrödinger Equation

Recall that the superposition principle states, in words, that if some stuff can be either in one state **or** in another, it can also be in the first state **and** in the second one, i.e., in superposition of the two states. The linearity of the Schrödinger equation implies then that such superposition lasts for ever. Consequently, in a theory without the measurement problem and in which everything (except "I") satisfies the superposition principle (and without hyperdeterminism) it must be the case that it is the Schrödinger equation that has to be modified. First attempts to modify the Schrödinger equation tried to extend it to some non-linear but still deterministic equation [43, 44, 45]. But this turned out to be hopeless, as could be expected from the discussions in the previous sections. A quite convincing argument came from the observation that any such deterministic nonlinear generalization of the Schrodinger equation activates non-locality, i.e., predicts the possibility of arbitrarily fast communication [1, 2, 3].

Hence, one has to go for a non-deterministic generalization.[18] Non-deterministic merely means not deterministic, that is it does not say how the equation should be, it only says how the equation should not be. However, assuming that the evolution is Markovian and the solution continuous in time, then—for those who know stochastic differential equations—possibilities are quite easy to find [48, 1, 49, 50]. Essentially, there is only one [51]. This solution depends on some operators, a

---

[17] I don't believe that everything is merely matter and energy as described by today's physics, not even stuff described by any physics theory at any given point in time. Though I believe in endless progress.

[18] Note that sticking to a linear equation is also hopeless, as the Schroödinger equation is the only linear equation that preserves the norm of the state-vector; see also [46, 47]

bit like the Schrödinger equation depends on the Hamiltonian. At this point, all that remains is to fix this operator of the new, non-linear and stochastic term of the hypothetically fundamental dynamical equation of the complemented quantum theory and look for the new predictions.

Let's be a bit more explicit. Consider the following Itô stochastic differential equation, see, e.g., [48, 52]:

$$|d\psi_t\rangle = -iH|\psi_t\rangle dt$$
$$+ \sum_j \left(2\langle L_j^\dagger \rangle_{\psi_t} L_j - L_j^\dagger L_j - \langle L_j^\dagger \rangle_{\psi_t} \langle L_j \rangle_{\psi_t} \right) |\psi_t\rangle dt$$
$$+ \sum_j \left(L_j - \langle L_j \rangle_{\psi_t} \right) |\psi_t\rangle d\xi_j \qquad (12.2)$$

where H is the usual Hamitonian, L_js are (Lindblad linear) operators, $\langle L_j \rangle_{\psi_t} = \frac{\langle \psi_t | L_j | \psi_t \rangle}{\langle \psi_t | \psi_t \rangle}$ are the expectation values of the operators L_j, and the $d\xi_j$s are independent complex Wiener processes satisfying:

$$M[d\xi_j] = 0 \qquad (12.3)$$
$$M[d\xi_j d\xi_k] = 0 \qquad (12.4)$$
$$M[d\xi_j d\xi_k^*] = \delta_{jk}\, dt \qquad (12.5)$$

where $M[\ldots]$ denotes the mean value. Note that eq. (12.2) preserves the norm of $|\psi_t\rangle$.

Equation (12.2) describes a sort of Brownian motion in Hilbert space of the state-vector $|\psi_t\rangle$. It is the analog of a stochastic description of Browian motion at the individual particle level. It is assumed that it is not merely an approximation but the foundamental dynamical law describing how isolated quantum systems evolve. Hence it predicts deviations from the standard Schrödinger dynamics, i.e., it predicts new physics. Consequently, at least, such modified dynamical laws could be wrong!

To illustrate eq. (12.2) and for simplicity, let's consider the case with a single operator L; furthermore, assume it is self-adjoint and commutes with the Hamiltonian H. Then, interestingly, the solution to (12.2) follows a sort of Brownian motion and eventually tends to an eigenstate $|l\rangle$ of L. Moreover, the probability to tend to a given $|l\rangle$ equals the quantum probability $|\langle l|\psi_0\rangle|^2$, see Fig. 12.1.

When averaging over all solutions of (12.2), i.e., averaging over all Wiener processes $d\xi_j$, one obtains a density matrix $\rho(t)$ that satisfies the linear evolution equation:

$$\frac{d\rho(t)}{dt} = -i[H, \rho(t)] \qquad (12.6)$$
$$- \sum_j \left(L_j^\dagger L_j \rho(t) + \rho(t) L_j^\dagger L_j - 2 L_j \rho(t) L_j^\dagger \right)$$

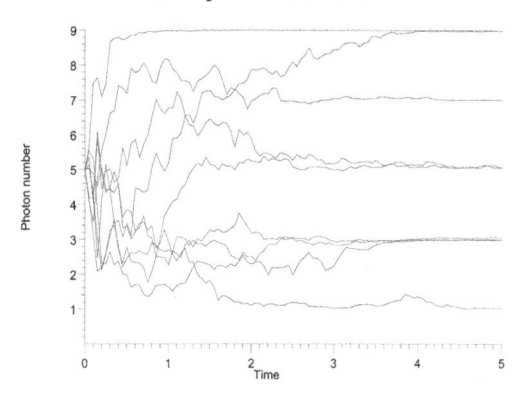

Figure 12.1 *Example of some solutions to eq. (12.2) in case of a photon-number measurements, i.e., $H = L = a^\dagger a$. The initial state is an equal superposition of odd photon-number states: $|1\rangle + |3\rangle + |5\rangle + |7\rangle + |9\rangle$. The convergence to the eigenstates can be clearly seen. Taken from [52].*

Equation (12.6) is the quantum analog of a classical Fokker-Planck equation describing the probability distribution of an ensemble of classical Browian particles.

Let us emphasize that since the density matrices at all times follow a closed-form equation, this modification of the Schrödinger equation does not lead to the possibility of faster-than-light communication [48, 1].

Remains to find what the operators L_j could be. Here comes the beautiful finding of Ghirardi, Rimini, and Weber [53].[19] Assume the L_j are proportional to the positions of all elementary particles, with a proportionality coefficient small enough that it barely affects the evolution of systems made out of one or only a few particles. Hence, microscopic systems would essentially not be affected by the modified Schrödinger equation (12.2). However, if a pointer is in superposition of pointing here and pointing there, then, since the pointer is made out of an enormous number of particles, let's say about 10^{20}, the modified Schrödinger equation predicts a quasi instantaneous collapse: it suffices that a single particle gets localized by the

---

[19] In 1988, Professor Alberto Rimini visited Geneva to present a colloquium. He presented the famous GRW paper [53] in the version Bell gave of it [54]. In the GRW theory, the non-linear stochastic terms added to the Schrödiger equation lead to solutions with discontinuous jumps of the wave-packet, i.e., to some sort of spontaneous collapses triggered by nothing but mere random chance, as time passes. Near the end of his colloquium, Rimini mentioned that an open question was to massage the stochastic modifications in such a way that the solutions would be continuous trajectories (in Hilbert space). He also emphasized the need for an equation that would preserve (anti-)symmetric states. He may have added that, with Philip Pearle [55], they have a solution, but for sure he had no time to explain it. Immediately after the colloquium, I went to Alberto and told him that I knew how to answer his questions. He encouraged me, and I immediately added a small section to a paper already quasi-finished [1]. There is no doubt that Philip Pearle found CSL independently. Lajos Diósi, by the way, did also find it [49]. Actually, everyone who, at that time, knew both GRW and Itô stochastic differential calculus would have found it, because it is quite trivial once you know the tools and the problem. Anyway, Ghirardi and Pearle got very angry that I published my result first, and I decided to leave that field. I didn't like fights and wanted a carrier.

stochastic nonlinear terms of equation (12.2) for the entire pointer to localize, i.e., the pointer localizes about 10^{20} times faster than individual particles.

I remain convinced that collapse models of the form sketched earlier, see [56] and references there in, is the best option we have today to solve the unacceptable quantum measurement problem. Note, however, the following two critical points.

First, one unpleasant characteristic of such a modified dynamics is that the very same equation (12.2) can also be derived within standard quantum theory by assuming some coupling between the quantum system and its environment and conditioning the system's state on some continuous measurement outcomes carried out on the environment [57]. This makes it highly non-trivial to demonstrate an evolution satisfying equation (12.2) as a fundamental evolution, as one would have to convincingly show that the system does not interact significantly with its environment. Note also that sufficient error corrections could hide the additional stochastic terms of eq. (12.2) and thus prevent that the developments of advanced quantum information processors reveals them.

A second delicate point about eq. (12.2) is that it is not relativistic, and it seems impossible to make it relativistic [58].

The previous two points are part of the reasons I left the field some 20 years ago.

12.8 Conclusion

I want to understand Nature. For me this requires that "I" exist and that there is something out there to be understood, in particular that there is a world out there. Physics is all about extracting information about How Nature Does It. For physicists, extracting information means performing measurements. Hence the measurement problem has to be taken seriously. It is a real physics problem, and its solution will provide new physics and new and testable predictions.

Taking standard Hilbert-space quantum theory at face value, without the vague collapse postulate, leads to the many-worlds: everything that can happen happens. The problem, besides hyper-determinism, is that "I" am excluded from the many-worlds. In order to re-introduce the "I," one has to introduce some effective collapses that happen when "I" interact with the world. Note that this step is usually not taken explicitly by the many-worlds followers, except when they compute predictions, i.e., when they do physics. This is a bit similar to the well-known Wigner friend story [59], though Wigner never presented it in a many-worlds context. Hence, it seems that in order to make physical sense of many-worlds, one needs some form of dualism: "I" trigger effective collapses. Before me, everything coexisted. Now that I am here, in order to make predictions, I have to condition these "coexisting things," on those that correlated to me, using some effective collapses. Since it is a fact that "I" exist, wouldn't it be much simpler and cleaner to assume that the effective collapses are truly real and to include them in our physics theory?

Bohmian mechanics is a nice and constructive existence proof of non-local hidden variables. But it suffers from similar drawbacks to the many-worlds. It is hyper-deterministic, and in order to make predictions, one has to introduce an "I" that does some conditioning by de facto effective collapses. Hence, again, it is cleaner to assume real collapses in our physics theory. Moreover, doing so we may at least be wrong, i.e., at least we may predict new phenomena.

Remains the question of what triggers the collapses. Should we formulate the measurement problem as a search for those configurations of atoms and photons that trigger a collapse, as formulated in section 12.1? This is an interesting line of experimental research.

Dualism is a very natural position in our culture. Actually, I don't see how to avoid it for our science to make sense. But I believe much premature to jump to the conclusion that it is the interaction between "I" and the outside world that triggers the collapses of the quantum states. Other forms of dualism, actually trialism: "I" plus two sorts of stuff out there are logical possibilities, but there are no good candidates and introducing some new stuff seems too high a price to pay, especially when it is not (yet?) needed.

Remains spontaneous collapses, described for instance by some modified Schödinger equation to which one adds some non-linear stochastic terms, as in eq. (12.2). These additional terms lead continuously and spontaneously, i.e., by mere random chance, to collapses that barely affect microscopic systems, but quickly localize macroscopic objects. It seems to me that the scientific method that has been that efficient so far tells us that this spontaneous collapse approach is by far the most promising one. For me, it is also the only one that is consistent with what I expect from physics. Indeed, at the end of the day, a theory without collapses, doesn't predict any events, hence has zero explanatory power.

One additional value of collapse theories is that they naturally incorporate the passage of time. I am well aware that it is fashion in physics to claim that time is an illusion [60]. Admittedly, time is a complex notion, or series of notions with many facets, time may be relative, difficult to grasp, etc. But time exists. Moreover, time passes [17, 18, 15].

With spontaneous collapse theories, time exists and passes, the world out there exists and undergoes a stochastic evolution. And "I" exist, outside the theory, able to contemplate it, to develop it and act as an observer.

Acknowledgment

This work profited from stimulating discussions with Florian Fröwis and Renato Renner. Financial support by the European ERC-AG MEC and the Swiss NSF is gratefully acknowledged.

References

[1] N. Gisin, Stochastic quantum dynamics and relativity, *Helv. Phys. Acta* **62**, 363 (1989).
[2] N. Gisin, Weinberg non-linear quantum-mechanics and superluminal communications, *Phys. Lett. A* **143**, 1 (1990).
[3] N. Gisin and M. Rigo, Relevant and irrelevant nonlinear Schrödinger equations, *J. Phys. A* **28**, 7375 (1995).
[4] D. Frauchiger and R. Renner, Single-world interpretations of quantum theory cannot be self-consistent, arXiv:1604.07422.
[5] J. Von Neumann, *Mathematical Foundations of Quantum Mechanics*, Beyer, R. T., 1932.
[6] T.J. Barnea, M.-O. Renou, F. Fröwis, and N. Gisin, Macroscopic quantum measurements of noncommuting observables, *Phys. Rev. A* **96**, 012111 (2017)
[7] http://www.iqoqi-vienna.at/blog/article/nicolas-gisin/
[8] A. Einstein, B. Podolsky, and N. Rosen, Can quantum-mechanical description of physical reality be considered complete?, *Phys. Rev.* **47**, 777 (1935).
[9] J. S. Bell, *Speakable and Unspeakable in Quantum Mechanics: Collected Papers on Quantum Philosophy* (Cambridge University Press, Cambridge, 1987).
[10] J.S. Bell, Against measurements, *Physics World*, pp. 33–40, August 1990.
[11] E. Schrödinger, *Mind and Matter*, Ch. 3, Cambridge Univ. Press, Cambridge, 1958).
[12] Ch. Fuchs, The perimeter of quantum Bayesianism, arXiv:1003.5209; N. D. Mermin, QBism puts the scientist back into science, *Nature*, **507**, 421–423 (2014).
[13] D. Deutsch, *The Fabric of Reality*, (The Penguin Press, 1997).
[14] *Many Worlds? Everett, Quantum Theory and Reality*, S. Saunders, J. Barrett, A. Kent and D. Wallace (eds.), (Oxford University Press, Oxford, 2010).
[15] N. Gisin, Time really passes, science can't deny that, arxiv/1602.01497.
[16] G. Chaitin, *The Labyrinth of the Continuum, in Meta Math!* (Vintage, 2008).
[17] J. Norton, Time really passes, *Journal of Philosophical Studies* **13**, 23–34 (2010), www.pitt.edu/~jdnorton/Goodies/passage, www.humanamente.eu/index.php/pages/36-issue13.
[18] L. Smolin, *Time Reborn* (Houghton Mifflin Harcourt, 2013).
[19] R. Griffiths, Consistent histories and the interpretation of quantum mechanics, *J. Stat. Phys.*, **36**, 219 (1984); R. Omnès, *Rev. Mod. Phys.*, **64**, 339 (1992); H.F. Dowker and J.J. Halliwell, Quantum mechanics of history: The decoherence functional in quantum mechanics, *Phys. Rev. D*, **46**, 1580 (1992); M. Gell-Mann and J.B. Hartle, Classical equations for quantum systems, *Phys. Rev. D*, **47**, 3345 (1993).
[20] D. Bohm, A suggested interpretation of the quantum theory in terms of "hidden" Variables, *Phys. Rev.*, **85**, 166 and 180 (1952).
[21] D. Bohm and B.J. Hiley, *The Undivided Universe* (Routledge, London and NY, 1993; p. 347 of the paperback edition).
[22] B.-G. Englert et al., Surrealistic Bohm trajectories, *Z. Naturforsch.* **47a**, 1175–1186 (1992)

[23] N. Gisin, Why Bohmian mechanics? One- and two-time position measurements, Bell inequalities, philosophy and physics, arXiv:1509.00767.
[24] B. Van Fraassen, Hidden variables and the modal interpretation of quantum theory, *Synthese* **42**, 155 (1979).
[25] https://plato.stanford.edu/entries/qm-modal/
[26] J.S. Bell, On the impossible pilot wave, *Found. Phys.* **12**, 989–999 (1982).
[27] A. Valentini, Signal-locality, uncertainty, and the sub-quantum H-theorem, *Physics Letters* A, **156**, 5–11 (1991); *Phys. Lett.* A **158**, 1–8 (1991).
[28] Popescu, S., and Rohrlich, D. Quantum nonlocality as an axiom, *Found. Phys.* **24**, 379–385 (1994).
[29] N. Gisin, Quantum cloning without signaling, *Phys. Lett.* A **242**, 1 (1998).
[30] Ch. Simon, V. Buzek, and N. Gisin, No-signaling condition and quantum dynamics, *Phys. Rev. Lett.* **87**, 170405 (2001).
[31] N. Brunner, D. Cavalcanti, S. Pironio, V. Scarani, and S. Wehner, Bell nonlocality, *Rev. Mod. Phys.* **86**, 419 (2014).
[32] N. Gisin, *Quantum Chance, Nonlocality, Teleportation and Other Quantum Marvels* (Springer, 2014).
[33] Isaac Newton Papers & Letters on Natural Philosophy and related documents. Under the direction of Bernard Cohen and Robert E. Schofield, Harvard University Press, 1958.
[34] B. Hensen et al., Loophole-free Bell inequality violation using electron spins separated by 1.3 kilometres, *Nature* **526**, 682 (2015).
[35] M. Giustina et al., Significant-loophole-free test of Bell's theorem with entangled photons, *Phys. Rev. Lett.* **115**, 250401 (2015).
[36] L. K. Shalm et al., Strong loophole-free test of local realism, *Phys. Rev. Lett.* **115**, 250402 (2015).
[37] N. Gisin, in *Le plus Grand des Hasards, Surprises Quantiques*, pp. 184–186, eds J.-F. Dars et A. Papillault (Belin, Paris, 2010).
[38] F. Karolyhazy, A. Frenkel, and B. Lukacs, in *Physics as Natural Philosophy*, edited by A. Shimony and H. Feschbach (MIT, Cambridge, MA, 1982), p. 204.
[39] L. Diósi, A universal master equation for the gravitational violation of quantum mechanics, *Phys. Lett.* A **120**, 377 (1987).
[40] R. Penrose, On gravity's role in quantum state reduction, *Gen. Rel. Gravit.* **28**, 581–600 (1996).
[41] W. Marshall, C. Simon, R. Penrose, and D. Bouwmeester, Towards quantum superpositions of a mirror, *Phys. Rev. Lett.* **91**, 130401 (2003).
[42] S. Adler, Comments on proposed gravitational modifications of Schrödinger dynamics and their experimental implications, *J. Phys.* A **40**, 755 (2007).
[43] M.D. Kostin, Friction and dissipative phenomena in quantum mechanics, *J. Stat. Phys.* **12**, 145 (1975).
[44] I. Bialynicki-Birula and J. Mycielski, Nonlinear wave mechanics, *Annals of Physics* **100**, 62 (1976).
[45] N. Gisin, A simple nonlinear dissipative quantum evolution equation, *J. Phys.* A **14**, 2259 (1981).
[46] I.E. P. Wigner, *Group Theory* (Academic, New York, 1959).
[47] N. Gisin, Irreversible quantum dynamics and the Hilbert space structure of quantum kinematics, *J. Math. Phys.* **24**, 1779 (1983).
[48] N. Gisin, Quantum measurements and stochastic processes, *Phys. Rev. Lett.* **52**, 1657–1660 (1984).

[49] L. Diósi, Continuous quantum measurement and Itô formalism, *Phys. Lett.* A **129**, 419 (1988).

[50] P. Pearle, Combining stochastic dynamical state-vector reduction with spontaneous localization, *Phys. Rev.* A **39**, 2277, (1989).

[51] I.C. Percival, *Quantum State Diffusion* (University Press, Cambridge, 1998).

[52] N. Gisin and I.C. Percival, The quantum-state diffusion model applied to open systems, *J. Phys.* A, **25**, 5677 (1992); Quantum state diffusion, localization and quantum dispersion entropy, *J. Phys.* A, **26**, 2233 (1993); The quantum state diffusion picture of physical processes, *J. Phys.* A, **26**, 2245 (1993).

[53] G.C. Ghirardi, A. Rimini, and T. Weber, Unified dynamics for microscopic and macroscopic systems, *Phys. Rev.* D **34**, 470–491 (1986).

[54] J. S. Bell, Are there quantum jumps?, in Schrödinger, *Centenary of a Polymath* (Cambridge University Press, Cambridge, 1987).

[55] G. C. Ghirardi, P. Pearle, and A. Rimini, Markov processes in Hilbert space and continuous spontaneous localization of systems of identical particles, *Phys. Rev.* A **42**, 78 (1990).

[56] A. Bassi et al., Models of wave-function collapse, underlying theories, and experimental tests, *Rev. Mod. Phys.* **85**, 471 (2013).

[57] H. M. Wiseman and Z. Brady, Robust unravelings for resonance fluorescence, *Phys. Rev.* A **62**, 023805 (2000).

[58] N. Gisin, Impossibility of covariant deterministic nonlocal hidden-variable extensions of quantum theory, *Phys. Rev.* A **83**, 020102 (2011).

[59] E.P. Wigner, in *Symmetries and Reflections*, chapter Remarks on the mind-body question, pp. 171–184 (Indiana University Press, Bloomington, 1967).

[60] J. Barbour, *The End of Time* (Oxford University Press, Oxford, 1999).

13

Three Arguments for the Reality of Wave-Function Collapse

SHAN GAO

In this chapter, I suggest three possible ways to examine the three main solutions to the measurement problem, namely Bohm's theory, Everett's theory, and collapse theories. The first way is to analyze whether the result assumptions and the corresponding forms of psychophysical connection of these theories satisfy certain principles or restrictions. My analysis supports the standard result assumption of quantum mechanics in collapse theories, namely that the measurement result is determined by the wave function. The second way is to analyze whether each solution to the measurement problem is consistent with the meaning of the wave function. It is argued that Bohm's and Everett's theories can hardly accommodate the recently suggested ontological interpretation of the wave function in terms of random discontinuous motion of particles, especially when considering the origin of the Born probabilities. While the suggested ontology may not only support the reality of the collapse of the wave function but also provide more resources for formulating a promising collapse theory. The third way is to analyze whether the solutions to the measurement problem are consistent with the principles in other fields of fundamental physics. It is argued that certain discreteness of space–time, which may be a fundamental postulate in the final theory of quantum gravity, may result in the dynamical collapse of the wave function, and the minimum duration and length may also yield a plausible collapse criterion consistent with experiments. These new analyses provide strong support for the reality of wave-function collapse and suggest that collapse theories are in the right direction to solve the measurement problem.

13.1 Introduction

Quantum mechanics is an extremely successful physical theory due to its accurate empirical predictions. The core of the theory is the Schrödinger equation and the Born rule. The Schrödinger equation is linear, and it governs the time evolution of the wave function assigned to a physical system. The Born rule says that the result of a measurement on a physical system is definite but generally random, and the probability is given by the modulus squared of the wave function of the system. However, when assuming the wave function of a physical system is a complete description of the system, the linear Schrödinger equation is apparently incompatible with the Born rule, in particular the appearance of definite results of measurements. This leads to the measurement problem. Maudlin (1995) gave a precise formulation of the problem in terms of the incompatibility. Correspondingly, the three approaches to avoiding the incompatibility lead to the three main solutions to the measurement problem: Bohm's theory, Everett's theory, and collapse theories. It is widely thought that these theories can indeed solve the measurement problem, although each of them still has some other problems.

Then, which solution is the right one or in the right direction? Although there have been many analyses of this issue, the investigation seems still not thorough and complete. In my view, there are three possible ways to examine these competing solutions before experiments can finally test them.

The first way is to analyze the representation of the measurement result, and in particular whether the representation satisfies certain principles or restrictions. In Bohm's theory, Everett's theory, and collapse theories, the measurement results are represented by different physical states. Then, which physical state represents the measurement result? There are at least two restrictions. The first one is the Born rule; the measurement result represented by a certain physical state should be consistent with the Born rule. It is not so obvious that each of these theories such as Everett's theory satisfies the Born rule. The second restriction concerns the psychophysical connection. It has been realized that the measurement problem is essentially the determinate-experience problem (see, e.g., Barrett, 1999). In the final analysis, the problem is to explain how the linear dynamics can be compatible with the existence of definite experiences of conscious observers. This necessitates the existence of a certain form of psychophysical connection, while the forms are different in Bohm's theory, Everett's theory, and collapse theories. A reasonable restriction is that the form of psychophysical connection required by a quantum theory should not violate the principle of psychophysical supervenience. It is not obvious either that these quantum theories all satisfy this restriction.

The second way to examine the solutions to the measurement problem is to analyze whether they are consistent with the meaning of the wave function. The conventional research program is to first find a solution to the measurement problem and then try to make sense of the wave function in the solution. By such an approach, the meaning of the wave function will have no implications for solving the measurement problem. However, this approach is arguably problematic. The reason is that the meaning of the wave function (in the Schrödinger equation) is independent of how to solve the measurement problem, while the solution to the measurement problem relies on the meaning of the wave function. For example, if assuming the operationalist ψ-epistemic view, then the measurement problem will be arguably dissolvedi.

There are two issues here. The first one concerns the nature of the wave function, and the second one concerns the ontology behind the wave function. Is the wave function ontic, directly representing a state of reality, or epistemic, merely representing a state of (incomplete) knowledge, or something else? If the wave function is not ontic, then what, if any, is the underlying state of reality? If the wave function is indeed ontic, then exactly what physical state does it represent? It has been argued that even when assuming the ψ-ontic view, the ontological meaning of the wave function may also have implications for solving the measurement problem (Gao, 2017a).

In my view, the underlying ontology and the psychophysical connection are the two extremes that should be understood in the first place when trying to solve the measurement problem; the underlying ontology is at the lowest quantum level, and the psychophysical connection is at the highest classical level. It is very likely that once we have found the underlying ontology and the psychophysical connection, we will know which solution of the measurement problem is in the right direction. Certainly, we still need to understand the dynamics bridging the quantum and classical worlds.

The third way to examine the solutions to the measurement problem is to analyze whether they are consistent with the principles in other fields of fundamental physics. Penrose's conjecture on gravity's role in wave-function collapse is a typical example (Penrose, 1996, 2004). According to Penrose (1996), there is a fundamental conflict between the superposition principle of quantum mechanics and the principle of general covariance of general relativity. As a result, the definition of the time-translation operator for a superposition of different space–time geometries involves an inherent ill-definedness, leading to an essential uncertainty in the energy of the superposed state. Then it is arguable that this superposition, like an unstable particle in quantum mechanics, is also unstable, and it will decay or collapse into one of the states in the superposition after a finite lifetime (see also Gao, 2013b). In this chapter, I will propose a new conjecture on the origin of wave-function collapse in terms of discreteness of space–time.

This chapter is organized as follows. In Section 13.2, I first introduce Maudlin's conventional formulation of the measurement problem, and then suggest two new formulations of the problem which lay more stress on the result assumption and the psychophysical connection, respectively. It is pointed out that the three main solutions to the measurement problem, namely Bohm's theory, Everett's theory, and collapse theories, make three different result assumptions and assume three different forms of psychophysical connection. In Section 13.3, I analyze the question of what physical state determines the measurement result. The analysis supports the standard result assumption of quantum mechanics, namely that the measurement result is determined by the wave function.

First, Bohm's theory is analyzed in Section 13.3.1. It is argued that the two suggested result assumptions and the corresponding two forms of psychophysical connection of the theory both have potential problems. In particular, the more popular result assumption (i.e., the assumption that the measurement result is determined by the relative configuration of Bohmian particles) may cause a real difficulty for Bohm's theory to satisfy the Born rule. Second, Everett's theory is analyzed in Section 13.3.2. The theory assumes that for a post-measurement wave function there are many equally real worlds, in each of which there is an observer who is consciously aware of a definite result. This means that each measurement result is determined by a certain branch of the wave function, and correspondingly, the mental state of each observer is determined not uniquely by her whole wave function, but only by the corresponding branch of the wave function. It is argued that this special form of psychophysical connection seems to violate the well-accepted principle of psychophysical supervenience. Last, I analyze collapse theories, in particular, how the mental state of an observer supervenes on her wave function in these theories in Section 13.3.3. It is argued that the analysis may help solve the tails problem of collapse theories.

In Section 13.4, I analyze whether the three main solutions to the measurement problem are consistent with the meaning of the wave function. It is argued that Bohm's and Everett's theories can hardly accommodate the recently suggested ontological interpretation of the wave function in terms of random discontinuous motion of particles, especially when considering the origin of the Born probabilities. The suggested ontology may not only support the reality of the collapse of the wave function but also provide more resources for formulating a promising collapse theory. In Section 13.5, I argue that certain discreteness of space–time, which may be a fundamental postulate in the final theory of quantum gravity, may result in the dynamical collapse of the wave function, and the minimum duration and length may also yield a plausible collapse criterion consistent with experiments. Conclusions are given in the last section.

13.2 The Measurement Problem

According to Maudlin's (1995) formulation, the measurement problem originates from the incompatibility of the following three claims:

(C1) the wave function of a physical system is a complete description of the system;

(C2) the wave function always evolves in accord with a linear dynamical equation, e.g., the Schrödinger equation;
(C3) each measurement has a definite result.

The proof of the inconsistency of these three claims is familiar. Suppose a measuring device M measures the x-spin of a spin one-half system S that is in a superposition of two different x-spins $1/\sqrt{2}(|up\rangle_S + |down\rangle_S)$. If (C2) is correct, then the state of the composite system after the measurement must evolve into the superposition of M recording x-spin up and S being x-spin up and M recording x-spin down and S being x-spin down:

$$1/\sqrt{2}(|up\rangle_S |up\rangle_M + |down\rangle_S |down\rangle_M). \tag{13.1}$$

The question is what kind of state of the measuring device this represents. If (C1) is also correct, then this superposition must specify every physical fact about the measuring device. But by symmetry of the two terms in the superposition, this superposed state cannot describe a measuring device recording either x-spin up or x-spin down. Thus if (C1) and (C2) are correct, (C3) must be wrong.

It can be seen that there are three direct solutions to the measurement problem thus formulated. The first solution is to deny the claim (C1) and add some hidden or additional variables and corresponding dynamics to explain the appearance of definite measurement results. A well-known example is Bohm's theory (Bohm, 1952). The question is: Can these hidden variables determine the measurement result?

The second solution is to deny the claim (C3) and assume the existence of many equally real worlds to accommodate all possible results of measurements (Everett, 1957; DeWitt and Graham, 1973). In this way, it may explain the appearance of definite measurement results in each world, including our world. This approach is called Everett's interpretation of quantum mechanics or Everett's theory. It can be seen that Everett's theory assumes a very special form of psychophysical connection. In this theory, a post-measurement wave function corresponds to many observers, and the mental state of each observer is not determined uniquely by her whole wave function but determined only by one branch of the wave function. The question is: Is this special form of psychophysical connection consistent with the well-accepted principle of psychophysical supervenience?

The third solution is to deny the claim (C2) and revise the Schrödinger equation by adding certain nonlinear and stochastic evolution terms to explain the appearance of definite measurement results. Such theories are called collapse theories (Ghirardi, 2011). A key issue of these theories is: What causes the collapse of the wave function? This is the legitimation problem called by Pearle (2004). It can be expected that it will be very helpful to solve the problem if we first understand the real meaning of the wave function.

Certainly, there are also other solutions to the measurement problem, such as a recently suggested solution that denies both the claims (C1) and (C2) (Gao, 2017a). In order to facilitate our later analysis of these solutions and in particular the above questions, I will give two other formulations of the measurement problem. In the first formulation, the measurement problem originates from the incompatibility of the following three assumptions about measurements:

(A1) the linear dynamics: the wave function of a physical system evolves in accord with a linear dynamical equation, e.g., the Schrödinger equation, during a measurement process;

(A2) the Born rule: after each measurement the probability of obtaining a particular result is given by the modulus squared of the wave function of the measured system;

(A3) the standard result assumption: the measurement result is represented by the wave function of the measuring device.[1]

It can be seen that by a similar proof as before these three assumptions are incompatible (see also Maudlin, 1995). This formulation is more precise than Maudlin's original formulation, and especially, it highlights the important role of the result assumption in causing the measurement problem. In order to lead to the measurement problem, the wave function of a physical system is not necessarily a complete description of the system, and it is only required that the wave function of the measuring device determines the measurement result. Besides, it is not required that the wave function always evolves in accord with the linear Schrödinger equation, and it is only required that during the measurement process the wave function evolves in accord with the linear Schrödinger equation.

However, this formulation of the measurement problem is still not precise. It has been realized that the measurement problem in fact has two levels, the physical level and the mental level, and it is essentially the determinate-experience problem (see, e.g., Barrett, 1999). The problem is not only to explain how the linear dynamics can be compatible with the appearance of definite measurement results obtained by physical devices but also, and more importantly, to explain how the linear dynamics can be compatible with the existence of definite experiences of conscious observers. After all, what we are sure of is that we as observers obtain a definite result and have a definite mental state after a measurement, while we are not sure of what physical state this mental state corresponds to. However, this mental aspect of the measurement problem is ignored in the physicalistic formulation, which defines the problem at the physical level (see Gao, 2017b for further discussion). This leads us to the second mentalistic formulation of the measurement problem, which

---

[1] This result assumption is supported by the standard formulation of quantum mechanics, and thus it may be called the standard result assumption.

defines the problem at the mental level and lays more stress on the psychophysical connection. In the formulation, the measurement problem originates from the incompatibility of the following three assumptions:

(A1) the linear dynamics: the wave function of a physical system evolves in accord with a linear dynamical equation, e.g., the Schrödinger equation, during a measurement process;

(A2) the Born rule: after each measurement, the probability of obtaining a particular result is given by the modulus squared of the wave function of the measured system;

(A3) the standard psychophysical connection: the mental state of an observer is determined by her wave function.

The proof of the inconsistency of these assumptions is also similar to the earlier proof. Suppose an observer M measures the x-spin of a spin one-half system S that is in a superposition of two different x-spins, $1/\sqrt{2}(|up\rangle_S + |down\rangle_S)$. If (A1) is correct, then the physical state of the composite system after the measurement will evolve into the superposition of M recording x-spin up and S being x-spin up and M recording x-spin down and S being x-spin down:

$$1/\sqrt{2}(|up\rangle_S|up\rangle_M + |down\rangle_S|down\rangle_M). \tag{13.2}$$

If (A3) is also correct, then the mental state of the observer M will be determined by this superposed wave function.[2] Since the mental states corresponding to the physical states $|up\rangle_M$ and $|down\rangle_M$ differ in their mental content, the observer M being in the superposition (13.2) will have a conscious experience different from the experience of M being in each branch of the superposition by the symmetry of the two branches. In other words, the record that M is consciously aware of is neither x-spin up nor x-spin down when she is physically in the superposition (13.2). While according to (A2), after the measurement, the record that M is consciously aware of is either x-spin up or x-spin down with the same probability $1/2$. Therefore, (A1), (A2) and (A3) are incompatible.

By these two new formulations of the measurement problem, we can look at the three main solutions of the problem from a new angle. Since the assumption (A2), namely the Born rule, has been verified by experiments with great precision, and even an approximate form of the Born rule is enough for leading to the incompatibility, denying (A2) is not an option. Then the solution to the measurement problem must deny either the assumption (A1) or the assumption (A3) or both. Denying the assumption (A1) is the same as denying the claim (C2) in Maudlin's

---

[2] Note that (A3) already excludes Everett's theory. As noted before, in Everett's theory, a post-measurement wave function corresponds to many observers, and the mental state of each observer is not determined by her wave function but determined only by one branch of the wave function.

(1995) original formulation, which means that the Schrödinger equation must be revised. This corresponds to collapse theories. In this case, the measurement result is determined by the wave function of the measuring device, and the mental state of an observer is determined by her wave function. Denying the assumption (A3) means that the mental state of an observer is determined not by her wave function but by another physical state. If the physical state is the so-called hidden variables, then the solution will be Bohm's theory. If the physical state is one branch of the post-measurement wave function, then the solution will be Everett's theory. Finally, denying both the assumptions (A1) and (A3) will lead to a particular form of collapse theories. I will not discuss it in this chapter.

To sum up, the three main solutions to the measurement problem, namely Bohm's theory, Everett's theory, and collapse theories, correspond to three different result assumptions and three different forms of psychophysical connection. In fact, there are only three types of physical states that may determine the measurement result and the mental state of an observer, which are (1) the wave function, (2) certain branches of the wave function, and (3) other hidden variables. The question is: Which one is the actual determiner? It can be expected that an analysis of this question may help solve the measurement problem.

13.3 What Determines the Measurement Result?

In this section, I will try to answer the question of what physical state determines the measurement result. As we will see, the analysis supports the standard result assumption of quantum mechanics, namely that the measurement result is determined by the wave function.

13.3.1 Bohm's Theory

I will first analyze Bohm's theory. In this theory, there are two suggested result assumptions and correspondingly two forms of psychophysical connection, in each of which the Bohmian particles determine the measurement result. The first one is that the measurement result is represented by the branch of the post-measurement wave function occupied by the Bohmian particles, and correspondingly, the mental state of an observer is also determined by the branch of her wave function occupied by her Bohmian particles. The second one is that the measurement result is directly represented by the relative configuration of Bohmian particles, and correspondingly, the mental state of an observer is also determined by the relative configuration of her Bohmian particles.[3]

---

[3] Note that the absolute configuration of Bohmian particles in an inertial frame, which is not invariant in all inertial frames, cannot represent the measurement result, since the representation of a measurement result should be independent of the selection of an inertial frame.

The first result assumption is also called Bohm's result assumption (Brown and Wallace, 2005). Indeed, Bohm initially assumed this result assumption and the corresponding form of psychophysical connection. He said, "the packet entered by the apparatus [hidden] variable... determines the actual result of the measurement, which the observer will obtain when she looks at the apparatus" (Bohm, 1952, p.182). In this case, the role of the Bohmian particles is to select the branch from amongst the other non-overlapping branches of the superposition, which represents the actual measurement result.

It has been widely argued that Bohm's result assumption is problematic (Stone, 1994; Brown and Wallace, 2005; Lewis, 2007a). For example, according to Brown and Wallace (2005), in the general case each of the non-overlapping branches in the post-measurement superposition has the same credentials for representing a definite measurement result as the single branch does in the predictable case (i.e., the case in which the measured system is in an eigenstate of the measured observable). The fact that only one of them carries the Bohmian particles does nothing to remove these credentials from the others, and adding the particles to the picture does not interfere destructively with the empty branches either.

In my view, the main problem with the first form of psychophysical connection is that the empty branches and the occupied branch have the same qualification to be the supervenience base for the mental state. Moreover, although it is imaginable that the Bohmian particles may have influences on the occupied branch, e.g., disabling it from being supervened by the mental state, it is hardly conceivable that the Bohmian particles have influences on all other non-overlapping empty branches, e.g., disabling them from being supervened by the mental state.

In view of the first form of psychophysical connection being problematic, most Bohmians today seem to support the second form of psychophysical connection (Lewis, 2007a). If assuming this form of psychophysical connection, namely assuming the mental state is determined by the relative configuration of Bohmian particles, then the listed problems can be avoided. However, it has been argued that this form of psychophysical connection is inconsistent with the popular functionalist approach to consciousness (Brown and Wallace, 2005; see also Bedard, 1999). The argument can be summarized as follows. If the functionalist assumption is correct, for consciousness to supervene on the Bohmian particles but not the wave function, the Bohmian particles must have some functional property that the wave function does not share. But the functional behaviour of the Bohmian particles is arguably identical to that of the branch of the wave function in which they reside.

Here one may respond, as Lewis (2007b) did, that all theories must give up some intuitive familiar theses, and functionalism is the one that Bohm's theory must give up. However, it has been argued that the second form of psychophysical connection also leads to another serious problem of allowing superluminal signaling (Brown and Wallace, 2005; Lewis, 2007a). If the mental state supervenes on the positions of Bohmian particles, then an observer can in principle know the configuration of the Bohmian particles in her brain with a greater level of accuracy than that defined by the wave function. This will allow superluminal signaling and lead to a violation of the no-signaling theorem (Valentini 1992).

In the following, I will argue that the earlier two result assumptions of Bohm's theory may have a few more potential problems. To begin with, for the first result assumption or Bohm's result assumption, it seems that there is in principle no way to determine which branch of the post-measurement wave function the Bohmian particles occupy. The reason is as follows. On the one hand, each result branch extends over the whole of configuration space in actual situations. On the other hand, the configuration of the Bohmian particles is just a location in configuration space, and it is influenced by all branches that are non-zero at that location according to the guiding equation of Bohm's theory. This means that the configuration of the Bohmian particles is not necessarily associated with any branch of the post-measurement wave function. As a result, it seems that Bohm's theory with this result assumption cannot tell us what the actual result of a measurement is, let alone satisfying the Born rule.

Certainly, one may add a practical rule of associating a configuration of Bohmian particles with a result branch of a post-measurement wave function to Bohm's theory so that the theory may yield a probability distribution of measurement results that may be consistent with the Born rule. But such a rule can hardly be taken as an essential assumption of a fundamental theory like Bohm's theory. A practical rule, which has been widely but implicitly used, is that a configuration of Bohmian particles belongs to the result branch of a post-measurement wave function that has the highest amplitude at the location of these Bohmian particles in configuration space. However, this rule also has some potential problems. For one, when the two result branches of a post-measurement wave function have the same amplitudes at the location of the Bohmian particles in configuration space, it cannot be determined by this rule that which result branch the relative configuration of the Bohmian particles belongs to. Note that the measure of the set of these locations may be not zero.

Next, it can be argued that the non-existence of a necessary association of Bohmian particles with any branch of the wave function may also cause a serious difficulty for Bohm's theory with the second result assumption (i.e., the assumption that the measurement result is represented by the relative configuration of Bohmian particles) to satisfy the Born rule. The reason is that the Born rule requires that there should exist a one-to-one correspondence from the relative configurations of the Bohmian particles to the result branches of the post-measurement superposition, while the non-existence of a necessary association of Bohmian particles with any branch of the wave function denies the existence of this one-to-one correspondence at the bottom.

The requirement of the Born rule can be argued as follows. According to the Born rule, the modulus squared of the amplitude of each result branch of a post-measurement superposition gives the probability of obtaining the measurement result corresponding to the branch. For example, the modulus squared of the amplitude of the branch $|up\rangle_S|up\rangle_M$ in the superposition (13.2), 1/2, gives the probability of obtaining the x-spin up result. Then, if relative particle configurations represent measurement results and Bohm's theory with this result assumption satisfies the Born rule, different result branches must correspond to different relative configurations of Bohmian particles. In other words, there must exist a one-to-one correspondence from the relative configurations of the Bohmian particles to the result branches of the post-measurement superposition.

There is a deeper basis for this requirement of the Born rule. A measurement is an interaction between the measured system and the measuring device. In Bohm's theory, this interaction is also described by the Schrödinger equation for the wave function as in standard quantum mechanics. As a result, the different values of the measured quantity are correlated with the different branches of the post-measurement wave function. Then different measurement results must be represented, first of all, by the different result branches of the post-measurement wave function. For example, in the Stern-Gerlach experiment, which measures the spin of a particle, the measurement is realized by the spin-magnetic field interaction, which is described by the Schrödinger equation for the wave function with a potential term describing the interaction. Thus the measurement result being spin-up or spin-down is encoded in one branch of the post-measurement wave function. Then, if there are other quantities that also represent the measurement results, they should be correlated with these result branches of the superposition. For Bohm's theory with the second result assumption, this is just the earlier requirement of the Born rule.

It seems also worth giving an example to illustrate that in some situations there is no one-to-one correspondence from the relative configuration of the Bohmian particles to the wave function branches in Bohm's theory (Gao, 2017b). Consider again the previous spin measurement, in which a measuring device or an observer M measures the x-spin of a spin one-half system S that is in a superposition of two different x-spins, $\frac{1}{\sqrt{2}}(|up\rangle_S + |down\rangle_S)$. The wave function of the composite system after the measurement evolves into the superposition of M recording x-spin up and S being x-spin up and M recording x-spin down and S being x-spin down:

$$\frac{1}{\sqrt{2}}(|up\rangle_S|up\rangle_M + |down\rangle_S|down\rangle_M). \tag{13.3}$$

According to Bohm's theory, although the post-measurement wave function is a superposition of two definite result branches, the configuration of the Bohmian particles of the device is definite after the measurement, being in one of the two branches with epistemic probability consistent with the Born rule (under the quantum equilibrium hypothesis). Now suppose the spatial part of $|down\rangle_S$ is $\psi(x_0, y_0, z_0, t)$, the spatial part of $|up\rangle_S$ is $\psi(x_0 - a_0, y_0, z_0, t)$, and the spatial part of $|down\rangle_M$ is $\phi(x_1, y_1, z_1, \ldots, x_N, y_N, z_N, t)$, the spatial part of $|up\rangle_M$ is $\phi(x_1 - a_1, y_1, z_1, \ldots, x_N - a_N, y_N, z_N, t)$, where $a_0, a_1, \ldots$ and a_N are large enough so that the two branches of the superposition (13.3) are non-overlaping in configuration space and the superposition may be a valid post-measurement state. When all a_i ($i = 0, \ldots N$) are different, and the difference between two of them is larger than the spreading size of the wave function $\psi(x_0, y_0, z_0, t)\phi(x_1, y_1, z_1, \ldots, x_N, y_N, z_N, t)$ in configuration space, then there is a one-to-one correspondence from the relative configurations of the Bohmian particles to the two branches of the post-measurement superposition (when applying this practical rule).

However, it can be seen that there are also situations in which the one-to-one correspondence does not exist. For example, consider the situations in which two a_i such as a_1 and a_2 are different, and the difference between them is smaller than the spreading size of the wave function $\psi(x_0, y_0, z_0, t)\phi(x_1, y_1, z_1, \ldots, x_N, y_N, z_N, t)$ in configuration space. In this case, if a relative configuration of the Bohmian particles appears in some experiments in the region of the first branch in configuration space which overlaps with the branch that is generated by spatially translating the second branch by a_1 or a_2, it may also appear in other experiments in the region of the second branch in configuration space that overlaps with the branch that is generated by spatially translating the first branch by $-a_1$ or $-a_2$. This means that the relative configurations of the Bohmian particles that (are assumed by the practical rule to) reside in different branches of the superposition may be the same, and there does not exist a one-to-one correspondence from the relative configurations of the Bohmian particles to the result branches of the post-measurement superposition. Although one may debate whether these situations are really valid measurement situations in Bohm's theory, it does illustrate that the one-to-one correspondence from the relative configuration of the Bohmian particles to the wave function branches does not always exist in Bohm's theory (see Gao, 2017b, for further discussion).

Finally, I note that if the measurement result is determined by the wave function or a certain branch of the wave function, not by the hidden variables such as the relative configuration of Bohmian particles, then this will raise serious concern about the whole strategy of hidden-variable theories to solve the measurement problem. Why add hidden variables such as positions of Bohmian particles to quantum mechanics? It is usually thought that adding these variables that have definite values at all times is enough to ensure the definiteness of measurement results and further solve the measurement problem. Indeed, the existing no-go theorems for hidden-variable theories, such as the Kochen-Specker theorem (Kochen and Specker, 1967), consider only whether observables can be assigned sharp values or whether there exist such hidden variables. However, if these hidden variables do not determine the measurement results, then even though they have definite values at all times, their existence does not help solve the measurement problem.

13.3.2 Everett's Theory

Let us now turn to Everett's theory. The theory assumes that the wave function of a physical system is a complete description of the system, and the wave function always evolves in accord with the linear Schrödinger equation. In order to solve the measurement problem, the theory further assumes that for the post-measurement state (13.3), namely $\frac{1}{\sqrt{2}}(|up\rangle_S|up\rangle_M + |down\rangle_S|down\rangle_M)$, there are two equally real worlds, in each of which there is an observer who is consciously aware of a definite record, either x-spin up or x-spin down.[4]

There are three ways of understanding the notion of multiplicity in Everett's theory: (1) measurements lead to multiple worlds at the fundamental level (DeWitt and Graham, 1973), (2) measurements lead to multiple worlds only at the non-fundamental "emergent" level (Wallace, 2012), and (3) measurements only lead to multiple minds (Zeh, 1981).[5] In either case, for the named post-measurement state (13.2), each measurement result is not determined uniquely by the whole wave function but determined only by a branch of the wave function. Correspondingly, the mental state of each observer is determined not uniquely by her wave function but only by a branch of the wave function.[6]

It is widely thought that this result assumption will meet a serious difficulty to account for the Born probabilities, and some opponents have argued that Everett's theory cannot even make sense of the probability itself (see Saunders et al, 2010 and references therein). On the other hand, some Everettians have argued that this problem can be solved in a satisfactory way (see, e.g., Wallace, 2012). I will not discuss this probability problem here. In the following, I will argue that even if the probability problem can be solved, it seems that the form of psychophysical

---

[4] In Wallace's (2012) formulation of Everett's theory, the number of the emergent observers after the measurement is not definite due to the imperfectness of decoherence. My following analysis also applies to this case.

[5] Albert and Loewer's (1988) many-minds theory does not assume the usual notion of multiplicity as listed earlier. It assumes the existence of infinitely many minds even for a post-measurement product state, and it already entails dualism and violates the principle of psychophysical supervenience. I will not discuss this theory here.

[6] Note that if the mental state of each observer is not determined by the corresponding branch of the post-measurement superposition, then the predictions of the theory will be not consistent with the predictions of quantum mechanics and experience for some unitary time evolution of the superposition.

connection assumed by Everett's theory also violates the well-accepted principle of psychophysical supervenience (see Gao, 2017c, for a more detailed analysis).

Consider a unitary time evolution operator, which changes the first branch of the superposition (13.3) to its second branch and the second branch to the first branch. It is similar to the NOT gate for a single q-bit and is permitted by the Schrödinger equation in principle. Then after the evolution the superposition does not change. According to Everett's theory, the wave function of a physical system is a complete description of the system. Therefore, the physical properties or physical state of the composite system does not change after the unitary time evolution.

On the other hand, after the evolution, the mental state of each observer that is determined by the corresponding branch of the superposition will change; the mental state determined by the first branch will change from being aware of x-spin up to being aware of x-spin down, and the mental state determined by the second branch will change from being aware of x-spin down to being aware of x-spin up.[7] Then the mental state of each observer does not supervene on the whole superposition or the physical state of the composite system. Since the mental states of the system are composed of the mental states of the two observers, they do not supervene on the physical state of the system either. Therefore, it seems that the psychophysical supervenience is violated by Everett's theory in this example. Note that supervenience is used here in its standard definition, and the principle of psychophysical supervenience requires that the mental properties of a system cannot change without the change of its physical properties (see McLaughlin and Bennett, 2014).

There are two possible ways to avoid the violation of psychophysical supervenience in the example. The first way is to deny that after the evolution, the physical state of the composite system has not changed. This requires that the wave function of a system is not a complete description of the physical state of the system, and additional variables are needed to be introduced to describe the complete physical state. However, this requirement is not consistent with Everett's theory. Moreover, it is worth noting that in order to save psychophysical supervenience, it is also required that the additional variables should be changed by the unitary time evolution of the wave function, and the mental state of an observer should also supervene on the additional variables; otherwise the introduction of these variables cannot help save psychophysical supervenience in the example.

The second way is to deny that after the evolution, the total mental properties of the composite system have changed. For example, one may argue that after the

---

[7] If this is not the case, then for other evolution or other post-measurement states such as those containing only one branch of the superposition, the predictions of the theory may be inconsistent with the predictions of quantum mechanics and experience (when assuming the mental dynamics is linear). See later discussion.

evolution, there remains a mental state corresponding to seeing a spin up result and a mental state corresponding to seeing a spin down result, and thus the total mental states of the composite system have not changed. However, this seems to require that each observer has no identity, while the absence of identities of observers is inconsistent with the predictions of quantum mechanics and our experience. If each observer has an identity (whether her identity changes or not after the evolution), then her mental properties including her identity and mental state will change after the evolution, and thus the total mental properties of the composite system, which are composed of the identities and mental states of these observers, also change after the evolution. On the other hand, if each observer has no identity in the example, then this seems equivalent to saying that there is only one observer with two mental contents, which are seeing a spin-up result and seeing a spin-down result. In this case, the total mental properties of the composite system will not change after the evolution, and the principle of psychophysical supervenience can be satisfied (see further discussion about collapse theories in the next subsection). But this is not consistent with Everett's theory, which assumes multiplicity.

In order to avoid the violation of psychophysical supervenience, one may even resort to a more complicated dynamics for the mental state, such as a mental dynamics that keeps the mental state of each observer unchanged for the evolution of the superposition (13.3). Certainly, for the evolution of a product state by the same unitary evolution operator, the mental dynamics must still change the mental state of the observer as usual so that it can be consistent with the predictions of quantum mechanics and experience. Thus, such a dynamics must be nonlinear.[8] Although a nonlinear dynamics for the physical state or the wave function is obviously inconsistent with Everett's theory, it seems that a nonlinear dynamics for the mental state is not prohibited by the theory; the many-minds theory is an example (Albert and Loewer, 1988; Barrett, 1999). However, the existence of a nonlinear dynamics for the mental state in Everett's theory already entails dualism. This is no better than the violation of psychophysical supervenience. Moreover, such a nonlinear dynamics seems very ad hoc, and it is also difficult to determine what the dynamics is for an arbitrary superposition such as $\alpha|up\rangle_S|up\rangle_M + \beta|down\rangle_S|down\rangle_M$, where α and β are not zero and satisfy the normalization condition $|\alpha|^2 + |\beta|^2 = 1$.

To sum up, I have argued that Everett's theory seems to violate the principle of psychophysical supervenience. The violation results from the three key

---

[8] A linear dynamics requires that the evolution of one branch of a superposition is independent of the evolution of other branches, as well as whether these branches exist. Thus, by the same unitary evolution operator, the evolution of one branch of the post-measurement superposition (13.2), such as the branch $|up\rangle_S|up\rangle_M$ in the sperposition, will be the same as the evolution of the post-measurement state containing only this branch, such as the product state $|up\rangle_S|up\rangle_M$. This is true for the evolution of both the physical state and the mental state. Otherwise the linearity of dynamics will be violated, and the resulting dynamics will be nonlinear.

assumptions of the theory: (1) the completeness of the physical description by the wave function, (2) the linearity of the dynamics for the wave function, and (3) multiplicity. It seems that one must go beyond Everett's theory in order to avoid the violation of psychophysical supervenience.

13.3.3 Collapse Theories

I have argued that one will meet some serious difficulties if assuming the measurement result is represented by hidden variables or a certain branch of the wave function (and correspondingly, the mental state of an observer is determined by the hidden variables or a certain branch of her wave function). This seems to suggest that the measurement result is represented directly by the wave function and the mental state of an observer is also determined by her wave function. As a result, by the previous analysis of how to solve the measurement problem (see Section 13.2), this means that the linear Schrödinger equation must be revised, and collapse theories may be in the right direction to solve the measurement problem.

However, collapse theories such as the GRW theory predict that the post-measurement state is still a superposition of different outcome branches with similar structure (although the modulus squared of the coefficient of one branch is close to one), and they need to explain why high modulus-squared values are macro-existence determiners. This is the tails problem (Albert and Loewer, 1996). It seems that this problem, especially the structured tails problem, has not been solved in a satisfactory way (see McQueen, 2015, and references therein). In my view, the key to solving the tails problem is not to analyze the connection between high modulus-squared values and macro-existence but to analyze the connection between these values and our experience of macro-existence, which requires us to further analyze how the mental state of an observer is determined by or supervenes on her wave function.[9]

Consider an observer M being in the following superposition:

$$\alpha|1\rangle_P|1\rangle_M + \beta|2\rangle_P|2\rangle_M, \tag{13.4}$$

where $|1\rangle_P$ and $|2\rangle_P$ are the states of a pointer being centered in positions x_1 and x_2, respectively, $|1\rangle_M$ and $|2\rangle_M$ are the physical states of the observer M who observes the pointer being in positions x_1 and x_2, respectively, and α and β, which are not zero, satisfy the normalization condition $|\alpha|^2 + |\beta|^2 = 1$. The question is: What does M observe when her mental state supervenes on the given superposition?

---

[9] Note that this issue is independent of whether the observer can correctly report her mental content, which is related to the bare theory (Albert, 1992; Barrett, 1999).

First of all, it can be seen that the mental content of the observer M is related to the modulus squared of the amplitude of each branch of the superposition she is physically in. When $|\alpha|^2=1$ and $|\beta|^2=0$, M will observe the pointer being only in position x_1. When $|\alpha|^2=0$ and $|\beta|^2=1$, M will observe the pointer being only in position x_2. When $\alpha = \beta = 1/\sqrt{2}$, by the symmetry of the two branches the mental content of M will be neither the content of observing the pointer being in position x_1 nor the content of observing the pointer being in position x_2.

In addition, it can be seen that when $|\alpha|^2 \neq 0$ and $|\beta|^2 \neq 0$, the mental content of the observer M is composed of the mental contents determined by the two terms in the superposition. The reason is as follows. When $|\alpha|^2=0$ or $|\beta|^2 = 0$, the mental content of the observer M does not contain the content of observing the pointer being in position x_1 or x_2. Similarly, the mental content of the observer M does not contain the content of observing the pointer being in another position x_3 that is different from x_1 and x_2, since the amplitude of the corresponding term $|3\rangle_P|3\rangle_M$ is exactly zero. On the other hand, when $|\alpha|^2 = 1$ or $|\beta|^2 = 1$, the mental content of the observer M is the content of observing the pointer being in position x_1 or x_2. Then, when $|\alpha|^2 \neq 0$ and $|\beta|^2 \neq 0$, the mental content of the observer M can only contain the content of observing the pointer being in position x_1 and the content of observing the pointer being in position x_2.

Next, it can be argued that the mental content of the observer M is not related to the phase of each branch of the superposition she is physically in. If the mental contents for $\alpha|1\rangle_P|1\rangle_M + \beta|2\rangle_P|2\rangle_M$ and $\alpha|1\rangle_P|1\rangle_M - \beta|2\rangle_P|2\rangle_M$ are different, then what is the difference? It seems that since the coefficients of the first terms of the two superpositions are the same, the mental contents determined by them should be the same. Then the mental contents determined by the second terms of the two superpositions will be different. On the other hand, an overall phase has no physical meaning, and two wave functions with only a difference of overall phase represents the same physical state. Since $\alpha|1\rangle_P|1\rangle_M+\beta|2\rangle_P|2\rangle_M$ and $-(\alpha|1\rangle_P|1\rangle_M+\beta|2\rangle_P|2\rangle_M)$ represent the same physical state, their mental content should be the same by psychophysical supervenience. Then by the stated reasoning, the mental contents determined by the first terms of $-(\alpha|1\rangle_P|1\rangle_M+\beta|2\rangle_P|2\rangle_M)$ and $\alpha|1\rangle_P|1\rangle_M-\beta|2\rangle_P|2\rangle_M$ should be different, as the second terms of the two superpositions are the same now. This leads to a contradiction. Therefore, it seems that the mental content of M is related not to the phases of the two branches of the superposition she is physically in.

Finally, I will analyze how the mental content of the observer M is determined by the modulus squared of the amplitude of each branch of the superposition she is physically in. This is a difficult task. And I can only give a few speculations here. When $|\alpha|^2 = 0$, the mental content of the observer M does not contain the content of observing the pointer being in position x_1. When $|\alpha|^2 = 1$, the mental content of the observer M contains only the content of observing the pointer being in position x_1. Similarly, when $|\beta|^2 = 0$, the mental content of the observer M does not contain the content of observing the pointer being in position x_2. When $|\beta|^2 = 1$, the mental content of the observer M contains only the content of observing the pointer being in position x_2. Then it seems reasonable to assume that the mental property determined by the modulus squared of the amplitude is a certain property of vividness of conscious experience. For example, when $|\alpha|^2$ is close to one, the conscious experience of M observing the pointer being in position x_1 is the most vivid, while when $|\alpha|^2$ is close to zero, the conscious experience of M observing the pointer being in position x_1 is the least vivid. In particular, when $|\alpha|^2 = |\beta|^2 = 1/2$, the conscious experience of M observing the pointer being in position x_1 and the conscious experience of M observing the pointer being in position x_2 have the same intermediate vividness.

To sum up, I have argued that the mental content of an observer is related to the amplitude of each branch of the superposition she is physically in, and it is composed of the mental content corresponding to every branch of the superposition. Moreover, the modulus squared of the amplitude of each branch may determine the vividness of the mental content corresponding to the branch.

This analysis of how the mental state of an observer supervenes on her wave function may help solve the structured tails problem of collapse theories. In particular, if assuming the modulus squared of the amplitude of each branch indeed determines the vividness of the mental content corresponding to the branch, then the structured tails problem may be solved. Under this assumption, when the modulus squared of the amplitude of a branch is close to zero, the mental content corresponding to the branch will be the least vivid. It is conceivable that below a certain threshold of vividness, an ordinary observer or even an ideal observer will not be consciously aware of the corresponding mental content. Then even though in collapse theories the post-measurement state of an observer is still a superposition of different outcome branches with similar structure, the observer can only be consciously aware of the mental content corresponding to the branch with very high amplitude, and the branches with very low amplitudes will have no corresponding mental content appearing in the whole mental content of the observer. This will solve the structured tails problem of collapse theories.

13.4 What Physical State Does the Wave Function Represent?

In this section, I will turn to the second way to examine the solutions to the measurement problem, namely analyzing whether these solutions are consistent with the meaning of the wave function. I will analyze the ontological meaning of the wave function and argue that Bohm's and Everett's theories can hardly be consistent with the recently suggested ontological interpretation of the wave function in terms of random discontinuous motion of particles, especially when considering the origin of the Born probabilities. The suggested ontology may not only support the reality of the collapse of the wave function but also provide more resources for formulating a promising collapse theory. For a more detailed analysis, see Gao (2017a).

13.4.1 The Wave Function as a Description of Random Discontinuous Motion of Particles

If the wave function in quantum mechanics indeed represents the physical state of a single system (Pusey, Barrett and Rudolph, 2012; Leifer, 2014; Gao, 2015), then what physical state does the wave function represent? The popular position among philosophers of physics and metaphysicians seems to be wave function realism, according to which the wave function represents a real, physical field on configuration space (Albert, 1996, 2013). This is the most straightforward way of thinking about the wave function realistically. However, it is well known that this interpretation is plagued by the problem of how to explain our three-dimensional impressions, while a satisfying solution to this problem seems still missing (see Albert, 2015, for a recent attempt). This motivates a few authors to suggest that the wave function represents a property of particles in three-dimensional space (see, e.g., Monton, 2002, 2006, 2013; Lewis, 2004, 2013, 2016), although they do not give a concrete ontological picture of these particles in space and time and specify what property the property is.

Recently, I have proposed a new ontological interpretation of the wave function in terms of random discontinuous motion of particles (Gao, 2017a). It can be regarded as a further development of the earlier suggestion, although it already came to my mind more than 20 years ago (Gao, 1993). According to this interpretation, a quantum system is a system of particles that undergo random discontinuous motion in our three-dimensional space. Note that the concept of particle here is used in its usual sense. A particle is a small localized object with mass and charge, and it is only in one position in space at each instant. The wave function of a quantum system then describes the state of the random discontinuous motion of the particles of the system and, in particular, the modulus squared of the wave function gives the probability density that these particles appear in every possible group of positions in space. At a deeper level, the wave function may represent the propensity property of the particles that determines their random discontinuous motion.

For a quantum system of N particles, the system can be represented by a point in a $3N$-dimensional configuration space. During an arbitrarily short time interval or an infinitesimal time interval around each instant, these particles perform random discontinuous motion in three-dimensional space, and correspondingly, this point performs random discontinuous motion in the configuration space, and visually speaking, it forms a "cloud" extending throughout the space. Then the state of the system can be described by the density $\rho(x_1, y_1, z_1, \ldots, x_N, y_N, z_N, t)$ and flux density $j(x_1, y_1, z_1, \ldots, x_N, y_N, z_N, t)$ of the cloud, which may further constitute the $3N$-dimensional wave function $\psi(x_1, y_1, z_1, \ldots, x_N, y_N, z_N, t)$. The density of the cloud in each position, $\rho(x_1, y_1, z_1, \ldots, x_N, y_N, z_N, t)$, represents the probability density that particle 1 appears in position (x_1, y_1, z_1) and particle 2 appears in position $(x_2, y_2, z_2) \ldots$ and particle N appears in position (x_N, y_N, z_N).

This picture of random discontinuous motion (RDM in brief) of particles clearly shows that interpreting the multi-dimensional wave function as representing the state of motion of particles in three-dimensional space *is* possible. In particular, the RDM of particles can explain quantum entanglement in a more vivid way, which thus reduces the force of the main motivation to adopt wave function realism. In any case, this new interpretation of the wave function in terms of RDM of particles provides an alternative to wave function realism.

Furthermore, the picture of RDM of particles has more explanatory power than wave function realism. It can help explain our three-dimensional impressions more readily (Gao, 2017a), which is still a difficult task for wave function realism. Moreover, it can also explain many fundamental features of the Schrödinger equation that governs the evolution of the wave function, which seem puzzling for wave function realism. For example, the existence of N particles in three-dimensional space for an N-body quantum system can readily explain why there are N mass parameters that are needed to describe the system and why each mass parameter is only correlated with each group of three coordinates of the $3N$ coordinates on the configuration space of the system and why each group of three coordinates of the $3N$ coordinates transforms under the Galilean transformation between two inertial coordinate systems in our three-dimensional space and so on.

In my view, the difference between particle ontology and field ontology may also result in different predictions that may be tested with experiments, and thus the two interpretations of the wave function can be distinguished in physics. The main difference between a particle and a field is that a particle exists only in one position in space at each instant, while a field exists throughout the whole space at each instant. It is a fundamental assumption in physics that a physical entity being at an instant has no interactions with itself being at another instant, while two physical entities may have interactions with each other. Therefore, a particle at an instant has no interactions with the particle at another instant, while any two parts of a field in space (as two local physical entities) may have interactions with each other. In particular, if a field is massive and charged, then any two parts of the field in space will have gravitational and electromagnetic interactions with each other.

Now consider a charged one-body quantum system such as an electron being in a superposition of two separated wavepackets. Since each wavepacket of the electron has gravitational and electromagnetic interactions with another electron, it is arguable that it is also massive and charged. Indeed, when Schrödinger introduced the wave function and founded his wave mechanics in 1926, he assumed that the charge of an electron is distributed in the whole space, and the charge density in position x at instant t is $-e|\psi(x,t)|^2$, where $\psi(x,t)$ is the wave function of the electron. Moreover, there is also a modern variant of Schrödinger's charge density hypothesis, which has been called mass density ontology (Ghirardi, Grassi and Benatti, 1995; Ghirardi, 1997, 2016). However, some people may be not convinced by this heuristic argument. Why? A common reason, I guess, may be that if each wavepacket of the electron is massive and charged, then the two wavepackets will have gravitational and electromagnetic interactions with each other, but this is inconsistent with the superposition principle of quantum mechanics and experimental observations. But this reason is not valid. For if each wavepacket of the electron is not massive and charged, then how can it have gravitational and electromagnetic interactions with another electron? Rather, there should be a deeper reason why there are no interactions between the two wavepackets of the electron.

Besides the given heuristic argument, protective measurement, a method to measure the expectation value of an observable on a single quantum system (Aharonov and Vaidman 1993; Aharonov, Anandan and Vaidman 1993) may provide a more convincing argument for the existence of the mass and charge distributions of an electron in space. Note that the first protective measurement has been recently realized in experiment (Piacentini et al, 2017). As I have argued in Gao (2015, 2017a), when combined with a reasonable connection between the state of reality and results of measurements, protective measurements imply the reality of the wave function. For example, for an electron whose wave function is $\psi(x)$ at a given instant, we can measure the density $|\psi(x)|^2$ in each position x in space by a protective measurement, and by the connection the density $|\psi(x)|^2$ is a physical property of the electron. Then what density is the density $|\psi(x)|^2$? Since a measurement must always be realized by a certain physical interaction between the measured system and the measuring device, the density must be, in the first place, the density of a certain interacting charge. For instance, if the measurement is realized by an electrostatic interaction between the electron and the measuring device, then the density multiplied by the charge of the electron, namely $-|\psi(x)|^2 e$, will be the charge density of the electron in position x. This means that an electron indeed has mass and charge distributions throughout space, and two separated wavepackets of an electron are both massive and charged.

Let us now see how the difference between particle ontology and field ontology may result in different empirical predictions. If the wave function of an electron represents the state of motion of a particle, then although two separated wavepackets of an electron are massive and charged, these mass and charge distributions are effective, formed by the motion of a particle with the mass and charge of the electron that moves throughout the two separated regions, and thus the two separated wavepackets will have no interactions with each other; a particle at an instant has no interactions with the particle at another instant. This is consistent with the superposition principle of quantum mechanics and experimental observations. On the other hand, if the wave function of an electron represents a physical field, then since this field is massive and charged, any two parts of the field will have gravitational and electromagnetic interactions with each other, which means that two separated wavepackets of an electron will have gravitational and electromagnetic interactions with each other. This is inconsistent with the superposition principle of quantum mechanics and experimental observations. Therefore, it is arguable that the interpretation of the wave function in terms of motion of particles, rather than wave function realism, is supported by quantum mechanics and experience.

13.4.2 Implications for Solving the Measurement Problem

An important aspect of the measurement problem is to explain the origin of the Born probabilities or the probabilities of measurement results.[10] According to the suggested interpretation of the wave function in terms of RDM of particles, the ontological meaning of the modulus squared of the wave function of an electron in a given position is that it represents the probability density that the electron as a particle appears in this position, while according to the Born rule, the modulus squared of the wave function of the electron in the position also gives the probability density that the electron is found there. It is hardly conceivable that these two probabilities, which are both equal to the modulus squared of the wave function, have no connection. On the contrary, it seems natural to assume that the origin of the Born probabilities is the RDM of particles.

There is a further argument supporting this assumption. According to the picture of RDM of particles, the wave function of a quantum system represents an instantaneous property of the particles of the system that determines their motion. However, the wave function is not a complete description of the instantaneous state of the particles. The instantaneous state of the particles at a given instant also

---

[10] The Born probabilities are usually given the frequency interpretation, which is neutral with respect to the issue whether the probabilities are fundamental or due to ignorance. Although there have been attempts to derive the Born rule from more fundamental postulates of quantum mechanics, it is still controversial whether these arguments really derive the Born rule or they are in fact circular (see Landsman, 2009, and references therein). In my opinion, only after finding the origin of the Born probabilities can one find the correct derivation of the Born rule (see Gao, 2017a).

includes their random positions, momenta, and energies at the instant. Although the probability of the particles being in each random instantaneous state is completely determined by the wave function, their stay in the state at each instant is independent of the wave function. Therefore, it seems natural to assume that the random states of the particles may have certain stochastic influences on the evolution of the system, which are manifested in the measurement process, e.g., generating the random result during a measurement.

If the assumption that the origin of the Born probabilities is the RDM of particles turns out to be true, then it will have significant implications for solving the measurement problem, because the existing solutions have not taken this assumption into account.[11] In Bohm's theory, the dynamics is deterministic, and the Born probabilities are epistemic in nature. In the latest formulation of Everett's theory (Wallace, 2012), the Born probabilities are subjective in the sense that it is defined via decision theory. In collapse theories, although the Born probabilities are objective, it is usually assumed that the randomness originates from a classical noise source independent of the wave function of the studied system. In short, none of these main solutions to the measurement problem assumes that the Born probabilities originate from the wave function itself. Therefore, if the Born probabilities originate from the objective probabilities inherent in the RDM of particles described by the wave function, then all these realistic alternatives to quantum mechanics need to be reformulated so that they can be consistent with this result. The reformulation may be easier for some alternatives but more difficult or even impossible for others.

In order that the Born probabilities originate from the RDM of particles, there must exist an additional random dynamics besides the linear, deterministic dynamics, which results from the RDM of particles and results in the appearance of a random measurement result. Since Everett's theory only admits the linear, deterministic dynamics, it cannot accommodate this assumption. In other words, the Born probabilities cannot originate from the RDM of particles in Everett's theory. In Bohm's theory, since the motion of the Bohmian particles is not ergodic, the Born probabilities cannot wholly originate from the RDM of particles either, and they must also depend on the initial probability distribution of the positions of the Bohmian particles. Moreover, if there is an additional random dynamics responsible for generating the measurement result, then the guiding equation, which is added for the same purpose, will be redundant. Last but not least, the existence of RDM of particles itself already reduces the necessity of introducing additional Bohmian particles in the first place; otherwise the theory will be clumsy and unnatural,

---

[11] Note that in Bell's Everett (?) theory, the Born probabilities directly come from the objective probabilities inherent in the random motion of particles.

since an electron will contain two particles, one undergoing random discontinuous motion and the other undergoing deterministic continuous motion.

Compared with Bohm's and Everett's theories, collapse theories seem to be the most appropriate framework to accommodate the assumption about the origin of the Born probabilities, since a random dynamics responsible for generating the measurement result is just what these theories need. Moreover, the RDM of particles may also provide more resources for formulating a dynamical collapse theory, e.g., it already provides an appropriate noise source to collapse the wave function. Certainly, the resulting theory will be different from the existing collapse theories, in which the Born probabilities do not originate from the RDM of particles. A concrete model of wave-function collapse in terms of RDM of particles has been proposed (Gao, 2013a, 2017a).

To sum up, I have argued that an analysis of the ontological meaning of the wave function and the origin of the Born probabilities favors collapse theories and disfavors Everett's theory and Bohm's theory.

13.5 Discreteness of Space–time May Result in the Collapse of the Wave Function

In this section, I will turn to the third way to examine the solutions to the measurement problem, namely analyzing whether these solutions are consistent with the principles in other fields of fundamental physics. As noted before, Penrose's conjecture on gravity's role in wave-function collapse is a typical example (Penrose, 1996, 2004). Here I will argue that certain discreteness of space–time, which may be a fundamental postulate in the final theory of quantum gravity, may result in the dynamical collapse of the wave function, and the minimum duration and length may also yield a plausible collapse criterion consistent with experiments. This analysis, if it is valid, will provide further support for collapse theories.

Although quantum mechanics and general relativity are both based on the concept of continuous space–time, it has been widely argued that their proper combination leads to the existence of the Planck scale, a lower bound to the uncertainty of distance and time measurements (Garay, 1995). For example, when we measure a space interval near the Planck length, the measurement will inevitably introduce an uncertainty comparable to the Planck length, and as a result, we cannot accurately measure a space interval shorter than the Planck length. This is clearly expressed by the generalized uncertainty principle (Adler and Santiago, 1999):

$$\Delta x = \Delta x_{QM} + \Delta x_{GR} \geq \frac{\hbar}{2\Delta p} + \frac{2L_p^2 \Delta p}{\hbar} \qquad (13.5)$$

where Δx is the total position uncertainty of the measured system, Δx_{QM} is the position uncertainty resulting from quantum mechanics, Δx_{GR} is the position uncertainty resulting from general relativity, $\hbar$ is the reduced Planck constant, Δp is the momentum uncertainty of the system, and L_p is the Planck length. Moreover, different approaches to quantum gravity also lead to the existence of a minimum length, a resolution limit in any experiment (Garay, 1995).

This result seems surprising. However, it should be pointed out that the existence of the minimum *observable* space and time intervals does not mean that space and time themselves are discrete or the (ontological) discreteness of space–time. The reason is obvious. The derivation of the result is based on quantum mechanics and general relativity, which are both physical theories in continuous space and time. If the result of the derivation indeed says that space and time are not continuous but discrete, then the derivation must be wrong. On the other hand, the existence of the minimum observable space and time intervals *does* suggest that space and time may be discrete at the Planck scale. Moreover, the black hole entropy formula and the hypothetical holographic principle also support the existence of certain discreteness of space–time.

In the following, I will analyze a possible implication of the discreteness of space–time for the evolution of the wave function. Since there are different understandings of the discreteness of space–time in the literature, I will first make clear what the discreteness of space–time means here. The discreteness of time means that there is no physical change during a time interval shorter than the Planck time. The discreteness of space means that there is no physical variation in a space interval shorter than the Planck length. Notwithstanding such discreteness of space–time, it seems that continuous space–time may be still used as a description framework, which, as we will see, is also necessary for my following analysis.[12]

It can be seen that the discreteness of time does not influence the slower time evolution of the wave function in a significant way. But for the faster time evolution of the wave function, it will have observable effects (Gao, 2014). Consider a quantum superposition of two energy eigenstates. Each eigenstate has a well-defined static mass distribution in the same spatial region with radius R. For example, they are rigid balls of radius R with different uniform mass density. The initial state is

$$\psi(x,0) = \frac{1}{\sqrt{2}}[\phi_1(x) + \phi_2(x)], \qquad (13.6)$$

---

[12] In my opinion, there are at least two reasons to support this view. First, it seems that there is no physical restriction on the difference of the happening times of two causally independent events, e.g., the difference is not necessarily an integral multiple of the Planck time. Next, it seems that continuous space–time is still needed to describe nonphysical superluminal motion, e.g., superluminal light pulse propagation in an anomalously dispersive media. During such superluminal propagations, moving a distance equal to the Planck length will correspond to a time interval shorter than the Planck time (Gao, 2014).

where $\phi_1(x)$ and $\phi_2(x)$ are two energy eigenstates with energy eigenvalues E_1 and E_2, respectively. By the linear Schrödinger evolution, we have:

$$\psi(x,t) = \frac{1}{\sqrt{2}}[e^{-iE_1 t/\hbar}\phi_1(x) + e^{-iE_2 t/\hbar}\phi_2(x)] \tag{13.7}$$

and

$$\rho(x,t) = |\psi(x,t)|^2 = \frac{1}{2}[|\phi_1(x)|^2 + |\phi_2(x)|^2 + 2\phi_1(x)\phi_2(x)\cos(\Delta E t/\hbar)]. \tag{13.8}$$

This result indicates that the density will oscillate with a period $h/\Delta E$ in each position of space, where $\Delta E = |E_2 - E_1|$ is the energy difference.[13]

If the energy difference is much smaller than the Planck energy as for usual quantum states of microscopic systems, the density will oscillate with a period much longer than the Planck time, and the discreteness of time will have no significant influences on the density oscillation. Concretely speaking, the influence of the discreteness of time is to remove the tiny change of the density during each Planck time, and thus on the whole it will only make the amplitude of the density oscillation a little bit smaller. However, if the energy difference exceeds the Planck energy, the density will oscillate with a period shorter than the Planck time, and the discreteness of time will have significant influences on the density oscillation.[14] Concretely speaking, the density oscillation within the Planck time is prohibited by the discreteness of time, which requires that there should be no real physical change during a time interval shorter than the Planck time. As a result, the superposition with an energy difference larger than the Planck energy cannot exist. This obviously leads to the violation of the superposition principle.

This result also means that before the formation of such a superposition, the wave function has already evolved to another state, and the evolution is not linear. The question is: Which state does the wave function evolve to? Suppose the energy difference of the above superposition is much smaller than the Planck energy initially. Then when the energy difference becomes larger and larger, the discreteness of time will make the amplitude of the density oscillation smaller and smaller. In the end, when the energy difference approaches the Planck energy, the period of the density oscillation will approach the Planck time, and the amplitude of the

---

[13] Here I omit the gravitational fields in the superposition, as their existence does not influence my conclusion. When the energy difference is very small such as for a microscopic particle, the corresponding gravitational fields in the superposition are almost the same and not orthogonal, and the interference effect or the density oscillation can be detected in experiment, while when the energy difference becomes larger and larger such as approaching the Planck energy, the gravitational fields in the superposition are not orthogonal either, and thus the density oscillation can also be detected in principle. Moreover, as I will argue later, the superposition can still be defined when considering the existence of the gravitational fields.

[14] Note that there is no restriction on the maximum value of the energy of each eigenstate in the superposition. For example, the energy of a macroscopic object in a stationary state can be larger than the Planck energy.

density oscillation will be close to zero. In other words, when the energy difference of the superposition approaches the Planck energy, the superposition will evolve to a final state without density oscillation. Since a solution of the Schrödinger equation which has no density oscillation is an energy eigenstate,[15] the final state will be an energy eigenstate.

Then which energy eigenstate? Since the energy difference increasing process may happen during a measurement of an initial superposition with small energy difference, and the result for an energy measurement satisfies the Born rule, the superposition can evolve only to one of the energy eigenstates in the superposition, and the probability of evolving to each energy eigenstate must also satisfy the Born rule. This means that the superposition will collapse to one of the energy eigenstates in the superposition when the energy difference of the superposition approaches the Planck energy.[16] Moreover, the collapse is instantaneous, or the collapse time is the Planck time when the energy difference is the Planck energy. This sets a criterion for the collapse of the wave function.

In addition, it can be seen that when the energy difference of the superposition is smaller than the Planck energy, the superposition will also undergo a dynamical collapse process. The reason is as follows. The discreteness of time will always decrease the amplitude of the density oscillation. Then when the energy eigenstates in the superposition remain unchanged as argued earlier, this requires that the product of the modula of the amplitudes of the two energy eigenstates in the superposition should be decreased during the evolution, as can be seen from (13.8). This further requires that during the evolution, the modulus of the amplitude of one energy eigenstate should become larger, and the modulus of the amplitude of the other energy eigenstate should become smaller. In other words, the superposition also undergoes a dynamical collapse process when the energy difference is smaller than the Planck energy, although the collapse process is slower.

Moreover, since the discreteness of time requires that there should be no real physical change during a time interval shorter than the Planck time, the collapse process will be a discrete process; each tiny collapse must happen during one Planck time unit or more. Note that such collapse evolution satisfies the principle

---

[15] Here it is implicitly assumed that an energy eigenstate does not undergo wave-function collapse, and thus the Schrödinger equation is still the dynamical equation that governs its time evolution. This assumption is consistent with the given analysis.

[16] It is worth pointing out that the discreteness of space will similarly require that the superpositions of two momentum eigenstates with momentum difference larger than the Planck energy divided by the speed of light should not exist, since these superpositions lead to spatial oscillation of density within the Planck length. As a result, such a superposition will have collapsed to one of the momentum eigenstates that has no such spatial oscillation of density before it forms. However, it is arguable that the discreteness of space may not lead to the collapse of an energy eigenstate such as the ground state of a hydrogen atom; although the spatial form of an energy eigenstate will be changed by the discreteness of space, the state may be still an energy eigenstate with the same energy eigenvalue, which has no temporal oscillation of density.

of conservation of energy (for an ensemble of identically prepared systems). It has been shown that such a discrete model of energy-conserved wave-function collapse that satisfies the given criterion is consistent with existing experiments and our macroscopic experience (Gao, 2013a, 2017a).[17]

Since there is a connection between the difference of energy distribution and the difference of space–time geometries according to general relativity, the result also suggests that quantum superpositions of two different space–time geometries cannot exist and must collapse to one of the definite space–time geometries in the superposition. In order to make this statement more precise, we need to define the difference between two space–time geometries here. As suggested by the generalized uncertainty principle (13.5), the energy difference ΔE corresponds to the space–time geometry difference $\frac{2L_p^2 \Delta E}{\hbar c}$, where c is the speed of light. The physical meaning of this quantity can be further clarified as follows. Let the two energy eigenstates in the superposition be limited in the regions with the same radius R (they may locate in different positions in space). Then the space–time geometry outside the region can be described by the Schwarzschild metric. By assuming that the metric tensor inside the region R is the same order as that on the boundary, the proper size of the region is

$$L \approx 2 \int_0^R \left(1 - \frac{r_s}{R}\right)^{-\frac{1}{2}} dr, \qquad (13.9)$$

where $r_s = \frac{2GE}{c^4}$ is the Schwarzschild radius, and G is Newton's gravitational constant. Then the spatial difference of the two space–time geometries in the superposition inside the region R can be characterized by

$$\Delta L \approx \int_0^R \frac{\Delta r_s}{R} dr = \frac{2L_p^2 \Delta E}{\hbar c}. \qquad (13.10)$$

where $L_p = \sqrt{\frac{\hbar G}{c^3}}$. This result is consistent with the generalized uncertainty principle. Therefore, for the two energy eigenstates in a superposition, we may define the difference of their corresponding space–time geometries as the difference of the proper spatial sizes of the regions occupied by these states. Such difference represents the fuzziness of the point-by-point identification of the spatial section of the two space–time geometries (cf. Penrose, 1996).

The space–time geometry difference defined here can be rewritten in the following form (when omitting the numerical factor):

$$\frac{\Delta L}{L_p} \approx \frac{\Delta E}{E_p}. \qquad (13.11)$$

---

[17] It seems that in order to satisfy the Born rule, the discreteness of space–time is also at the ensemble level, like the principle of conservation of energy.

This relation indicates one kind of equivalence between the difference of energy and the difference of space–time geometries for superpositions of two energy eigenstates. Therefore, we can also give a collapse criterion in terms of space–time geometry difference. If the difference of the space–time geometries in a superposition is close to the Planck length, the superposition will collapse to one of the definite space–time geometries within one Planck time unit. If the difference of the space–time geometries in a superposition is smaller than the Planck length, the superposition will collapse after a finite time interval longer than the Planck time. As a result, a superposition of space–time geometries can only possess space–time uncertainty smaller than the minimum length. If the uncertainty limit is exceeded, the superposition will have collapsed to one of the definite space–time geometries. This will ensure that the wave function and its evolution can still be consistently defined during the process of wave-function collapse, as the space–time geometries with a difference smaller than the Planck size may be regarded as physically identical in discrete space–time.

To sum up, I have argued that the discreteness of space–time (as defined at the beginning of this section) may result in the dynamical collapse of the wave function and prohibit the existence of quantum superpositions of different space–time geometries. This may not only provide further support for collapse theories but also suggest a possible road to quantum gravity. In this theory, there is no quantized gravity in its usual meaning; space–time geometry is not a pure quantum dynamical entity, but it is not wholly classical either. In contrast to the semiclassical theory of quantum gravity, the theory may naturally include the back-reactions of quantum fluctuations to space–time, e.g., the influences of wave-function collapse on the geometry of space–time, and thus it might be able to provide a consistent framework for a fundamental theory of quantum gravity.

13.6 Conclusions

The measurement problem is the most important and hardest problem of quantum mechanics. It has been debated which solution to the problem is right or in the right direction due to the absence of experimental tests. In this chapter, I suggest three possible ways to examine the three main solutions to the measurement problem, namely Bohm's theory, Everett's theory, and collapse theories.

The first way is to analyze whether the result assumptions and the corresponding forms of psychophysical connection of these theories satisfy certain principles or restrictions. It is argued that the result assumptions of Bohm's theory may cause a serious difficulty for the theory to satisfy the Born rule, and the special form of psychophysical connection required by Everett's theory seems to violate the well-accepted principle of psychophysical supervenience. This analysis supports the standard result assumption of quantum mechanics in collapse theories, namely that the measurement result is determined by the wave function. Moreover, an analysis of how the mental state of an observer supervenes on her wave function in collapse theories may help solve the tails problem of these theories.

The second way is to analyze whether each solution to the measurement problem is consistent with the meaning of the wave function. It is argued that Bohm's and Everett's theories can hardly accommodate the recently suggested ontological interpretation of the wave function in terms of random discontinuous motion of particles, especially when considering the origin of the Born probabilities. The suggested ontology may not only support the reality of the collapse of the wave function but also provide more resources for formulating a promising collapse theory.

The third way is to analyze whether the solutions to the measurement problem are consistent with the principles in other fields of fundamental physics. It is argued that certain discreteness of space–time, which may be a fundamental postulate in the final theory of quantum gravity, may result in the dynamical collapse of the wave function, and the minimum duration and length may also yield a plausible collapse criterion consistent with experiments.

These new analyses may provide strong support for the reality of wave-function collapse and suggest that collapse theories are in the right direction to solve the measurement problem.

Acknowledgments

This work is partly supported by the National Social Science Foundation of China (Grant No. 16BZX021).

References

Adler, R. J. and Santiago, D. I. (1999). On gravity and the uncertainty principle. *Mod. Phys. Lett.* A 14, 1371.

Albert, D. Z. (1992). *Quantum Mechanics and Experience.* Cambridge, MA: Harvard University Press.

Albert, D. Z. (1996). Elementary quantum metaphysics. In J. Cushing, A. Fine and S. Goldstein (eds.), *Bohmian Mechanics and Quantum Theory: An Appraisal.* Dordrecht: Kluwer, pp. 277–284.

Albert, D. Z. (2013). Wave function realism. In Ney, A. and D. Z. Albert (eds.) (2013). *The Wave Function: Essays on the Metaphysics of Quantum Mechanics.* Oxford: Oxford University Press, pp. 52–57.

Albert, D. Z. (2015). *After Physics.* Cambridge, MA: Harvard University Press.

Albert, D. Z. and B. Loewer. (1988). Interpreting the many worlds interpretation. *Synthese* 77, 195–213.

Albert, D. Z. and B. Loewer (1996). Tails of Schrödinger's cat. In *Perspectives on Quantum Reality*, ed. R. Clifton. Dordrecht: Kluwer Academic Publishers, pp. 81–92.

Barrett, J. A. (1999). *The Quantum Mechanics of Minds and Worlds.* Oxford: Oxford University Press.

Bedard, K. (1999). Material objects in Bohm's interpretation. *Philosophy of Science* 66, 221–242.

Bohm, D. (1952). A suggested interpretation of quantum theory in terms of "hidden" variables, I and II. *Physical Review* 85, 166–193.

Brown, H. R. and D. Wallace (2005). Solving the measurement problem: de Broglie-Bohm loses out to Everett. *Foundations of Physics* 35, 517–540.

DeWitt, B. S. and N. Graham (eds.). (1973). *The Many-Worlds Interpretation of Quantum Mechanics.* Princeton: Princeton University Press.

Everett, H. (1957). "Relative state" formulation of quantum mechanics. *Rev. Mod. Phys.* 29, 454–462.

Gao, S. (1993). A Suggested Interpretation of Quantum Mechanics in Terms of Discontinuous Motion (unpublished manuscript).

Gao, S. (2013a). A discrete model of energy-conserved wavefunction collapse. *Proceedings of the Royal Society* A 469, 20120526.

Gao, S. (2013b). Does gravity induce wavefunction collapse? An examination of Penrose's conjecture. *Studies in History and Philosophy of Modern Physics* 44, 148–151.

Gao, S. (2014). Three possible implications of spacetime discreteness. In *Space–Time Geometry and Quantum Events*, Ignazio Licata (ed.). New York: Nova Science Publishers, 197–214.

Gao, S. (2015). An argument for ψ-ontology in terms of protective measurements. *Studies in History and Philosophy of Modern Physics* 52, 198–202.

Gao, S. (2017a). *The Meaning of the Wave Function: In Search of the Ontology of Quantum Mechanics.* Cambridge: Cambridge University Press.

Gao, S. (2017b). The measurement problem revisited. *Synthese*. First Online: 27 June 2017.

Gao, S. (2017c). Failure of psychophysical supervenience in Everett's theory. http://philsci-archive.pitt.edu/12954/.

Garay, L. J. (1995). Quantum gravity and minimum length. *Int. J. Mod. Phys.* A 10, 145.

Ghirardi, G. C. (2011). Collapse theories. *The Stanford Encyclopedia of Philosophy* (Fall 2008 Edition), Edward N. Zalta (ed.), http://plato.stanford.edu/archives/win2011/entries/.

Leifer, M. S. (2014). Is the quantum state real? An extended review of ψ-ontology theorems. *Quanta* 3, 67–155.

Lewis, P. J. (2004). Life in configuration space. *British Journal for the Philosophy of Science* 55, 713–729.

Lewis, P. J. (2007a). How Bohm's theory solves the measurement problem. *Philosophy of Science* 74, 749–760.

Lewis, P. J. (2007b). Empty waves in Bohmian quantum mechanics. *British Journal for the Philosophy of Science* 58, 787–803.

Lewis, P. J. (2013). Dimension and illusion. In Ney, A. and D. Z. Albert (eds.) (2013). *The Wave Function: Essays on the Metaphysics of Quantum Mechanics.* Oxford: Oxford University Press, pp. 110–125.

Lewis, P. J. (2016). *Quantum Ontology: A Guide to the Metaphysics of Quantum Mechanics.* Oxford: Oxford University Press.

Maudlin, T. (1995a). Three measurement problems. *Topoi* 14, 7–15.

Maudlin, T. (1995b). Why Bohm's theory solves the measurement problem. *Philosophy of Science* 62, 479–483.

McLaughlin, B. and Bennett, K. (2014). Supervenience. *The Stanford Encyclopedia of Philosophy* (Spring 2014 Edition), Edward N. Zalta (ed.), https://plato.stanford.edu/archivessupervenience/.

McQueen, K. J. (2015). Four tails problems for dynamical collapse theories. *Studies in History and Philosophy of Modern Physics* 49, 10–18.

Monton, B. (2002). Wave function ontology. *Synthese* 130, 265–277.

Monton, B. (2006). Quantum mechanics and 3N-dimensional space. *Philosophy of Science* 73, 778–789.

Monton, B. (2013). Against 3N-dimensional space. In Ney, A. and D. Z. Albert (eds.) (2013). *The Wave Function: Essays on the Metaphysics of Quantum Mechanics.* Oxford: Oxford University Press pp. 154–167.

Ney, A. and D. Z. Albert (eds.) (2013). *The Wave Function: Essays on the Metaphysics of Quantum Mechanics.* Oxford: Oxford University Press.

Penrose, R. (1996). On gravity's role in quantum state reduction. *Gen. Rel. Grav.* 28, 581.

Penrose, R. (2004). *The Road to Reality: A Complete Guide to the Laws of the Universe.* London: Jonathan Cape.

Piacentini, F. et al. (2017). Determining the quantum expectation value by measuring a single photon. arXiv:1706.08918 [quant-ph].

Pusey, M., Barrett, J. and Rudolph, T. (2012). On the reality of the quantum state. *Nature Phys.* 8, 475–478.

Stone, A. D. (1994). Does the Bohm theory solve the measurement problem? *Philosophy of Science* 62, 250–266.

Valentini, A. (1992). *On the Pilot-Wave Theory of Classical, Quantum and Subquantum Physics.* Ph.D. Dissertation. Trieste, Italy: International School for Advanced Studies.

Wallace, D. (2012). *The Emergent Multiverse: Quantum Theory According to the Everett Interpretation.* Oxford: Oxford University Press.

Zeh, H. D. (1981). *The Problem of Conscious Observation in Quantum Mechanical Description, Epistemological Letters of the Ferdinand-Gonseth Association in Biel (Switzerland), 63.* Also Published in *Foundations of Physics Letters* 13 (2000) 221-233.

14

Could Inelastic Interactions Induce Quantum Probabilistic Transitions?

NICHOLAS MAXWELL

What are quantum entities? Is the quantum domain deterministic or probabilistic? Orthodox quantum theory (OQT) fails to answer these two fundamental questions. As a result of failing to answer the first question, OQT is very seriously defective: it is imprecise, ambiguous, ad hoc, non-explanatory, inapplicable to the early universe, inapplicable to the cosmos as a whole, and such that it is inherently incapable of being unified with general relativity. It is argued that probabism provides a very natural solution to the quantum wave/particle dilemma and promises to lead to a fully micro-realistic, testable version of quantum theory that is free of the defects of OQT. It is suggested that inelastic interactions may induce quantum probabilistic transitions.

14.1 Two Fundamental Questions

What are quantum entities in view of their baffling wave and particle properties? Is the quantum domain deterministic or probabilistic? Even though quantum theory was invented more than 90 years ago, there are still no agreed answers to these two simple and fundamental questions.

In a series of papers published some years ago,[1] I have argued that the answer to the second question is that the quantum domain is fundamentally probabilistic, this answer having the great virtue that it provides a very natural solution to the first question—the baffling quantum wave/particle dilemma.[2] A few words of explanation.

We need to appreciate that there is a rough-and-ready one-to-one correspondence between physical properties of entities on the one hand and dynamical laws of the

---

[1] See Maxwell (1972; 1976; 1982; 1988; 1994; 1998, ch. 7; 2004; 2011). See also Maxwell (2017b, pp. 135–151).
[2] At the time of my first publication on fully micro-realistic, fundamentally probabilistic quantum theory, in 1972, there were very few others working on this approach. Since then, theoretical and experimental work on probabilistic collapse has become a growing field of research. For a recent survey of approaches and literature, see Bassi et al. (2013).

relevant theory on the other hand.³ Thus, in speaking of the physical properties of physical entities—electric charge, mass, wavelength, and so on—we are, in effect, speaking of dynamic laws that govern the way these entities evolve and interact. And vice versa: in speaking of dynamic laws, we are, in effect, speaking of physical properties that entities, postulated by the theory, will possess. Thus, if we change, more or less radically, the *laws* that govern the behaviour of entities, we thereby change, more or less radically, the *physical properties* of these entities, the kind of physical entities that they are.

The classical particle and the classical wave (or field) belong to classical, deterministic physics. If, now, we move from the deterministic laws of classical physics to fundamentally probabilistic laws of quantum theory, we must thereby change radically the physical nature of the entities that probabilistic quantum theory would be about.

Granted that quantum theory is indeed fundamentally probabilistic, so that the physical entities of the quantum domain interact with one another *probabilistically*, then the entities of the quantum domain must be radically different from anything encountered within deterministic, classical physics. It would indeed be a disaster, as far as the comprehensibility of the quantum domain is concerned—assuming it is fundamentally probabilistic—if quantum entities turned out to be classical particles or classical waves (or fields), as both are to be associated with *determinism*. That quantum entities—electrons, photons, atoms, and the rest—turn out to be very different from deterministic classical particles or waves is thus excellent news from the standpoint of the intelligibility of the quantum domain.

The founding fathers of quantum theory struggled to find an answer to the question "Are quantum entities particles or waves?" In the end, most physicists despaired of finding an answer, came to the conclusion that the question of what sort of entities quantum entities are could not be answered, and adopted the orthodox interpretation of quantum theory, which evades the whole issue by restricting quantum theory to being a theory about the results of performing *measurements*.⁴

---

[3] This point is spelled out in much greater detail in Maxwell (1976; 1988); see also Maxwell (1968; 1998, pp. 141–155; 2017a, pp. 116–118).

[4] This was the situation up to around 1990 perhaps. Increasingly, in the last two decades or so, physicists have come to appreciate that the orthodox interpretation of quantum theory is unsatisfactory, even unacceptable. It is now acceptable for physicists to publish research on the subject. This is strikingly borne out by the present volume and by Shan Gao's *Meaning of the Wave Function* (2017), which provides a sustained exploration of interpretative problems of quantum theory and gives an indication of the vast and growing literature on the subject. For a recent survey of fundamentally probabilistic or "collapse" versions of quantum theory, see Bassi et al. (2013).

What everyone failed to appreciate—both those who accepted orthodox quantum theory (OQT) following Bohr and Heisenberg and those who were critical of OQT following Einstein and Schrödinger—is that the key question that most physicists despaired of answering, namely "Are quantum entities particles or waves?" is the *wrong* question to ask. The *right* questions, granted the crucial assumption that the quantum domain is fundamentally probabilistic, are:

(1) What sort of *unproblematic*, fundamentally probabilistic entities are there as possibilities?
(2) Are quantum entities some variety of a kind of unproblematic, fundamentally probabilistic entity?

In short, what everyone still tends to take for granted—namely that the quantum domain is inherently baffling and incomprehensible because it cannot be made sense of in terms of the classical particle or the classical wave (or field)—is actually very good news indeed as far as the intelligibility of the quantum domain is concerned. The quantum domain would have been incomprehensible indeed if composed of classical, deterministic particles or waves. That it is *not* composed of such entities is very encouraging from the standpoint of the quantum domain being thoroughly intelligible.[5]

14.2 Is the Quantum Domain Fundamentally Probabilistic?

The foregoing considerations depend, crucially, on the quantum domain being fundamentally probabilistic. What grounds do we have for holding this to be true?

There are two possibilities: determinism and probabilism. Both should be explored.[6] Here, I explore some consequences of adopting the probabilistic hypothesis. Not only does this hypothesis imply that the traditional "Wave or particle?" question is the wrong question to ask. It has, as I hope to make clear in a moment, the great virtue of rendering the behaviour and properties of quantum entities thoroughly comprehensible in an entirely natural way.

Probabilism has additional virtues. It must lead, relentlessly, to a version of quantum theory that makes some empirical predictions that differ from those of OQT. It leads to testable physics, in other words.

---

[5] For a brief discussion of the point that, though hints of probabilism can be found in the development of quantum theory from the outset, nevertheless probabilistic quantum realism was long ignored, see Maxwell (2011, sections 12 and 13).

[6] At the time of writing (2016), the most popular deterministic version of quantum theory seems to be the Everett or many-worlds interpretation: for a recent exposition see Wallace (2012); for discussion see Bacciagaluppi and Ismael (2015). For my criticisms see Maxwell (2017b, pp. 148–151).

The point is this. OQT, in general, makes probabilistic predictions about the results of measurement. Measurement is, thus, a *sufficient* condition for probabilistic events to occur (granted that we are not interpreting such measurements in deterministic terms). What is not acceptable, however, is that measurement should be a *necessary* condition for probabilistic events. If nature is fundamentally probabilistic, then probabilistic events will occur whether or not there are physicists around to make measurements. They will have occurred long before life began on earth. Probabilistic quantum theory (PQT) is thus committed to holding that there must be fully *quantum mechanical* physical conditions for probabilistic transitions to occur, conditions which will obtain in nature entirely in the absence of measurement. Any version of PQT which specifies precisely what these conditions are must make predictions that differ from those of OQT (which restricts probabilism to measurement). The major task confronting the development of PQT is to specify precisely what the physical conditions for probabilistic events to occur are.

PQT has further potential advantages. It leads to a fully micro-realistic theory, one which specifies how quantum entities evolve and interact physically, in physical space and time, entirely independently of preparation and measurement. It is thus able to explain how the classical world emerges, as an approximation, from the quantum world. OQT, because it fails to solve the wave/particle problem and thus fails to be a fully micro-realistic theory about quantum entities (there being no coherent idea as to what such entities are) must, as a result, be a theory about the results of measurement. This means some part of classical physics (or some other theory) must be used to describe the measuring instrument. This in turn means that OQT suffers from the very serious defects that it is imprecise, ambiguous, ad hoc, non-explanatory, inapplicable to the early universe, inapplicable to the cosmos as a whole, and such that it is inherently incapable of being unified with general relativity.[7] PQT is free of these seven very serious defects precisely because it has its own micro-quantum ontology (it is about "beables" as John Bell called them), and thus makes no reference to measurement in its basic postulates. All these seven defects of OQT arise because the basic postulates of OQT refer to observables and thus to measurement. And PQT is free of the key eighth defect of OQT, responsible for all the others, namely the failure of OQT to provide a coherent micro-ontology for the quantum domain.

---

[7] These points are argued for in some detail in Maxwell (1988, pp. 1–8; 1993, section iv; 2004, section 1). Some of these points have been argued for even earlier, in Maxwell (1972; 1976; 1982). Bell (1973; 1987) also argued that the concept of measurement needs to be eliminated from the basic postulates of quantum theory, the theory being interpreted to be about "beables" rather than "observables." Bell tended to restrict himself, however, to the point that, if quantum theory is about measurement, it is inherently *imprecise*. I discuss Bell's contribution in Maxwell (1992).

There are, in short, decisive grounds for developing a better version of quantum theory than OQT, whether deterministic or probabilistic.

14.3 What Kinds of Possible Fundamentally Probabilistic Entity Are There?

How, in general terms, is the fundamentally probabilistic physical entity to be understood? Just as probabilism generalizes determinism, so too the idea of the fundamentally probabilistic object generalizes the notion of the deterministic object, familiar from common sense and classical physics. Objects of common sense and of classical deterministic physics have dispositional or necessitating properties which determine how the object in question behaves in certain circumstances.[8] A ball that is *elastic* bounces when dropped onto the floor. An *inflammable* object catches fire when exposed to a naked flame. A body with *Newtonian gravitational charge* obeys Newton's law $F = Gm_1m_2/d^2$. Fundamentally probabilistic objects, which interact with one another probabilistically, and which I shall call *propensitons*, have physical properties which generalize the familiar, deterministic notion of physical property, and which determine how objects behave *probabilistically* in appropriate circumstances. An example of such a probabilistic property, or *propensity*,[9] is what may be called the "bias" of a die—the property of the die which determines the probabilities of the outcomes 1 to 6 when the die is tossed onto a table. A value of bias assigns a probability to each of the six possible outcomes. We can even imagine that the value of the bias of the die itself changes: there is, perhaps, a tiny magnet imbedded in the die and an electromagnet under the table. As the strength of the magnet beneath the table varies, so the value of the bias of the die will change.

The idea, then, is to interpret the ψ-function of quantum theory as attributing values of physical *propensities* to quantum propensitons, thus specifying the real physical states of propensitons in physical space and time. The ψ-function carries implications about how quantum propensitons interact with one another probabilistically should appropriate quantum conditions for such interactions to occur arise.

Granted we adopt probabilism as our working hypothesis, the crucial question to ask is: What different kinds of propensiton are there as possibilities? There are two cases to consider. Probabilistic events might occur (i) continuously in time or (ii) discontinuously, only when specific physical conditions arise.

---

[8] For this account of "necessitating" physical properties, see Maxwell (1968; 1998, pp. 141–155).
[9] Popper introduced the idea of propensities in connection with interpretative problems of QT; see Popper (1957; 1967; 1982), although, as Popper (1982, pp. 130–135) has pointed out, Born, Heisenberg, Dirac, Jeans, and Landé have all made remarks in this direction. The version of the propensity idea employed here is, however, in a number of respects, different from and an improvement over the notion introduced by Popper: see Maxwell (1976, pp. 284–286; 1985, pp. 41–42). For a discussion of Popper's contributions to the interpretative problems of quantum theory, see Maxwell (2016, section 6, especially note 19).

The structure of OQT suggests that we should initially favour (ii). Given the astonishing empirical success of OQT—the wealth and variety of phenomena it predicts, the accuracy of its predictions, and the lack of empirical refutations—it is clear that OQT has got quite a lot right. All the defects of OQT stem from the failure to solve the wave/particle problem and the resulting need to formulate the theory as being about measurement. This suggests that we should, initially at least, keep as close to OQT as possible and modify the theory just sufficiently to remove these defects. According to OQT, probabilistic events only occur when measurements are made; quantum states evolve deterministically, in accordance with the Schrödinger equation, when no measurement is made. In developing PQT, then, we should, initially, favour (ii), hold onto the Schrödinger equation when no probabilistic events occur, and replace measurement with some physical condition more plausible to be the condition for the occurrence of probabilistic events. Let us call a physical entity that evolves deterministically until precise physical conditions arise for probabilistic transitions to occur an "intermittent propensiton."

Here is a very simple-minded model of a world made up of such intermittent propensitons. The world is composed of spheres, which expand deterministically at a steady rate. The instant two spheres touch, they collapse to a pre-determined minute sphere, each within its spherical volume, the precise location only probabilistically determined. This, one might say, is the simplest, the most elementary, and the most comprehensible example of the intermittent propensiton.

A slightly more complicated model is the following. Each deterministically expanding sphere is filled with a "stuff," which varies in density, from place to place. Where the "stuff" is dense, it is correspondingly more probable that it is there that the tiny sphere will be located when the probabilistic collapse occurs; where the "stuff" is sparse, it is correspondingly less probable.

We can now imagine five further modifications. First, the "stuff"—which we may call "position probability density"—may vary in a wavelike fashion within the sphere. Second, the propensiton ceases to be a sphere and may acquire any shape and size whatsoever within physical space. Third, we suppose that the physical state of the propensiton, from moment to moment in physical space, is specified by a complex function $\psi(r,t)$. Fourth, the evolution of the propensiton state is determined by Schrödinger's time-dependent equation. And fifth, $|\psi|^2 dv$ specifies the density of the "stuff" of the propensiton in volume element dv—that is, the probability of the propensiton undergoing a probabilistic interaction in dv should the physical conditions for such an interaction to occur obtain.

We began with the simplest, the most elementary, the most intelligible example of an intermittent propensiton conceivable. We have added a few details, and we have come up with a physical entity which behaves in space and time almost precisely in accordance with quantum theory. The quantum intermittent propensiton may seem mysterious, but that is just because ordinarily we have no experience of such entities. The quantum propensiton just indicated is in fact a special case of the simplest, the most intelligible intermittent propensiton conceivable, in turn a probabilistic generalization of the ordinary, quasi-deterministic objects we are familiar with: footballs, tables, pencils.

In the next section, I reply to objections to my claim that the foregoing provides us with a fully intelligible, micro-realistic version of quantum theory. In the section after that, I put forward my proposal about what it is, physically, that induces probabilistic events to occur.

14.4 Can the ψ-function be Interpreted to Specify the Physical State of the Propensiton in Physical Space?

Objection (1): The ψ-function is complex and hence cannot be employed to describe the physical state of an actual physical system.

Reply: The complex ψ is equivalent to two interlinked real functions, which can be regarded as specifying the propensity state of quantum systems. In any case, as Penrose (2004, p. 539) reminds us, complex numbers are used in classical physics, without this creating a problem concerning the reality of what is described. The complex nature of ψ has to do, in part, with the fact that the wave-like character of ψ is not in physical space, except when interference leads to spatio-temporal wave-like variations in the intensity of ψ and thus in $|\psi(x,t)|^2$ as well.

Objection (2): Given a physical system of N quantum entangled systems, the ψ-function is a function of 3N-dimensional configuration space and not 3-dimensional physical space. This makes it impossible to interpret such a ψ-function as specifying the physical state of N quantum entangled physical systems in physical space when $N \geq 2$.

Reply: $\psi(r_1, r_2 \ldots r_N)$ can be regarded as assigning a complex number to any point in 3N-dimensional configuration space. Equally, however, we can regard $\psi(r_1, r_2 \ldots r_N)$ as assigning the complex number to N points in 3-dimensional physical space. Suppose $\psi(r_1, r_2 \ldots r_N)$ is the quantum entangled state of N distinct kinds of particle. Then $\psi(r_1, r_2 \ldots r_N)$, in assigning a complex number to a point in configuration space, can be interpreted as assigning this number to N points in physical space, each point labelled by a different particle. The quantum propensiton state in physical space will be multi-valued at any point in physical space and also highly non-local, in that its values at any given point cannot be dissociated from values at N–1 other points. If we pick out N distinct points in physical space, there will be, in general, N! points in configuration space which assign different values of ψ to these N physical points, corresponding to the different ways the

N particles can be reassigned to these N points. If we pick out just one point in physical space (x_o, y_o, z_o), the ψ-function will in general assign infinitely many different complex numbers to this point (x_o, y_o, z_o), corresponding to different locations of the particles in physical space—there being infinitely many points in configuration space that assign a complex number to this point (x_o, y_o, z_o) in physical space. The N-particle, quantum-entangled propensiton is, in physical space, a complicated, non-local, multi-valued object, very different from anything found in classical physics. Its physical nature in 3-dimensional physical space is, nevertheless, precisely specified by the single-valued $\psi(r_1, r_2 \ldots r_N)$ in 3N-dimensional configuration space.[10]

Objection (3): The ψ-function is highly non-local in character. This, again, makes a realistic interpretation of it impossible.

Reply: As my reply to objection (2) indicates, quantum propensitons of the type being considered here, made up of a number of quantum entangled "particles," are highly non-local in character, in that one cannot specify what exists at one small region of physical space without simultaneously taking into account what exists at other small regions. Propensitons of this type seem strange because they are unfamiliar—but we must not confuse the unfamiliar with the inexplicable or impossible. Non-local features of the ψ-function do not prevent it from specifying the actual physical states of propensitons; propensitons just are, according to the version of PQT being developed here, highly non-local objects in the sense indicated.

Objection (4): If the ψ-function is interpreted realistically, it follows that when a position measurement is made, a quantum system that has a state spread throughout a large volume of space collapses instantaneously into a minute region where the system is detected. Such an instantaneous collapse, possibly across vast distances of space, becomes wholly implausible when regarded as a real physical process.

Reply: Such instantaneous probabilistic collapse is an inherent feature of the intermittent propensiton. There is nothing implausible or inexplicable about such probabilistic transitions.[11] To suppose otherwise is to be a victim of deterministic prejudice—a victim of the dogma that nature must be deterministic, the more general idea of probabilism being excluded a priori, without any valid reason.

---

[10] This solution to the problem was outlined in Maxwell (1976, pp. 666–667; and 1982, p. 610). Albert (1996) has proposed that the quantum state of an N-particle entangled system be interpreted to exist physically in 3N-dimensional configuration space. But configuration space is a mathematical fiction, not a physically real arena in which events occur. Albert's proposal is untenable, and in any case unnecessary.

[11] Instantaneous probabilistic collapse is, however, highly problematic the moment one considers developing a Lorentz-invariant version of the theory. For my views about this problem, see Maxwell (1985; 2006; 2011; 2017c). For a fascinating discussion of the problem of reconciling special relativity and the non-locality of quantum theory, see Maudlin (2011).

I conclude that there are no valid objections to interpreting ψ as specifying the actual physical states of propensitons in physical space.

14.5 Do Inelastic Interactions Induce Probabilistic Collapse?

In order to specify the precise nature of the quantum intermittent propensitons under consideration and at the same time give precision to the version of PQT being developed here, we need now to specify (a) the precise quantum conditions for a probabilistic transition to occur in a quantum system, (b) what the possible outcome quantum states are, given that the quantum state at the instant of probabilistic transition is ψ, and (c) how ψ assigns probabilities to the possible outcomes. No reference must be made to measurement, observables, macroscopic systems, irreversible processes, or classically described systems.

One possibility is the proposal of Ghirardi, Rimini, and Weber (1986)—see also Ghirardi (2002)—according to which the quantum state of a system such as an electron collapses spontaneously, on average after the passage of a long period of time, into a highly localized state. When a measurement is performed on the electron, it becomes quantum entangled with millions upon millions of quantum systems that go to make up the measuring apparatus. In a very short time, there is a high probability that one of these quantum systems will spontaneously collapse, causing all the other quantum-entangled systems, including the electron, to collapse as well. At the micro level, it is almost impossible to detect collapse, but at the macro level, associated with measurement, collapse occurs very rapidly all the time.

Another possibility is the proposal of Penrose (1986, 2004, ch. 30), according to which collapse occurs when the state of a system evolves into a superposition of two or more states, each state having, associated with it, a sufficiently large mass located at a distinct region of space. The idea is that general relativity imposes a restriction on the extent to which such superpositions can develop, in that it does not permit such superpositions to evolve to such an extent that each state of the superposition has a substantially distinct space–time curvature associated with it.

The possibility that I favour, put forward before either Ghirardi, Rimini, and Weber's proposal or Penrose's proposal, is that probabilistic transitions occur whenever, as a result of inelastic interactions between quantum systems, new "particles," new bound, stationary, or decaying systems, are created (Maxwell, 1972, 1976, 1982, 1988, 1994). A little more precisely:

(I) Whenever, as a result of an inelastic interaction, a system of interacting "particles" creates new "particles," bound, stationary, or decaying systems, so that the state of the system goes into a superposition of states, each state having associated with it different particles or bound, stationary, or decaying systems, then, when the interaction is nearly at an end, spontaneously and probabilistically, entirely in the absence of measurement, the superposition collapses into one or other state.

An example of the kind of inelastic interaction that is involved is the following:

$$e^- + H$$
$$e^- + H \rightarrow e^- + H^*$$
$$e^- + H + \gamma$$
$$e^- + e^- + p$$

Here e^-, H, H^*, γ, and p stand for electron, hydrogen atom, excited hydrogen atom, photon, and proton, respectively.)

What exactly does it mean to assert that the "interaction is very nearly at an end" in the previous postulate? My suggestion, here, is that it means that forces between the "particles" are very nearly zero except for forces holding bound or decaying systems together. In order to indicate how this can be formulated precisely, consider the toy interaction:

$$a + b + c \text{ (A)}$$
$$a + b + c \rightarrow$$
$$a + (bc) \text{ (B)}$$

Here, a, b, and c are spinless particles, and (bc) is the bound system. Let the state of the entire system be $\Phi(t)$, and let the asymptotic states of the two channels (A) and (B) be $\psi_A(t)$ and $\psi_B(t)$ respectively. Asymptotic states associated with inelastic interactions are fictional states towards which, according to OQT, the real state of the system evolves as $t \rightarrow +\infty$. Each outcome channel has its associated asymptotic state, which evolves as if forces between particles are zero, except where forces hold bound systems together.

According to OQT, in connection with the toy interaction, there are states $\phi_A(t)$ and $\phi_B(t)$, such that:

(1) For all t, $\Phi(t) = c_A \phi_A(t) + c_B \phi_B(t)$, with $|c_A|^2 + |c_B|^2 = 1$;
(2) as $t \rightarrow +\infty$, $\phi_A(t) \rightarrow \psi_A(t)$ and $\phi_B(t) \rightarrow \psi_B(t)$.

According to the version of PQT under consideration here, at the first instant t for which $\phi_A(t)$ is very nearly the same as the asymptotic state $\psi_A(t)$, or $\phi_B(t)$ is very nearly the same as $\psi_B(t)$, then the state of the system, $\Phi(t)$, collapses spontaneously either into $\phi_A(t)$ with probability $|c_A|^2$ or into $\phi_B(t)$ with probability $|c_B|^2$. Or, more precisely:

(II) At the first instant for which $| \langle \psi_A(t) | \phi_A(t) \rangle |^2 > 1 - \varepsilon$ or $| \langle \psi_B(t) | \phi_B(t) \rangle |^2 > 1 - \varepsilon$, the state of the system collapses spontaneously into $\phi_A(t)$ with probability

$|c_A|^2$ or into $\phi_B(t)$ with probability $|c_B|^2$, ε being a universal constant, a positive real number very nearly equal to zero.[12]

According to (II), if $\varepsilon = 0$, probabilistic collapse occurs only when $t = +\infty$ and the corresponding version of PQT becomes equivalent to the many-worlds, or Everett, interpretation of quantum theory. As ε is chosen to be closer and closer to 1, so collapse occurs more and more rapidly, for smaller and smaller times t—and the corresponding versions of PQT become more and more readily falsifiable experimentally.

The evolutions of the actual state of the system, $\Phi(t)$, and the asymptotic states, $\psi_A(t)$ and $\psi_B(t)$, are governed by the respective channel Hamiltonians, H, H_A, and H_B, where:

$$H = -\left(\frac{\hbar^2}{2m_a}\nabla_a^2 + \frac{\hbar^2}{2m_b}\nabla_b^2 + \frac{\hbar^2}{2m_c}\nabla_c^2\right) + V_{ab} + V_{bc} + V_{ac}$$

$$H_A = -\left(\frac{\hbar^2}{2m_a}\nabla_a^2 + \frac{\hbar^2}{2m_b}\nabla_b^2 + \frac{\hbar^2}{2m_c}\nabla_c^2\right)$$

$$H_B == -\left(\frac{\hbar^2}{2m_a}\nabla_a^2 + \frac{\hbar^2}{2m_b}\nabla_b^2 + \frac{\hbar^2}{2m_c}\nabla_c^2\right) + V_{bc}$$

Here, m_a, m_b, and m_c are the masses of "particles" a, b, and c, respectively, and $\hbar = h/2\pi$ where h is Planck's constant.

This condition for probabilistic collapse can readily be generalized to apply to more complicated and realistic inelastic interactions between quantum systems.

According to this fully micro-realistic, fundamentally probabilistic version of quantum theory, the state function, $\Phi(t)$, describes the actual physical state of the quantum system—the propensiton—from moment to moment. The physical (quantum) state of the propensiton evolves in accordance with Schrödinger's time-dependent equation as long as the condition for a probabilistic transition to occur does not arise. The moment it does arise, the state jumps instantaneously and probabilistically, in the manner indicated, into a new state. (All but one of a superposition of states, each with distinct "particles" associated with them, vanish.) The new state then continues to evolve in accordance with Schrödinger's equation until conditions for a new probabilistic transition arise. Quasi-classical objects arise as a result of the occurrence of a rapid sequence of many such probabilistic transitions.

---

[12] The basic idea of (II) is to be found in Maxwell (1982 and 1988). It was first formulated precisely in Maxwell (1994).

14.6 Propensiton PQT Recovers All the Empirical Success of OQT

The version of propensiton quantum theory (PQT) just indicated recovers—in principle—all the empirical success of orthodox quantum theory (OQT). In order to see this, it is crucial to take note of the distinction between *preparation* and *measurement* (Popper, 1959, pp. 225–226; Margenau, 1958, 1963). A preparation is some physical procedure which has the consequence that if a quantum system exists (or is found) in some predetermined region of space, then it will have (or will have had) a definite quantum state. A measurement, by contrast, actually detects a quantum system and does so in such a way that a value can be assigned to some quantum "observable" (position, momentum, energy, spin, etc.). A measurement need not be a preparation. Measurements of photons, for example, far from preparing the photons to be in some quantum state, usually *destroy* the photons measured! On the other hand, a preparation is not in itself a measurement, because it does not *detect* what is prepared. It can be converted into a measurement by a subsequent detection.

From the formalism of OQT, one might well suppose that the various quantum observables are all on the same level and have equal status. This is in fact not the case. Position is fundamental,[13] and measurements of all other observables are made up of a combination of preparations and position measurements.[14] PQT, in order to do justice to quantum measurements, need only do justice to *position* measurements.

It might seem, to begin with, that PQT, based on the two postulates (I) and (II), which say nothing about position or localization, cannot predict that unlocalized systems become localized, necessary, it would seem, to predict the outcome of position measurements. The version of propensiton PQT just formulated does, however, predict that localizations occur. If a highly localized system, S_1, interacts inelastically with a highly unlocalized system, S_2, in such a way that a probabilistic transition occurs, then S_1 will localize S_2. If an atom or nucleus emits a photon or other "particle" which travels outwards in a spherical shell and which is subsequently absorbed by a localized third system, the localization of the photon or

---

[13] The possibility of formulating quantum theory so that it makes predictions exclusively about positions is made by Feynman and Hibbs, who write, "all measurements of quantum mechanical systems could be made to reduce eventually to position and time measurements..." Because of this possibility a theory formulated in terms of position measurements is complete enough in principle to describe all phenomena: see Feynman and Hibbs (1965, p. 96).

[14] Popper distinguished preparation and measurement in part in order to make clear that Heisenberg's uncertainty relations prohibit the simultaneous *preparation* of systems in a precise state of position and momentum but place no restrictions whatsoever on the simultaneous *measurement* of position and measurement. One needs, indeed, to measure position and momentum simultaneously well within the Heisenberg uncertainty relations simply to check up experimentally on the predictions of these relations: see Popper (1959, pp. 223–236). See Maxwell (2016, section 6) for a discussion of the important contribution that Popper makes, here, to the interpretation of orthodox quantum theory.

"particle" will localize the emitting atom or nucleus with which it was quantum entangled.[15]

That PQT recovers (in principle) all the empirical success of OQT is a consequence of the following four points.[16]

First, OQT and PQT use the same dynamical equation, namely Schrödinger's time-dependent equation.

Second, whenever a position measurement is made and a quantum system is detected, this invariably involves an inelastic interaction and the creation of a new "particle" (bound or stationary system, such as the ionisation of an atom or the dissociation of a molecule, usually millions of these). This means that whenever a position measurement is made, the conditions for probabilistic transitions to occur, according to PQT, are satisfied. PQT will reproduce the predictions of OQT (given that PQT is provided with a specification of the quantum state of the measuring apparatus). As an example of PQT predicting, probabilistically, the result of a position measurement, consider the following. An electron in the form of a spatially spread-out wave packet is directed towards a photographic plate. According to PQT, the electron wave packet (or propensiton) interacts with billions of silver bromide molecules spread over the photographic plate: these evolve momentarily into superpositions of the dissociated and undissociated states until the condition for probabilistic collapse occurs, and just one silver bromide molecule is dissociated, and all the others remain undissociated. When the plate is developed (a process which merely makes the completed position measurement more visible), it will be discovered that the electron has been detected as a dot in the photographic plate.

Third, all other observables of OQT, such as momentum, energy, angular momentum, or spin, always involve (i) a preparation procedure which leads to distinct spatial locations being associated with distinct values of the observable to be measured and (ii) a position measurement in one or other spatial location. This means that PQT can predict the outcome of measurements of all the observables of OQT.

Fourth, insofar as the predictions of OQT and PQT differ, the difference is extraordinarily difficult to detect and will not be detectable in any quantum measurement so far performed.

---

[15] If postulates (I) and (II) procure insufficient localization, (II) could be modified so that the outcome states are those that would have evolved from a spatial region characteristic of the size of the "particles" involved or characteristic of the distance associated with the inelastic interaction. Postulate (II), modified in this way, would certainly procure sufficient localization to ensure that the classical world does not gradually become increasingly unlocalized.

[16] In fact, from a formal point of view (ignoring questions of interpretation), PQT has exactly the same structure as OQT with just one crucial difference: the generalized Born postulate of OQT is replaced by postulate (II) of section 5. (The generalized Born postulate specifies how probabilistic information about the results of measurement is to be extracted from the ψ-function.)

14.7 Crucial Experiments

In principle, however, OQT and PQT yield predictions that differ for experiments that are extraordinarily difficult to perform and which have not yet, to my knowledge, been performed. Consider the following evolution:

```
              collision   superposition   reverse collision
                           a + b + c
  a + b + c    ⟶                            ⟶       a + b + c
                            a + (bc)
     (1)        (2)           (3)              (4)      (5)
```

Suppose the experimental arrangement is such that, if the superposition at stage (3) persists, then interference effects will be detected at stage (5). Suppose, now, that at stage (3), the condition for the superposition to collapse into one or other state, according to PQT, obtains. In these circumstances, OQT predicts interference at stage (5), whereas PQT predicts no interference at stage (5) (assuming this evolution is repeated many times). PQT predicts that in each individual case, at stage (3), the superposition collapses probabilistically into one or other state. Hence there can be no interference.

OQT and PQT make different predictions for decaying systems. Consider a nucleus that decays by emitting an α-particle. OQT predicts that the decaying system goes into a superposition of the decayed and undecayed state until a measurement is performed, and the system is found either not to have decayed or to have decayed. PQT, in appropriate circumstances, predicts a rather different mode of decay. The nucleus goes into a superposition of decayed and undecayed states, which persists for a time until, spontaneously and probabilistically, in accordance with the postulate (II), the superposition jumps into the undecayed or decayed state entirely independent of measurement. The decaying system will continue to jump, spontaneously and probabilistically, into the undecayed state until, eventually, it decays.

These two processes of decay are, on the face of it, very different. There is, however, just one circumstance in which these two processes yield the same answer, namely if the rate of decay is exponential. Unfortunately, the rate of decay of decaying systems, according to quantum theory, *is* exponential. It almost looks as if nature is here maliciously concealing the mode of her operations. It turns out, however, that for long times, quantum theory predicts departure from exponential decay (Fonda et al., 1978). This provides the means for a crucial experiment. OQT predicts that such long-time departure from exponential decay will, in appropriate circumstances, obtain, while PQT predicts that there will be no such departure.

The experiment is, however, very difficult to perform because it requires that the environment does not detect or "measure" decay products during the decay process. For further suggestions for crucial experiments, see Maxwell (1988, pp. 37–38).

There is a sense, it must be admitted, in which PQT is not falsifiable in these crucial experiments. If OQT is corroborated and PQT seems falsified, the latter can always be salvaged by letting ε, the undetermined constant of PQT, be sufficiently minute. Experiments that confirm OQT only set an upper limit to ε. There is always the possibility, however, that OQT will be refuted and PQT will be confirmed.

It would be interesting to know what limit present experiments place on the upper bound of ε.

14.8 The Potential Achievements of PQT

PQT provides a very natural possible solution to the quantum wave/particle dilemma. The theory is fully micro-realistic; it is, in the first instance, exclusively about "beables" to use John Bell's term. It makes sense of the mysterious quantum world. There is no reference to observables, to measurement, to macroscopic, quasi-classical, or irreversible phenomena or processes, or to the environment, whatsoever. As a result, PQT does not suffer from the eight defects, indicated in section 14.2, which beset OQT. The theory is restricted, in the first instance, to specifying how quantum micro-systems—quantum propensitons—evolve and interact with one another deterministically and probabilistically. But despite eschewing all reference to observables or measurement in its basic postulates, the theory nevertheless in principle recovers all the empirical success of OQT. At the same time, it is empirically distinct from OQT for experiments not yet performed and difficult to perform.

It is quite possible, of course, that the general propensiton solution to the wave/particle problem that I have outlined here is correct but the specific proposal that probabilistic transitions are to be associated with inelastic interactions is false. The falsity of the specific proposal does not mean that the general propensiton idea is false as well.

My chief objective in developing the version of PQT outlined here, from 1972 onwards, was to provoke physicists better qualified than I *to tackle the problem by doing straightforward physics*! I hoped physicists would put forward testable conjectures concerning the quantum conditions for probabilistic transitions to occur and then put these conjectures to the test of experiment. Straightforward physics, theoretical and experimental, is needed if we are to develop a more adequate, intelligible, and explanatory version of quantum theory than that which we possess today. This research is now, it seems, underway: see Bassi (2013).

References

Albert, D. Z., 1996, Elementary Quantum Metaphysics, in *Bohmian Mechanics and Quantum Theory*, edited by J. Cushing, A. Fine, and S. Goldstein, Kluwer Academic, Dordrecht, pp. 277–284.

Bacciagaluppi, G. and J. Ismael, 2015, Review of *The Emergent Multiverse* by David Wallace, *Philosophy of Science*, vol. 82, no. 1, pp. 129–148.

Bassi, A., K. Lochan, S. Satin, T. P. Singh, and H. Ulbricht, 2013, "Models of Wave-Function Collapse, Underlying Theories, and Experimental Tests, *Reviews of Modern Physics*, vol. 85, pp. 471–527.

Bell, J. S., 1973, Subject and Object, in *The Physicist's Conception of Nature*, edited by J. Mehra, Reidel, Dordrecht, pp. 687–690.

—, 1987, *Speakable and Unspeakable in Quantum Mechanics*, Cambridge University Press, Cambridge.

Feynman, R. P. and A. R. Hibbs, 1965, *Quantum Mechanics and Path Integrals*, McGraw-Hill, New York.

Gao, S., 2017, *Meaning of the Wave Function: In Search of the Ontology of Quantum Theory*, Cambridge University Press, Cambridge.

Ghirardi, G. C., 2002, Collapse Theories, in *Stanford Encyclopedia of Philosophy*, available online at http://plato.stanford.edu/achives/spr2002/entries/qm-collapse.

Ghirardi, G. C., Rimini, A., and Weber, T., 1986, Unified Dynamics for Microscopic and Macroscopic Systems, *Physical Review D*, vol. 34, pp. 470–491.

Margenau, H., 1958, Philosophical Problems Concerning the Meaning of Measurement in Physics, *Philosophy of Science*, vol. 25, pp. 23–33.

—, 1963, Measurements and Quantum States: Part I. *Philosophy of Science*, vol. 30, pp. 1–16.

Maudlin, T., 2011, *Quantum Non-Locality & Relativity*, Wiley-Blackwell, Chichester.

Maxwell, N., 1972, A New Look at the Quantum Mechanical Problem of Measurement, *American Journal of Physics*, vol. 40, pp. 1431–145.

—, 1976, Towards a Micro Realistic Version of Quantum Mechanics, *Foundations of Physics*, vol. 6, 1976, pp. 275–292 and 661–676.

—, 1982, Instead of Particles and Fields: A Micro Realistic Quantum "Smearon" Theory, *Foundations of Physics*, vol. 12, pp. 607–631.

—, 1985, Are Probabilism and Special Relativity Incompatible? *Philosophy of Science*, vol. 52, pp. 23–43.

—, 1988, Quantum Propensiton Theory: A Testable Resolution of the Wave/Particle Dilemma, *The British Journal for the Philosophy of Science*, vol. 39, pp. 1–50.

—, 1992, Beyond Fapp: Three Approaches to Improving Orthodox Quantum Theory and an Experimental Test, in *Bell's Theorem and the Foundations of Modern Physics*, edited by A. van der Merwe, F. Selleri and G. Tarozzi, World Scientific, Singapore, pp. 362–370.

—, 1993, Does Orthodox Quantum Theory Undermine, or Support, Scientific Realism? *The Philosophical Quarterly*, vol. 43, pp. 139–157.

—, 1994, Particle Creation as the Quantum Condition for Probabilistic Events to Occur, *Physics Letters A*, vol. 187, pp. 351–355.

—, 1998, *The Comprehensibility of the Universe: A New Conception of Science*, Clarendon Press, Oxford.

—, 2004, Does Probabilism Solve the Great Quantum Mystery?, *Theoria*, vol. 19/3, no. 51, pp. 321–336. https://www.academia.edu/2975222/Dispositions_Causes_and_Propensities_in_Science?auto_accept_coauthor=true

—, 2006, Special Relativity, Time, Probabilism and Ultimate Reality, in *The Ontology of Spacetime*, edited by D. Dieks, Elsevier, B. V., pp. 229–245.

—, 2011, Is the Quantum World Composed of Propensitons?, in *Probabilities, Causes and Propensities in Physics*, edited by Mauricio Suárez, Synthese Library, Springer, Dordrecht, pp. 221–243.

—, 2016, Popper's Paradoxical Pursuit of Natural Philosophy, in *Cambridge Companion to Popper*, edited by Jeremy Shearmur and Geoffrey Stokes, Cambridge University Press, Cambridge, pp. 170–207.

—, 2017a, *Understanding Scientific Progress*, Paragon House, St. Paul, Minnesota.

—, 2017b, *In Praise of Natural Philosophy: A Revolution for Thought and Life*, McGill-Queen's University Press, Montreal.

—, 2017c, Relativity Theory May Not Have the Last Word on the Nature of Time: Quantum Theory and Probabilism, in *Space, Time and the Limits of Human Understanding*, edited by G. Ghirardi and S. Wuppuluri, Springer, Dordrecht, pp. 109–124.

Penrose, R., 1986, Gravity and State Reduction, in *Quantum Concepts of Space and Time*, edited by R. Penrose and C. J. Isham, Oxford University Press, Oxford, pp. 129–146.

—, 2004, *The Road to Reality*, Jonathan Cape, London.

Popper, K., 1957, The Propensity Interpretation of the Calculus of Probability and the Quantum Theory, in *Observation and Interpretation*, edited by S. Körner, Butterworth, London, pp. 65–70.

—, 1959, *The Logic of Scientific Discovery*, London: Hutchinson (first published in 1934).

—, 1967, Quantum Mechanics Without "The Observer," in *Quantum Theory and Reality*, edited by M. Bunge, Springer, Berlin, pp. 7–44.

—, 1982, *Quantum Theory and the Schism in Physics*, Hutchinson, London.

Wallace, D., 2012, *The Emergent Multiverse*, Oxford University Press, Oxford.

15

How the Schrödinger Equation Would Predict Collapse: An Explicit Mechanism

ROLAND OMNÈS

One will sketch here a project of deriving wave function collapse, directly and only from the Schrödinger equation. This is deemed possible, with the help of properties of local entanglement between two quantum systems, which this equation implies when one of the systems is macroscopic. Although these properties are in agreement with the algebra of observables, they do not identify with it and are essentially topological: They express a growth of clustering, as an aspect of quantum dynamics, and measure it by new local probabilities, which evolve nonlinearly and are distinct from the usual quantum probabilities although compatible with them.

One can also apply local entanglement to a macroscopic system and fluctuations in its environment. This result is new and striking, because it implies that these fluctuations can inject, into the quantum state of this system, a form of incoherence, which would have remained ignored till now. This incoherence would not only be strong, but it would also be inaccessible to any form of direct experimental observation.

A conjugation of these two effects could bring out collapse, as the outcome of many tiny "slips in coherence," which would break at atomic level the algebraic entanglement between a measured system and a macroscopic measuring one. This mechanism can be made explicit, but it remains conjectural: real proofs, which could ascertain it, would need at least a new axiom, which would give sense to local entanglement and place it among the foundations, in the interpretation of quantum mechanics.

How can it be that a unique real fact emerges from a quantum measurement? This perplexing question came to light in 1935, together with negative answers, in two historic papers, respectively by Schrödinger [1] and by Einstein, Podolsky, and Rosen [2]. Although quantum physics has made outstanding progress since that time, no universally agreed answer exists yet for this problem, which remains a matter of great concern [3] and is still presently a subject of much research [4]. It has also become a major theme in the philosophy of science [5], and the present anthology shows the great variety of directions, along which tentative answers are still hunted for.

The approach, which is proposed in the present chapter, is very simple from the standpoint of a philosophy of science. It takes for granted that quantum dynamics—and especially the Schrödinger equation—is perfectly exact, if only because of its countless successes and of its own fecundity and depth. Its interpretation (when meant as its place in the field of knowledge) remains on the contrary unfinished from this new standpoint and should be completed, not by new laws, but by some more consequences of the Schrödinger equation: consequences, which were known but could have remained little noticed, or would be new. One will try to sketch the last viewpoint here.

A direct derivation of collapse from the Schrödinger equation will be shown a likely consequence of this approach, and proposed here as a conjecture. A real proof would certainly require however a significant extension and even a deep revision of the standard interpretation of quantum mechanics (expressed and used, for instance, in two basic books by Dirac [6] and by Von Neumann [7]). Some hints along that direction will be proposed, hither and thither, in this chapter.

15.1 Local Entanglement

Many thorough experiments led, during the last thirty years or so, to an empirical conclusion of great significance: "Wave function collapse" is strictly restricted to macroscopic measuring systems and is never observed at microscopic scale [8]. One will pay a special attention to this conclusion in this chapter and, as a starting point, one will return to the famous example of Schrödinger's cat, with special attention for everything in it regarding macroscopic features.

15.1.1 A Classical Exploration

Schrödinger divided the content of a box, in which a measurement was taking place, into a set of quantum subsystems including a radioactive source, a Geiger detector, a hammer, a bottle of poison, and a cat. The fact that he made no difference between a single radioactive nucleus and a macroscopic cat, when envisioned by quantum theory, will turn out essential in this chapter. Although these two aspects are obviously correlated by the atomic structure of matter, one will first stress here a significant conceptual difference between what is seen as macroscopic and what is thought as microscopic. One will look at that by using another model.

This model involves a charged particle, denoted by A, which crosses a very simple Geiger counter, in which the detecting part is essentially a gas (which one takes as made of argon, because that makes its constituents very simple). One cannot avoid introducing also a solid box (for containing the gas, from a practical standpoint, and for marking a frontier of the SYSTEM, conceptually). One denotes this macroscopic system by B. Anyway, one will approach it first in a completely classical way.

One assumes, that every particle in the AB system obeys classical dynamics and carries furthermore a color, which marks a memory of earlier actions from the small A-system. This color is either white or red: Before entry of the particle A into the gas, this particle is red and every atom in the system B is white. One assumes that the red color is conveyed by "contagion" so that when a red particle interacts with a white particle, both of them come out red from their interaction. Moreover, when a particle has become red, it keeps that color forever. Finally, when two white atoms collide, they keep their color after having interacted. The particles move classically and their collisions are random.

At some time t, after entry of the particle A into the box, one expects that the probability for an atom, near a point x in the box, has a probability for having become red before a time t, is some positive function $f_1(x,t)$. Its probability for having remained white is then the function $f_0(x,t)$, such that

$$f_1(x) + f_0(x) = 1. \tag{15.1}$$

Since the red color of an atom is conserved under its collisions with other atoms, one can assert that t the red color diffuses into the box, according to a standard diffusion equation

$$\partial f_1/\partial t_{\text{diffusion}} = D\Delta f_1, \tag{15.2}$$

where D is a diffusion coefficient (which one expresses approximately, in terms of the mean free path of atoms λ and of their mean free time τ, by $D = \lambda^2/6\tau$).

Because a contagion of the red color occurs when a locally entangled atom (red) collides with a non-locally entangled one (white), the associated increase in $f_1(x,t)$ is given by

$$\partial f_1/\partial t_{\text{contagion}} = f_1 f_0/\tau. \tag{15.3}$$

Using (15.2) and (15.3) give a nonlinear equation for the evolution and propagation of the red color which is

$$\partial f_1/\partial t = D\Delta f_1 + f_1(1-f_1)/\tau. \tag{15.4}$$

When looking at this equation, one finds that it cannot be satisfied by a function f_1, which would be everywhere positive and non-vanishing (as it does in the case of the diffusion equation (15.3)). Non-vanishing values of the probability of influence $f_1(x, t)$ must be therefore restricted behind a front, which starts initially from the trajectory of Particle A. (Such a behavior is not surprising, since one knows that a generation of sharp fronts is frequent in nonlinear wave equations such as (15.4) [9]). A model of random walk for the motion of atoms suggests then that the velocity of this front is close to the ratio $v = \lambda/\tau$ in one-dimensional space, and $v' = 3^{-1/2}v$ in three dimensions (this is close to the velocity of sound in a dilute gas, and this relation is probably not a coincidence).

This behavior shows that the influence of the incoming particle A progresses much more rapidly than standard diffusion would do, since diffusion expands only at time t to a distance of order $(Dt)^{1/2}$, whereas diffusion, when acting together with contagion in Equation (15.4) yields an expansion of influence at the much larger distance $v't$, which defines the position of the moving front boundary S at that time. Numerical solutions of the propagation equation (15.4) confirm this motion of a wave front S at the velocity v'. One will often make use in this chapter of the behavior of the probability of influence $f_1(x, t)$, just behind the front, where it increases rapidly from 0 to 1 over a distance of order the mean free path λ.

15.1.2 A Quantum Approach: Local Entanglement

Eliot Lieb and Derek Robinson discovered in the seventies a quantum form of these phenomena, where a microscopic system influences locally a macroscopic one. They called it "local entanglement" [10], and one will keep that name and sometimes abridge by "LE,". The adjective "local" in this name came in the Lieb-Robinson approach from its association with a spin-lattice model, where a particle, which carries a spin, is completely specified by its "location."

Rather than speaking of two colors, "red" and "white," to mean LE and non-LE, one can also use two indices, 1 and 0 [11]. These labels stand only as indices and do not refer to different values of an observable, as would be the case for the values of a spin component, for instance. These indices of local entanglement have no relation with any quantum observable and the index 1 marks only the memory of a past interaction of some atom with another one, which already carried that index. This notion of local entanglement makes sense only for macroscopic systems and it has no analog in microscopic systems.

One can build up easily mathematical expressions for local entanglement and its evolution. To do so, one can introduce three 2×2 matrices, in which the indices 0 and 1 denote rows and columns, namely:

$$P_0 = \begin{pmatrix} 0 & 0 \\ 0 & 1 \end{pmatrix}, P_1 = \begin{pmatrix} 1 & 0 \\ 0 & 0 \end{pmatrix}, S = \begin{pmatrix} 0 & 1 \\ 0 & 0 \end{pmatrix}. \quad (15.5)$$

P_0 can be interpreted as a projection matrix, which picks up an atom with *LE* index 0 and keeps this index unchanged. The same behavior holds for P_1, which picks up Index 1 and conserves it. The matrix S picks up an *LE* index 0 (which indicates an absence of previous influence of A) and brings it to local entanglement, shown by the index 1 (so that the influence of A becomes then imprinted on this atom).

One recalls that these matrices are not supposed to act on a state vector in a two-dimensional Hilbert space but only on a conventional index, which an atom carries as a kind of memory. An important feature of this family of matrices is that the adjoint matrix $S^\dagger$, which could be formally associated with S and would bring back local entanglement to no local entanglement, does not belong to the construction. This absence imposes an irreversible character to local entanglement, in accordance with its representation as a contagion.

One can make these rules of contagion mathematically explicit in a quantum way. To do so, one replaces the potential U_{Aa} for the interaction between the particle A and an atom a by a 2×2 matrix

$$U_{Aa} = U_{Aa}(P_{1a} + S_a) \tag{15.6a}$$

Similarly, the potential V_{ab} for interaction between two atoms a and b is replaced by

$$V_{ab} = V_{ab}(P_{0a} \otimes P_{0b} + P_{1a} \otimes P_{1b} + P_{1a} \otimes S_b + S_a \otimes P_{1b}), \tag{15.6b}$$

which describes adequately the rules of contagion. A kinetic energy term for the free propagation of an atom can be extended in the same way to a propagation of *LE* under the extended free Hamiltonian

$$H_{0a} = -(\nabla^2/2m)(P_{0a} + P_{1a}). \tag{15.6c}$$

15.1.3 Dynamics of Local Entanglement

The standard wave function ψ of the composite AB system evolves according to the Schrödinger equation

$$i\hbar \partial \psi / \partial t = H\psi. \tag{15.7}$$

When the various interaction terms and kinetic terms in the Hamiltonian H are replaced by the matrix expressions (15.6), equation (15.7) becomes a set of coupled equations for a set of locally entangled wave functions $\{\psi_s\}$, which involves a definite set of *LE* indices for all atoms. If N denotes the number of atoms in B, the index s of an *LE* function ψ_s is a sequence of N indices of local entanglement, equal either to 1 or 0.

All the atoms are initially non-locally entangled and there is a unique component ψ_0 where all the *LE* indices are equal to 0. One may notice that the Bose-Einstein or Fermi-Dirac symmetry of the wave function ψ remains valid in every ψ_s, because two atoms carry the same index, 0 or 1, after an interaction.

If one denotes by ψ' this set $\{\psi_s\}$ and one considers it as a vector with 2^N components, one gets a linear equation of evolution with the same abstract form as a Schrödinger equation, namely

$$i\hbar \partial \psi' / \partial t = H' \psi'. \tag{15.8}$$

The operator H' is a $2^N \times 2^N$ matrix. Its matrix elements involve differential operators representing kinetic energy and potentials (U, V) representing interactions. Before interaction between the two systems A and B, the vector ψ' has only one component in which all the *LE* indices are 0. This unique component coincides then with the standard wave function ψ, and one can show easily that, by means of (15.6) and (15.7), that the standard wave function coincides at all times with the sum

$$\psi = \sum_s \psi_s. \tag{15.9}$$

Several other properties of ψ' imply however significant differences between standard quantum dynamics and local entanglement (although the second one amounts only to rewriting the first one with an introduction of memory). The main one is that the operator H' in (15.8) is *not* self-adjoint. As a consequence, local entanglement is *irreversible* under time reversal. It always leads to a final situation in which all atomic states have become locally entangled and there remains a, unique component in the set $\{\psi_s\}$: the one where all the *LE* indices are the index 1, which indicates local entanglement.

A direct consequence is that local entanglement stands out of the standard interpretation: No quantum observable can extract an *LE* component $\psi_s(t)$ Local entanglement belongs nevertheless to the Schrödinger equation, since it results directly from a refined rewriting of this equation.

15.1.4 Probabilities of Local Entanglement and Their Evolution

One introduced earlier a local probability of contagion, $f_1(x, t)$, and one turns now to its derivation from the quantum framework of local entanglement. The relevant basis is a "cluster decomposition principle," which Steven Weinberg showed necessary for founding quantum field theory on a complete set of principles ([12], volume I, chapter 4). In addition to other principles, which mostly belong to algebra [7], this one is concerned with topological properties of connectedness, in the evolution operator of a quantum system. A remarkable consequence of this principle is a derivation of Feynman histories from this form of the principles of quantum field theory ([12], volume II). A connectedness between Feynman paths of atoms in a Feynman history, which occurs when two-body potentials have a finite range, is then very close to the one occurring in locally entangled atomic states.

When using this cluster principle under these conditions, one can derive the following properties:

One can represent the creation of locally entangled states of atoms and the existence of non-locally entangled ones by means of creation fields, $\phi_1^\dagger(x)$ and $\phi_0^\dagger(x)$. The standard creation field $\phi^\dagger(x)$ is then the sum $\phi_1^\dagger(x) + \phi_0^\dagger(x)$.

Average values of the products $\phi_1^\dagger(x)\phi_1(x)$ and $\phi_0^\dagger(x)\phi_0(x)$ in a macroscopic state can be used to define local densities of *LE* and no-*LE*, $n_1(x)$ and $n_0(x)$, like a product of standard fields $\phi^\dagger(x)\phi(x)$ defines a local density $n(x)$. One can then define the previous local probabilities by the positive ratios, $f_r(x) = n_r(x)/n(x)$, with $r = 1$ or 0.

The sum property (15.1) is, however, not trivial This construction of probabilities, for local entanglement and no local entanglement, remains doubtful and a deeper approach, by methods belonging to the field theory of many-body systems, would be necessary for appreciating the validity of results using the local probabilities $f_r(x)$. This limitation will not be considered however to a perilous impairment for the present approach: One will see that the present theory does not require absolute exactness and usual approximate methods can be fair enough (at least apparently),

15.2 The Environment and Its Action

A second major point in this study is concerned with the action of environment on the quantum state of a macroscopic system. One can express its main effect by the following proposition:

Proposition 1 *Fluctuations in the action of environment can inject a specific form of incoherence into a macroscopic system and its quantum state, in which associated random phases propagate along waves of local entanglement.*

A key point in this proposition is its mention of incoherence, which means in the present case an occurrence of intrinsically random phases, somewhat like in classical optics, and not a consequence of the unique and universal source of randomness in quantum mechanics, with holds in Born's axiom.

There are no random phases in the standard interpretation but only "arbitrary" ones, which affect the wave functions of an isolated system [6, 7]. When the standard interpretation deals for instance with the diffuse light coming from the sky, it considers that the photons came from the sun and underwent on their way some scattering with molecules in the atmosphere. The standard interpretation attributes then the impossibility of showing interference patterns with these photons to the unique accepted form of randomness, which this interpretation recognizes: the one in Born's axiom. The photons from the sky do not show interferences because different every one of them followed its own random way and these ways are uncorrelated.

The notion of waves of local entanglement, which occurs in Proposition 1, is also foreign to the standard interpretation, as well as some other elements in this proposition, like the mention of an "environment," its "action," "fluctuations" in this action, which need more precision.

It will be convenient to use a specific example and one will deal again with the case of the previous Geiger counter B, where an atomic gas is contained in a solid box. Its environment is an ordinary atmosphere, under standard conditions of temperature and pressure.

But it turns out that the consideration of local entanglement, as mentioned in Proposition 1, takes one immediately out of the security of abstraction: Local entanglement, as one means it here, is not a pure quantum effect, except in lattice models [10]). It is closely linked with *reality* and it shows, for instance, significant differences when it crosses the solid box and when it progresses into the gas: Phonons carry it through the box, and argon atoms inside the gas. No "observable" of the "quantum system," which unifies these components, enters in local entanglement. Too many things in it are not yet clearly understood (if it is not an illusion).

It does not matter much, moreover, whether one considers or not the environment as a quantum system, because Proposition 1 is concerned with a specific action of *fluctuations* in this environment, and with waves of local entanglement carrying these fluctuations. These notions about the physical system B are foreign to the mathematical concept of a quantum system [6]. They do not negate this concept however. They deal with another concept, which allows other properties (like the ones of local entanglement), but remains concerned with the same physical system (or real object). There is much of Bohr's deep idea of complementarity in this approach [3, 4].

The same considerations would apply to Everett's conception of a quantum state of the universe [13] (if that were useful for cosmological purposes). If local entanglement is acting somewhere in that universe, it can produce collapse there (or, at least, that will be the conclusion of the present chapter). One could say then that this collapse effect is a consequence of the Schrödinger equation, but it would not be a consequence resulting from the algebraic formulation of this equation. It would need a conception of the Schrödinger equation, which would fit with irreversibility (since local entanglement is irreversible). This is conceivable, for instance, by thinking of a Turing machine, endowed with an infinite memory...

Are not we saying there that one does not understand well enough all the aspects of the Schrödinger equation for understanding how it implies collapse? Maybe, but these are deep waters and that is not new [5].

15.2.1 A Theoretical Framework

The theoretical framework, which fits best Proposition 1, relies on quantum field theory, in Weinberg's version of its axioms involving a "cluster decomposition principle" and deriving from this principle a legitimate use of Feynman histories [12]. Local entanglement is an example of cluster decomposition and Feynman histories provide the best lighting for it.

Proposition 1 means from this standpoint that every Feynman path, which represents an incoming external molecule (which belongs itself to fluctuations) injects some random phase, say ϕ, into a Feynman history describing the system B. This random phase is transmitted then to more and more atoms in B, through growing interconnections in their Feynman paths inside a global Feynman history describing all the particles involved.

When the interactions between atoms have a finite range, a Feynman history involves, after some time, individual paths for the various atoms among which some paths carry that phase ϕ, whereas other paths do not carry it.

This behavior is not only uncommon, but it appears completely new because ignores the notion of an absolutely random phase escapes to the standard interpretation of quantum mechanics. When considered in a wave function of B, it implies that this wave function would be actually a sum of many components. Every one of these components would be a product of sub-wave functions involving different atoms. Every sub-wave function would involve locally entangled atoms and would carry the phase ϕ, whereas another sub-wave function would involve all the other non-locally entangled atoms and would not carry ϕ.

This behavior, where a phase factor affects a term in a product and not the other term seems meaningless. But mathematics can provide a proper framework: A mathematician would say that the standard Hilbert space of quantum mechanics becomes replaced by a fibered space. This is a plunge into deep waters, but it seems necessary for reinterpreting quantum mechanics, in such a way that local entanglement (and collapse), would have a place in it. Such a project requires a significant enlargement of the mathematical framework of quantum mechanics.

Needless to say, the present author stopped at that. One will proceed here from there on by assuming that the necessary mathematical foundations will be reached by further research. A derivation of heir physical consequences is easier, and one turns now to it.

15.2.2 A Model for an Injection of Incoherence

One deals again with the previous example of a macroscopic system, and one leaves aside the question regarding the transport of random phases through the solid box.

One introduces a reference length L, which is supposed to represent an order of magnitude for the size of this system, and a time Δt for the time that a wave of local entanglement takes for filling up the whole of B (their ratio is close to the velocity of sound in the gas).

Reminding that local entanglement "sees" only events at scales (λ, τ), one covers the boundary ∂B of the box by square "tiles" with side λ and one divides time into intervals with duration τ. One calls "event" the set of external molecules arriving on such a tile, during such a time interval. One also denotes a specific tile by T_q and the beginning of a time interval by t_n, so that an event is associated with a double index qn.

All events involve the same average number of molecules, but the associated average effect is trivial, because it exerts only an uninteresting pressure on the box. There are also fluctuations around this average in various events and one associates with every one of them a sign ε_{qn}, which equal to $+1$ for an excess above average and to -1 for a deficiency below. The individual excesses and deficiencies in various events are random, but one associates with every one of them, whatever its sign, the same number of molecules N_f (f for "fluctuation").

One might have doubts regarding this idea of attributing molecules to an event where they are missing, but one assumes that a more clever approach would make it sensible.

A next step is concerned with phases: One would like to deal with the phases of wave functions, which would be associated with ideal molecules belonging to fluctuations. The main character, which one can see regarding these molecules, is that they are abstract and impossible to define individually. One might also say that they are arbitrary and one attributes accordingly an arbitrary phase to every one of them. Every event injects accordingly a number N_f of arbitrary phases into a wave function of B. One goes next one step further and, judging that N_f is sufficiently large, one assimilates it with a *random* phase, which one denotes by ϕ_{qn} for an event qn.

There is a last question, which is: Can one associate a quantum probability with this phase ϕ_{qn} and, if so, what would be its value? This is again a deep question, and its answer would require a consistent interpretation, allowing for local entanglement. Let one look however at this question itself.

One is considering in the present case, an event regarding one tile T_q on the boundary of the measuring device, during a time τ. The average number of external molecules, which are associated with this event, has some definite value, which one can denote by N_t. The previous number N_f is then equal to $N_t^{1/2}$. But one must be careful again regarding the meaning of the question, which one asks: One is considering a unique *random* phase, ϕ_{qn}, which is meant to replace a set of $N_t^{1/2}$ *arbitrary* phases, one for each molecule in a fluctuation. The relevant question is concerned with the quantum probability (let one call it p) which one must associate with this random phase.

The answer, which one proposes here, relies on the relevant character of the environment, with which one is dealing here: This character is an absolute limitation of local entanglement, regarding its own precision: It cannot reach atoms below a distance of order λ, during at least a time τ, and that is the reason why one is dealing with a finite tile and one splits time into intervals τ. This question is whether one should reduce the probability p to the one of a unique phase, *viz.* $1/N_t$, or is it $1/N_f$, i.e., $1/N_t^{1/2}$. One recalls that

$$1/N_t \approx (n_e \lambda^3 v_e / v_a)^{-1}, \qquad (15.10)$$

The real point in this discussion is not so much the value, which one attributes to this probability of a random phase, but the fact that a reference to local entanglement raises new questions, for which the standard interpretation provides no answer. This is not far, moreover, from questions about the sex of angels, because the answer would only fix the time rate of collapse; which is inaccessible.

One can summarize this construction by the following set of characteristics for an event (in which the last one, p, remains problematic):

$$(T_q, t_n, N_f, \varepsilon_{qn}, \phi_{qn}, p). \qquad (15.11)$$

15.2.3 Incoherent Density Matrices

There was a great step from Feynman paths to density matrices, but one makes one more by asking now the question: What happens to the density matrix ρ of the gas, under the strong incoming flux of incoherent phases, which it receives from a multitude of external "events"?

Many phases ϕ_{qn} are injected into this density matrix. They enter through the boundary ∂B expand then deeper into the system B, under local entanglement: More precisely, when the quantum state of an atom carries a random phase ϕ_{qn}, and this atom collides with another one, in a state which does not carry that phase, both outgoing states of these atoms carry that phase after their collision. The same thing happens when an eternal molecule brings that phase. This process implies obviously a contagion of the random phase ϕ_{qn} into the density matrix $\rho(t)$, with its propagation by a wave of local entanglement. The probability for an atom to carry the phase ϕ_{qn}, near a point x at time t, must be therefore a local probability of *LE*, $f_{1qn}(x, t)$, similar to the ones in Section 1.

This result might look perplexing, if only because the density matrix $\rho(t)$ becomes random! One may wonder how that could be compatible with the standard notion, according to which the density matrix predicts definite average values for every observable?

One needs some notations for setting this question: One denotes accordingly by $\rho_0(t)$ the *bona fide* matrix of the standard interpretation and, after introducing the difference $\rho(t) - \rho_0(t)$, one splits it into a part involving its positive eigenvalues, and a second part involving its negative ones. One gets thus the following proposition:

Proposition 2 *When a macroscopic system interacts with a fluctuating environment, its density matrix splits into a sum*

$$\rho(t) = \rho_0(t) + \rho_+(t) - \rho_-(t), \tag{15.12}$$

where $\rho_0(t)$ is the usual density matrix in the standard interpretation. Both matrices $\rho_+(t)$ and $\rho_-(t)$ are self-adjoint and their traces are exactly equal. The main effect of external fluctuations is to make them completely random and full of random phases, which are various sums of the phases ϕ_{qn}, which are brought by the external events (15.11)

This proposition becomes more manageable when one uses an approximate form for $\rho_0(t)$, which is familiar in statistical physics and has the following form [14]:

$$\rho_0 = Z^{-1} \exp\left\{-H/k_B T - \sum_k \mu_k \Omega_k\right\}. \tag{15.13}$$

The notation H stands here for the Hamiltonian and some observables, which one denotes by Ω_k, are supposed to characterize some significant quantities in the system. The associated factors μ_k are then Lagrange parameters, which fix the average values of these observables. The Ω_k's are usually macroscopic, but they can also be microscopic in some cases. The notation Z is the one of a "partition function," which fixes the trace of ρ_0 to the value 1.

When an "event" (*i.e.*, a fluctuation in the flux of external molecules) acts on the boundary ∂B, Equation (15.12) splits it into three parallel events, where this fluctuation interacts simultaneously with three states of the gas, which are expressed respectively by ρ_0, ρ_+, and ρ_-.

When a fluctuation acts on ρ, it has various effects but the most interesting one is a transfer of some quantum probabilities between the three matrices in the right-hand side of (15.12): When $\varepsilon_{qn} = +1$, one can think of the collision as due to a "formal molecule," say M_{qn}, in an ordinary quantum state with probability (square norm or trace) p, which carries moreover a phase ϕ_{qn} and interacts with the system B. When one considers only the part ρ_0 of the state of B, one finds that the outgoing state of the collision consists in one or a few atoms in some quantum state, which "caught" the random phase ϕ_{qn}. In that sense, the injection of this random phase resulted in picking up the previous tiny quantity p from the trace of ρ_0 and adding it to the trace of ρ_+.

There are three matrices in the right-hand side of (15.12) and two possible signs for ε_{qn}, which make six cases for every index qn For a given value of the index n (i.e. a definite time interval $[t_n, t_n + \tau]$) the total number of tiles in which similar events occur is of order $(L/\lambda)^2$, with an equal number of positive and negative signs ε_{qn} (on average).

Perpetual small changes of order p occur everywhere in the traces of the matrices (ρ_0, ρ_+, ρ_-). Something systematic in these changes is that they always go from ρ_0 towards ρ_+ or ρ, with injection of random phases: To say otherwise, random phases are never destroyed (at least in this type of process), and they are perpetually created (on every tile, during every time interval τ). Each one of them only dies from exhaustion, when it becomes shared by every atom and merges into the arbitrary phases of eigenfunctions in ρ_+ or ρ_-.

15.2.4 The Trace of the Two Incoherent Matrices

One would need some estimates regarding these transfers of quantum probabilities, which one considers now: The two matrices $\rho_+(t)$ and $\rho_-(t)$ in (15.12) have the same trace, which one will denote by W and which will be one of the main parameters in this theory (as a matter of fact, it would control the rate of collapse)

A simple way for computing W uses the following academic model: One assumes that the environment does not act on the counter B before some time 0 (or it acted only by its average effect). The fluctuations act after that time 0. One knows however that a random phase, which is created by these fluctuations, can last only for a time Δt of order L/c_s: A calculation of W can be made therefore by computing the growth of $Tr(\rho_+(t))$ from time 0 to time Δt.

The number of tiles on the boundary is of order $(L/\lambda)^2$, half of them involving a positive ε_{qn} and half of them a negative one. One saw that every positive "event" brings a small probability p into ρ_+, for instance. Since two successive events are separated by a time τ, one finds that the trace of $\rho_+(\Delta t)$, when saturation of the growth of incoherence occurs, has reached the value $p(L/\lambda)^3$.

The previous problem regarding the relevant value of p comes however again. One will adopt the provisional value $1/N_f$ if only for illustration, together with the upper estimate

$$W \approx (n_e \lambda^3 v_e/v_a)^{-1/2} L^3/\lambda^3. \tag{15.14}$$

15.2.5 A Ballet of LE Waves

Whereas one found a very high number of active random phases ϕ_{qn} in ρ_+ and ρ_- their incoherence is certainly much stronger. More precisely, the number of different random phases is certainly much higher.

One can show that by noticing that there is an active part in a wave of local entanglement: a zone, just behind its moving front, where the probability of local entanglement, $f_1(x)$, goes from zero to come near 1 (numerical calculations give a width about 3λ for this zone).

Simple calculations show then that many active regions (from various elementary events) always overlap everywhere in the gas. In many of the local collisions between atoms, the two colliding atoms carry two different phases before the collision: say ϕ_1 and ϕ_2. Both of them carry the phase $\phi_1 + \phi_2$ in the final state of their collision.

When closing one's eyes, one can dream of a permanent ballet of waves of local entanglement, with innumerable shimmering colors of phases. Even if it does not exists, it warrants a glance.

One can look at the wave functions, which carry these phases. This analysis, which is partly given elsewhere (in a previous incomplete version of this theory [15]), leads to more explicit propositions. The first one is:

Proposition 3 *Every eigenfunction ψ of the matrix $\rho_{B+}(t)$ (for instance) splits under the effect of fluctuations in external collisions into a sum of component wave functions*

$$\psi = \sum_n \psi_n, \qquad (15.15)$$

where every component ψ_n carries a specific random phase and does not extend in space over a distance larger than the mean free path λ of atoms. It involves at most a limited number of atoms, of order of the ones in a region with size λ:

$$N_a \approx n_a \lambda^3. \qquad (15.16)$$

A direct consequence of Proposition 3 is striking, because one may say that it means that the incoherent parts (ρ_{B+} and ρ_{B-}) in the state of the system B become pulverized by incoherence into a random aggregate of independent parts with size λ: As an example, one may consider two arbitrary regions, say R and R', which are not smaller than λ and are distant by more than λ. If one defines a localized density matrix ρ_{R+} as the partial trace of ρ_{B+} over all the atoms outside of the region R, and one does the same for another regions R', as well as for the union $R'' = R \bigcup R'$, one gets a second useful proposition, which is:

Proposition 4

$$\rho_{R''+} = \rho_{R+} \otimes \rho_{R'+}. \qquad (15.17)$$

This proposition is remarkable because it implies the possibility that some events, in a small region (e.g. the one, denoted here by R) can sometimes affect immediately the quantum state of much larger regions (viz. R''). This behavior opens wide possibilities of very local effects, which would take the government of big global ones *(in the present case a small region R and a very large one, R″, almost as big as the whole system).*

15.3 Slips in Coherence

One begins with a brief formal description of a quantum measurement. One denotes by A a microscopic system involving an observable X, which is measured by the counter B. the measured system A can be for instance a particle, in an initial state

$$|A\rangle = \sum_k c_k |k\rangle. \tag{15.18}$$

One will denote the associated quantum probabilities by $p_k = |c_k|^2$.

The measuring system B can consist in several parts, like in a Stern–Gerlach experiment, for instance. One will not introduce however a specific notation for such a case, because the previous use of a point x in space is enough for specifying which apparatus is concerned

In place of the previous local probability for local entanglement, $f_1(x)$, one must deal now with as many quantities $f_k(x)$ as there are measurement channels k (in every part of the measuring device). The probability for an atom to be locally entangled with the channel k near a point x is given then by $p_k f_k(x)$, and its probability for no local entanglement is

$$f_0(x) = 1 - \sum_k p_k f_k(x). \tag{15.19}$$

15.3.1 Introducing Slips in Coherence

There exists in the present approach an elementary mechanism, which acts at atomic levels and works as the microscopic agent of collapse. One calls it a "slip in coherence," with the meaning of "slip" for a small mishap. This event happens in the present case in a collision between two atoms. It obeys quantum probabilities, according to the standard interpretation. The mishap is in the present case a slip in the quantum probabilities of some channels, which happens in some atomic collisions under some specific conditions.

One will take as an example the previous case of a Geiger counter. One can find in it various particles: argon atoms, electrons, ions, excited atoms ...It does not matter and a slip effect can involve any pair of them. One will therefore speak of a pair of "atoms," with no more precision..

A slip in coherence consists then by definition in a collision between two "atoms" (a, b) under the following conditions:

(*i*) The collision is incoherent.
(*ii*) The state of Atom a is locally entangled with a state $|j\rangle$ of the system A.
(*iii*) The state of Atom b is non-locally entangled with the system A.

Condition (*i*), which requires incoherence, requires also that one should only pay attention to the contribution of the matrix ρ_{AB+}, or of $-\rho_{AB-}$, to the collision. One will consider only the first case, in which an easy use of probabilities of local entanglement and of incoherence for initial states) involve respectively some probabilities of local entanglement, and the trace W of the incoherent matrices $\rho_{B\pm}$. A straightforward computation gives then the following result:

Proposition 5 *When a collision under Conditions (i–iii) is governed by the matrix ρ_{AB+}, and occurs near a point x, it produces changes in the quantum probabilities of various channels, explicitly given by*

$$\delta p_j = +Wp_j(1-p_j)f_j(x)f_0(x)/2N_a, \tag{15.20a}$$

$$\delta p_{j'} = -Wp_jp_{j'}f_j(x)f_0(x)/2N_a, \quad \text{for } j \neq j', \tag{15.20b}$$

The signs become opposite when the matrix $-\rho_{AB-}$ governs the collision.

The derivation of this proposition is essentially obvious: Condition (*i*) in the definition of a slip implies for instance the factors W, Condition (*ii*) and (*iii*) the factors $f_j(x)$ and $f_0(x)$ in (5.3b). The factor $1/N_a$ comes from (15.16): It expresses that all the atoms in a wave function feel a change of that wave function in the same way: Here a unique new atom has brought an individual increase (or decrease) in probability, and this increase must be shared by all atoms. (These indications are of course only hand waving, but they can be helpful).

The result warranting emphasis is anyhow the prediction of small switches between the quantum probabilities of various measurement channels. This is precisely what the standard interpretation deemed impossible [1]: Although this is not yet the death of Schrödinger's cat, it means that one hair has been taken at least out of his fur.

15.4 A Quantum Mechanism of Collapse

An explicit mechanism of collapse results almost directly from (15.20). It goes through from an accumulation of all the transitions in quantum probabilities, which result from all the slip events, during a short time interval δt. One will describe it again in the case of the matrix ρ_{AB+} and one will use the same name, "atom," for every type of particle entering in the process: This is a manner of mentioning in these results a kind of universality.

When Atom a is locally entangled with Channel j, one found that there are transfers of quantum probabilities from all other channels (with index $j' \neq j$) toward this channel j. But there are other slips, where the atom playing the part of a is locally entangled with a channel j', and the resulting average variations, $<\delta p_j>$ and $<\delta p_{j'}>$, cancel each other, in view of symmetry of the right-hand side of (15.20b) in j and j'. When one averages over all the slips, which occur during the time interval δt, the average variations $<\delta p_k>$ for all channels with every index k, must therefore vanish. Nevertheless, the standard deviations $<(\delta p_j)^2>$ and the quadratic correlation coefficients $<\delta p_j \delta p_{j'}>$ do not vanish but they add together.

When one looks then at the matrix $-\rho_{AB-}$ after these considerations on ρ_{AB+}, one finds identical results and they add up together (since the minus sign in $-\rho_{AB-}$ occurs twice: in the expression of δp_j under the transitions $j' \to j$ and of $\delta p_{j'}$ under $j \to j'$).

One gets thus finally

$$<(\delta p_j)^2> = W p_j (1 - p_j)(\delta t / N_c^2 \tau) \int n_a f_j(x) f_0(x) dx, \qquad (15.21)$$

$$<\delta p_j \delta p_k> = -W p_j p_k (\delta t / N_c^2 \tau) \int n_a [f_j(x) + f_k(x)] f_0(x) dx, \quad \text{for } j' \neq k. \qquad (15.22)$$

where $f_0(x)$ is given by (15.19).

One may notice that these results, which are obtained from a definite sample of fluctuations in external collisions, have exactly the same form for every sample. As for the local probabilities $f_k(x)$ and $f_0(x)$, they have a purely kinematical origin and they are therefore also independent of the sample of fluctuations, which is acting. Within the framework of interpretation, which one proposed for the effects of local entanglement with external fluctuations, one can hold the results (15.21-2) as having a wide generality.

15.4.1 From Fluctuations to Collapse

The linear behavior of the correlations (15.21-2) in δt implies that the quantum probabilities $\{p_i\}$ have a Brownian motion. Philip Pearle suggested this idea somewhat long ago [16] and the more recent theory of "continuous spontaneous localization" [17] extended this approach, by combination with an inspiring assumption by Ghirardi, Rimini, and Weber of extraneousl effects, which would perturb randomly the evolution under a Schrödinger equation [18]. In the present chapter, one encounters also Brownian variations of the quantum probabilities, but as a consequence of Schrödinger's equation itself, with no violation (in it but with an account of local entanglement, which is (was) its (overlooked) prediction

Pearle's theorem, which stands at the foundations of these CSL theories, is a refined version of an old "gambler's ruin theorem going back to Huygens. Pearle's one states that, if the quantum probabilities of the various channels undergo a Brownian motion, the quantum probabilities of the various channels must vanish, one by one, until a unique one remains. The theorem asserts that the Brownian probability, for this outcome to happen in some channel j, coincides with its initial quantum probability, namely the value $|c_j|^2$ in Born's fundamental axiom.

One might expect that an application of this theorem would imply immediately collapse, since Equations (15.21-2) define a Brownian process. One must take care however of the assumptions in this theorem. One of them is formal: It assumes that the fluctuations can be considered infinitesimal, so that the Brownian process is governed by a Fokker–Planck–Ito equation. A boundary condition, which insures that a quantity p_j can vanish under the motion, motion, can be expressed algebraically by the inequality

$$\partial <(\delta p_j)^2>/\partial p_j \neq 0, \text{ when } p_j = 0. \tag{15.23}$$

Pearle's theorem and its assumptions would then bring the conclusion that collapse occurs *after an infinite time*.

This conclusion is not however under the present *physical* conditions. The key point is that the variations δp_k for a slip, in (15.20), are small, but *finite*. But the generation of slips in a finite region (for instance with size λ) is a discrete Poisson process, not an infinitesimal Fokker-Planck-Ito Process. The Poisson law has a well-known property, when it deals with deals with random integers (the number of slips in a small region): the fluctuations are much larger than the average values when these average values are much smaller than 1 (*viz.* $x^{1/2} \gg x$, when $x \ll 1$). This property (which makes the Poisson law often called the "law of small numbers") becomes enforced by a another factor, of order $\pm(L/\lambda)^{3/2}$ from all the cells and the various signs of their contributions. One expects that it would be sufficient for giving a high chance of going to zero to a small p_k, which will never reappear.

Another character of these fluctuations is that the slips, which generate them, occur only in the two highly incoherent matrices of great significance, (ρ_{AB-}, ρ_{AB-}). They are independent in different cells and Proposition 5 increases highly their effects on the quantum probabilities of channels, by elevating every local change of channel probabilities in an incoherent density matrix of a small region to almost the same global change in the full matrix $\rho_{AB}(t)$.

Aside with these results, one must stress that even if these perspectives could look attractive, *no rigorous proof justifies them*, for the time being. They would need a much more thorough work, which could give them substance or lead to other ones.

Dreamers could a mysterious "veiled" incoherence, which would be inaccessible by all means. Poets could make its hidden depth a Chaos, mother of Reality. Philosophers of science could envision that this approach indicates a possible self-consistency in quantum mechanics, because of its agreement with Born's axiom. One will only close this chapter by the following final conjecture:

Proposition 6 *Within a proper new interpretation of quantum mechanics, the Schrödinger-Von Neumann equation for the evolution of quantum states can predict a random outcome of collapse under the conditions of quantum measurements, and an agreement of the relevant probabilities with Born's probability rule.*

References

[1] Schrödinger, E. (1935). Die gegenwärtige Situation in der Quantenmechanik. *Naturwissensschaften* **23**, 807, 823, 844. Reprinted with English translation in [3].

[2] Einstein, A., Podolsky, B. Rosen, N. (1935). Can quantum-mechanical description of physical reality be considered complete? *Phys. Rev.* **47**, 777.

[3] Wheeler, A., Zurek, W. H. (1983). *Quantum Theory and Measurement*, Princeton, Princeton University Press.

[4] Laloë, F. (2012). *Do We Really Understand Quantum Mechanics?* Cambridge, Cambridge University Press.

[5] d'Espagnat, B. (1995). *Veiled Reality*, Addison Wesley; (2006). *On Physics and Philosophy*, Princeton University Press (2006).

[6] Dirac, P.A.M. (1930). *The Principles of Quantum Mechanics*, Cambridge University Press.

[7] Von Neumann, J. (1932). *Mathematische Grundlagen der Quantenmechanik*, Berlin, Springer, Berlin. English translation by R. T. Beyer, *Mathematical Foundations of Quantum Mechanics*, Princeton University Press (1955).

[8] Haroche, S. Raimond, J. M. (2006). *Exploring the Quantum*, Oxford University Press.

[9] Dautray, R., Lions, J.-L. (2003). *Mathematical Analysis and Numerical Methods for Science and Technology. Evolution Problems, I, II*, Berlin, Springer.

[10] Lieb, E. M. Robinson, D. W. (1972). The finite group velocity of quantum spin systems. *Commun. Math. Phys.* **28**, 251.

[11] Omnès, R. (2013). in E. Agazzi (ed.), *Turing's Legacy,* Proc. 2012 Meeting of the International Academy of Philosophy of Science, Milano, FrancoAngeli; (2015). in *The Message of Quantum Science*, P. Blanchard, J. Fröhlilch (eds.), Berlin, Springer, 257–271; (2013). *Found. Phys.* **43**, 1339–1368.

[12] Weinberg, S. (2011). *The Quantum Theory of Fields, I, Foundations, II, Modern Applications*, Cambridge, Cambridge University Press.

[13] Everett, H. (1957). "Relative state" formulation of quantum mechanics. *Rev. Mod. Phys.* **29**, 454–462.

[14] Balian, R. (2006). *From Microphysics to Macrophysics*, Berlin, Springer.

[15] Omnès, R. (2016). *Sketch of a Derivation of Collapse From Quantum Dynamics II*, arxiv.org, quant-ph: 1611.06731.

[16] Pearle, P. (1976). Reduction of the state vector by a nonlinear Schrodinger equation *Phys. Rev.* **D13**, 857; (2007). How stands collapse I. *J. Phys. A, Math.-Theor.* **40**, 3189.

[17] Ghirardi, G. C., Pearle, P., Rimini, A. (1990). Markov processes in Hilbert space and continuous spontaneous localization of systems of identical particles. *Phys. Rev.* **A42**, 78.

[18] Ghirardi, G. C., Rimini, A., Weber, T. (1986). Unified dynamics for microscopic and macroscopic systems. *Phys. Rev.* **D34**, 470.

Part IV

Implications

16

Wave-Function Collapse, Non-locality, and Space–time Structure

TEJINDER P. SINGH

Collapse models possibly suggest the need for a better understanding of the structure of space–time. We argue that physical space and space–time are emergent features of the Universe, which arise as a result of dynamical collapse of the wave function. The starting point for this argument is the observation that classical time is external to quantum theory, and there ought to exist an equivalent reformulation which does not refer to classical time. We propose such a reformulation, based on a non-commutative special relativity. In the spirit of Trace Dynamics, the reformulation is arrived at as a statistical thermodynamics of an underlying classical dynamics in which matter and non-commuting space–time degrees of freedom are matrices obeying arbitrary commutation relations. Inevitable statistical fluctuations around equilibrium can explain the emergence of classical matter fields and classical space–time, in the limit in which the universe is dominated by macroscopic objects. The underlying non-commutative structure of space–time also helps us understand better the peculiar nature of quantum non-locality, where the effect of wave-function collapse in entangled systems is felt across space-like separations.

16.1 Introduction

Dynamical collapse models explain the collapse of the wave function and provide a solution of the quantum measurement problem by proposing that the Schrödinger equation is approximate. It is an approximation to a stochastic nonlinear dynamics, with the stochastic nonlinear aspect becoming more and more important as one progresses from microscopic systems to macroscopic ones. They are the only concrete class of models whose predictions differ from those of quantum theory in the mesoscopic and macroscopic domain. These predictions are currently being tested in a variety of experiments which are putting ever tighter constraints on departure from the standard linear theory [1]. Yet a large part of the parameter space remains to be ruled out. In Section II we give a very brief description of these models and then provide an overview of the status of their experimental tests.

The Schrödinger equation is of course non-relativistic, and so are the collapse models described in Section II. According to theory, collapse is an instantaneous process, and experiments also confirm that collapse of the wave function at one location in space produces an influence at another space-like separated location. This suggests that making a relativistic theory of collapse will be challenging, and indeed this turns out to be so. In Section II we briefly describe the current status of various attempts to make a relativistic theory of collapse.

It is possible that sometime in the future we will have a convincing relativistic theory of collapse of the wave function. In the present work though, we attempt to make a case that there is a fundamental reason why collapse and relativity are incompatible. The reason is what we call the 'problem of time' in quantum theory. As we argue in Section III, time being a classical concept, it is external to quantum theory. There should exist an equivalent reformulation of quantum theory which does not refer to an external classical time. Such a reformulation suggests space–time is fundamentally not classical but perhaps has a noncommutative structure. In the absence of collapse, as we shall argue, there is a map from non-commutative space–time to ordinary space–time, so that linear quantum theory is compatible with relativity.

However, when collapse is included, this map from non-commutative to ordinary space–time is not possible, and relativistic collapse can be described meaningfully only in a non-commutative space–time. This explains why collapse appears instantaneous and non-local, when seen from the (inadmissible) vantage point of ordinary space–time. We wish to suggest that the apparently acausal nature of collapse is incompatible with special relativity but compatible with the non-commutative space–time implied by a resolution of the problem of time in quantum theory. This aspect is discussed in Section IV.

Our work described in Sections III and IV represents a partial mathematical development of the proposed physical ideas. Considerable more work remains to be done in order to arrive at a complete mathematical model.

16.2 Collapse Models and Their Experimental Tests

The most advanced dynamical collapse model is known as Continuous Spontaneous Localization (CSL) [2, 3, 4, 5]. Its 'mass proportional' version is described by the following stochastic nonlinear modification of the Schrödinger equation, in Fock space:

$$d\psi_t = \left[-\frac{i}{\hbar} H dt + \frac{\sqrt{\gamma}}{m_0} \int d\mathbf{x} (M(\mathbf{x}) - \langle M(\mathbf{x}) \rangle_t) dW_t(\mathbf{x}) \right.$$
$$\left. - \frac{\gamma}{2m_0^2} \int d\mathbf{x} \, (M(\mathbf{x}) - \langle M(\mathbf{x}) \rangle_t)^2 dt \right] \psi_t$$

The first term describes standard Schrödinger evolution, with H being the quantum Hamiltonian. The second and third terms are the new terms which induce dynamical collapse. We note that the new terms are non-unitary, yet they maintain the norm-preserving nature of the evolution. γ is a (positive) new constant of nature which determines the strength of collapse—it must be detected in experiments if CSL is to be confirmed. m_0 is a reference mass, conventionally chosen to be the mass of the nucleon. $M(\mathbf{x})$ is the mass density operator, defined as

$$M(\mathbf{x}) = \sum_j m_j N_j(\mathbf{x}), \tag{16.1}$$

$$N_j(\mathbf{x}) = \int d\mathbf{y}\, g(\mathbf{y} - \mathbf{x}) \psi_j^\dagger(\mathbf{y}) \psi_j(\mathbf{y}), \tag{16.2}$$

Here $\psi_j^\dagger(\mathbf{y})$ and $\psi_j(\mathbf{y})$ are the creation and annihilation operators, respectively, for a particle j at the location $\mathbf{y}$. The smearing function $g(\mathbf{x})$ is given by

$$g(\mathbf{x}) = \frac{1}{\left(\sqrt{2\pi} r_c\right)^3} e^{-\mathbf{x}^2/2r_c^2}, \tag{16.3}$$

where r_c is the second new phenomenological constant of the model, which also should be detected experimentally. This is the length scale to which the collapsed wave function is localized. Stochasticity in the CSL model is described by $W_t(\mathbf{x})$, which is an ensemble of independent Wiener processes, one for each point in space.

The dynamical collapse of the wave function is understood in a straightforward manner by constructing the master equation for the density matrix, where the CSL terms cause decay of the off-diagonal elements. The decay constant is given by

$$\lambda_{\text{CSL}} = \frac{\gamma}{(4\pi r_c^2)^{3/2}}. \tag{16.4}$$

The CSL model assumes the value

$$\lambda_{\text{CSL}} \approx 10^{-17} \text{s}^{-1}. \tag{16.5}$$

This is the most conservative (i.e., smallest) value consistent with phenomenology—it gives a sufficiently high value for the superposition lifetime of a nucleon in two different position states and causes macroscopic superpositions to collapse sufficiently fast. The model proposes a value of about 10^{-5} cm for r_c—values much smaller or much larger than this would contradict observations.

Collapse models are phenomenological, designed for the express purpose of solving the measurement problem and explaining the Born probability rule and explaining why macroscopic objects behave classically. Essentially, because of the nonlinearity, the superposition principle becomes approximate: the lifetime of a quantum superposition (infinite according to the linear theory) is now finite, becoming increasingly smaller with increasing mass. Experiments test CSL by looking for this spontaneous breakdown of superposition, and since violation of superposition has not been observed, this puts bounds on λ_{CSL} and on r_c—essentially the allowed region in the $\lambda_{CSL} - r_c$ plane gets restricted.

The presence of the stochastic noise field in the theory results in a tiny violation of energy-momentum conservation. This leads to an energy gain and heating, which for a particle of mass M is given by the rate [6, 7]

$$\frac{dE}{dt} = \frac{3\lambda_{CSL}}{4} \frac{\hbar^2}{r_c^2} \frac{M}{m_N^2}. \tag{16.6}$$

A class of experiments put important bounds on CSL parameters by looking for this anomalous heating in controlled systems.

16.2.1 Experimental Tests

Collapse models such as CSL make predictions for experiments; these are different from the predictions made by quantum theory because of the introduction of the new non-linear stochastic terms and the two new constants of nature, λ_{CSL} and r_C. The most direct consequence is that the principle of quantum linear superposition is not an exact principle of nature but an approximate one. What this means is that a quantum superposition of states of a microscopic system such as an electron lasts for an astronomically long time, comparable to the age of the Universe. However, the quantum superposition of states of a macroscopic system lasts for an extremely short time interval, and this is what provides a resolution of the quantum measurement problem and explains the quantum-classical transition. Thus if one does a matter-wave interferometry experiment with progressively larger objects, going from electrons to neutrons, atoms, molecules, and ever larger objects, then the fact that one continues to observe the interference pattern (which of course is a consequence of superposition being valid) puts an upper bound on λ_{CSL}. Absence of the interference pattern would confirm CSL and the value of λ_{CSL}, provided all other known sources of decoherence have been ruled out (such as collisional and thermal decoherence). The largest molecule for which an interference experiment has been done to date has a mass of about 10^4 amu, which implies an upper bound $\lambda_{CSL} \sim 10^{-5}$, assuming the plausible value of $r_C \sim 10^{-5}$ cm. It has been suggested that if interferometry experiments can be carried out for a mass all the way up to 10^9 amu, and interference continues to be observed, that will push the upper bound on λ_{CSL} all the way down to (16.5) and essentially rule out the spontaneous collapse model. The state of the art in these experiments and their technological challenges and future prospects are discussed for instance in [1, 8, 9].

Optomechanical experiments of massive objects, typically in the nano-range and with a mass of about 10^{15} amu, aim to create position superpositions of mesoscopic objects by entangling them with light. A challenge for these experiments is that the position separation they achieve in the superposition is much smaller than the favoured value of 10^{-5} cm for r_C [10].

Very promising progress is currently being made on testing CSL through non-interferometric tests, using the heating effect described by Eqn. (16.6). The idea being that if one observes an isolated system for a certain length of time, then the CSL heating effect will leave an imprint, such as an anomalous rise in temperature or an anomalous random walk of an isolated nanosphere or an anomalous noise for which there is no other plausible explanation. There are various ways in this effect can be used to put bounds on CSL. First, the heating would contribute to results of laboratory experiments and astronomical observations that have already been carried out and which are in agreement with the predictions of quantum theory. In order that CSL should not disagree with the already observed result, one puts an upper bound on λ_{CSL}. These bounds come from ionisation of the intergalactic medium, decay of supercurrents, excitation of bound atomic and nuclear systems, absence of proton decay, heating of ultra-cold atoms in Bose-Einstein condensates, and rate of spontaneous 11 keV emission from germanium [7, 11]. This last effect provides the strongest bound to date on the CSL parameter: $\lambda_{CSL} < 10^{-11}$. Very impressive bounds have recently been put from the knowledge of the thermal noise in LIGO and the highly sensitive LISA pathfinder. Proposed tests of anomalous heating include observing the random walk of a micro-object at low temperatures, the anomalous motion of a levitated nanosphere, and spectral broadening [12, 13, 14, 15, 16, 17, 18, 19, 20]. The current bounds on λ_{CSL} and r_C can be found in Fig. 2 of [21].

A landmark recent experiment reports the strongest direct upper bound obtained by measuring the thermal noise in a ultra-cold nano-cantilever cooled to milli-kelvin temperatures [22]. (For the exclusion plot in the $\lambda - r_C$ plane, see Fig. 6.3 of this chapter.) A more refined version of the experiment reports the first possible observation and detection of a 'non-thermal force noise of unknown origin' which is compatible with previously known bounds on CSL [23]. This is perhaps the first time ever that a possible departure from quantum theory has been reported. It remains to be seen if the result will survive further similar tests of CSL.

16.2.2 Relativistic Collapse Models

A major remaining challenge is to construct a satisfactory relativistic quantum field theory of spontaneous collapse. Many serious attempts have been made in this direction [24, 25, 26, 27, 28, 29, 30, 31, 32, 33, 34]. The first such attempt was to convert the CSL equation into a Tomanaga-Schwinger type generalisation, with evolution being described with respect to an arbitrary space-like hypersurface. The problem arises in choosing a Lorentz invariant stochastic noise which couples locally to the quantum fields—typically such a choice leads to divergences, in the form of an infinite rate of energy production. Changing the coupling to a non-local coupling creates other difficulties. Various attempts have been made, and work is still in progress, to resolve the problem posed by divergences. Some of the other chapters in this volume discuss some of these developments in greater detail. In our present chapter, we take the stance that there are fundamental reasons (problem of time, quantum non-locality) because of which we possibly cannot have a relativistic generalisaton of CSL and a revised picture of space–time structure is essential, to make collapse consistent with relativity. We attempt to make the case that collapse is consistent with a non-commutative generalisation of special relativity but not with ordinary special relativity.

16.3 The Problem of Time in Quantum Theory

As in most other physical systems, evolution in time is central to the understanding of quantum systems. The time that is used to define evolution in quantum theory is obviously part of the classical space–time manifold, and the space–time geometry could be either Minkowski or curved—in the latter case there is a non-trivial metric overlying the manifold. One could take these classical aspects (manifold and geometry) as external givens in a quantum theory. However, doing so leads to a problem of principle, which suggests that the present formulation of quantum theory is incomplete.

According to the Einstein hole argument, the physical metric is required in order to give an operational meaning to the space–time manifold. The metric is determined, according to the laws of general relativity, by the classical distribution of macroscopic material bodies and matter fields. But these material bodies are a limiting case of quantum theory. Thus through its need for time, whose operational definition requires a space–time metric and hence the presence of classical objects, the formulation of quantum theory depends on its own classical limit (self-reference). From a fundamental viewpoint, this cannot be considered satisfactory. The classical-quantum divide reminds us of the Copenhagen interpretation, which we know has its shortcomings.

In particular, we could envisage a universe, or an epoch in the history of the universe (such as soon after the big bang), where classical material bodies are absent and matter is entirely microscopic in nature. Such matter fields would not give rise to a classical metric (unless one asserts as a matter of principle that gravity *is* classical no matter what, and/or one invokes a semiclassical theory of gravity, which by itself is problematic). Gravity produced by microscopic matter fields will possess fluctuations, as a result of which one can no longer give physical meaning to the underlying space–time manifold.

We may conclude from the reasoning in the previous two paragraphs that there ought to exist an equivalent reformulation of quantum theory which does not refer to classical time. The reformulation would reduce to ordinary quantum theory as and when the universe is dominated by classical macroscopic objects, so that a classical space–time manifold maybe meaningfully defined [35].

In order to arrive at such a reformulation, we ask as to what kind of space–time geometry might be compatible with quantum theory. The choices are undoubtedly vast, but as a starting point we propose that in analogy with the generic non-commutative nature of observables in quantum theory, space–time coordinates do not commute with each other in the sought-for reformulation. This seems a reasonable demand, if quantum space–time is to be produced by quantum fields, and if the canonical degrees of freedom in the latter have a non-commutative structure. We are thus led to a non-commutative geometry, with its own metric. We try to arrive at a model by suggesting that the reformulation should be invariant under general coordinate transformations of non-commuting coordinates, thus taking a clue from the principle of general covariance.

Ideally, a successful reformulation should also address those other aspects of quantum theory which maybe considered its shortcomings. We have in mind two aspects in particular. The first is that the theory is constructed not from first principles but by 'quantizing' its own classical limit. Thus for instance Poisson brackets from the classical theory are replaced by quantum canonical commutation relations, essentially as an ad hoc recipe. A more fundamental description of quantum theory would motivate the commutation relations in a manner independent of its classical limit. The second has to do with the phenomenological nature of collapse models: the origin of the noise field must ultimately be understood at a fundamental level. Here it is noteworthy that the standard quantum theory part of the collapse model is noise-free, and the fluctuations serve to modify the quantum evolution (negligibly in the micro-limit and significantly in the macro-world). This division of the collapse dynamics between a noise-free part and a noisy part is suggestive of statistical thermodynamics, as if the phenomenological collapse model is the thermodynamic limit of an underlying microscopic theory. From the underlying theory one may derive, after statistical averaging, quantum theory as the state of thermodynamic equilibrium; and the noise part is Brownian motion fluctuations around equilibrium. The theory of Trace Dynamics, developed by Adler and collaborators, goes a long way in providing such an underlying theory. We have made some preliminary attempts in generalizing the ideas of Trace Dynamics to achieve a reformulation of quantum theory without classical time. Much remains to be done, and at this stage we cannot claim that we have necessarily found the right direction for our purpose.

This is not the appropriate place for describing Trace Dynamics (TD) in detail - adequate details are available elsewhere [36, 1, 37, 38]. In brief, TD is the classical dynamics of matrices that exist on a background space–time, and whose elements are complex Grassmann numbers. The matrices could be 'bosonic (B) / fermionic (F),' i.e., their elements are even-grade / odd-grade elements of the Grassmann algebra, respectively. One can construct a polynomial P from these matrices, obtain its trace $\mathbf{P} = TrP$, and define a so-called Trace derivative of $\mathbf{P}$ with respect to a matrix. Given this, a trace Lagrangian can be constructed from the matrices (i.e., operators) $\{q_r\}$ and their time derivatives $\{\dot{q}_r\}$. Then one can develop a classical Lagrangian and Hamiltonian dynamics for the system of matrices, in the conventional way. The configuration variables and their canonical momenta all possess arbitrary commutation relations with each other. However, as a result of the global unitary invariance of the trace Hamiltonian, the theory possesses a remarkable conserved charge made up from the commutators and anti-commutators.

$$\tilde{C} = \sum_B [q_r, p_r] - \sum_F \{q_r, p_r\} \tag{16.7}$$

This charge, which has the dimensions of action, and which we call the Adler-Millard charge, is what makes TD uniquely different from the classical mechanics of point particles. It plays a key role in the emergence of quantum theory from TD. Assuming that one wishes to observe the dynamics at a coarse-grained level, one constructs the statistical mechanics of TD in the usual way, by defining a canonical ensemble and then determining the equilibrium state by maximising the entropy, subject to conservation laws. The canonical average of the Adler-Millard charge takes the form

$$\langle \tilde{C} \rangle_{AV} = i_{eff} \hbar; \qquad i_{eff} = diag(i, -i, i, -i, \ldots) \tag{16.8}$$

where $\hbar$ is a real positive constant with dimensions of action, which is eventually identified with Planck's constant.

Having found the equilibrium state, one can draw important conclusions from it. A general Ward identity (analog of the equipartition theorem in statistical mechanics) is derived as a consequence of invariance of canonical averages under constant shifts in phase space. Its implications are closely connected with the existence of the Adler-Millard charge and its canonical average (16.8). After applying certain realistic assumptions, the canonical averages of the TD operators are shown to obey Heisenberg equations of motion and the canonical commutation relations of quantum theory. In this sense, upon the identification of canonical averages of TD with Wightman functions in quantum field theory, one finds quantum theory emergent as a thermodynamic approximation to TD. The passage from the Heisenberg picture to the Schrödinger picture is made in the standard manner to arrive at the non-relativistic Schrödinger equation.

Next, one takes account of the ever-present thermal fluctuations around equilibrium; in particular there are fluctuations in the Adler-Millard charge about its canonical value (16.8). As a result, the Schrödinger equation picks up (linear) stochastic correction terms. If one invokes the assumptions that evolution should be norm-preserving in spite of the stochastic corrections and that there should be no superluminal signalling, the modified Schrödinger equation becomes non-linear and has the generic structure of a spontaneous collapse model. One can demonstrate dynamic collapse which obeys the Born probability rule. Of course the theory is not well-developed enough at this stage to uniquely pick out the CSL model or to predict the numerical values of the constants λ_{CSL} and r_c. Moreover, the assumption of norm preservation in the presence of stochastic corrections is ad hoc. Norm preservation should follow from a deeper principle, yet to be discovered. Another noteworthy feature is that in order to define a thermal equilibrium state, one is compelled to pick out a special frame of reference, possibly the cosmological rest frame of the cosmic microwave background.

In spite of important issues which remain to be resolved, Trace Dynamics is a significant example of an underlying theory from which quantum theory and collapse models are emergent. We have attempted to extend the ideas of TD to arrive at a reformulation of quantum theory without classical time by raising time and space coordinates to the level of non-commuting matrices, as is done in TD for the canonical degrees of freedom. This program has met with partial success, in that it could be implemented for Minkowski space–time. The case for a curved space–time remains to be developed, although a heuristic outline is available [39].

On a non-commutative Minkowski space–time, with $(\hat{t}, \hat{x}, \hat{y}, \hat{z})$ as non-commuting operators having arbitrary commutation relations, we define a trace proper time as follows [40]:

$$ds^2 = Tr d\hat{s}^2 \equiv Tr[d\hat{t}^2 - d\hat{x}^2 - d\hat{y}^2 - d\hat{z}^2]. \quad (16.9)$$

This line-element can be shown to be invariant under Lorentz transformations. Matter degrees of freedom 'live' on this non-commutative space–time and one can define a Poincarè invariant dynamics, by first introducing a four-vector $\hat{x}^\mu = (\hat{t}, \mathbf{x})$ and defining four velocity as $\hat{u}^\mu = d\hat{x}^\mu/ds$. Lagrangian and Hamiltonian dynamics can then be constructed in the spirit of Trace Dynamics, using the trace proper time to define evolution. As before, there exists in the theory, as a consequence of global unitary invariance, a conserved Adler-Millard charge for the matter degrees of freedom $\hat{y}^\mu$

$$\hat{Q} = \sum_{r \in B}[\hat{y}_r, \hat{p}_r] - \sum_{r \in F}\{\hat{y}_r, \hat{p}_r\} \qquad (16.10)$$

The important generalisation is that there is associated, with every degree of freedom, apart from the canonical pair $(\hat{q}, \hat{p})$, a conjugate pair $(\hat{E}, \hat{t})$, where $\hat{E}$ is the energy operator conjugate to $\hat{t}$. The metric given, although it is Lorentz invariant, does not admit a light-cone structure or point structure of ordinary space–time. This is what allows the recovery of a quantum theory without classical time.

From this point on, the construction parallels that in TD, namely an equilibrium statistical thermodynamics of the underlying classical theory is constructed, with coordinate time $\hat{t}$ now an operator and evolution being described with respect to trace proper time s. The quantum commutators emerge at the thermodynamic level, with the added feature that there is now an energy-time commutator as well. In the non-relativistic limit, one obtains the generalised Schrödinger equation

$$i\hbar \frac{d\Psi}{ds} = H\Psi(s) \qquad (16.11)$$

where the configuration variables now include also the time operator $\hat{t}$. It is important to emphasize that the configuration variables commute with each other. We call this a Generalised Quantum Dynamics (GQD), in which there is no classical space–time background, and all matter and space–time degrees of freedom have operator status. This is the sought-for reformulation of quantum theory which does not refer to classical time [41].

We now provide a heuristic picture as to how standard quantum theory on a classical space–time background is recovered from here [39]. For that, we need to first understand how GQD can give rise to a classical space–time. As we argued earlier, we expect a classical space–time to exist only when the universe is dominated by classical macroscopic objects. Consider now the role of statistical fluctuations of the Adler-Millard charge about equilibrium, which modify the Schrödinger equation into a non-linear equation, as discussed. And we consider the epoch of the very early universe, when tiny primordial matter density perturbations are present, which subsequently grow into large-scale structure. As these perturbations grow in magnitude, the associated statistical fluctuations of the non-linear stochastic Schrödinger equation become more and more significant, thereby eventually bringing about the classicalisation of these density perturbations, in the spirit of collapse models. The localization takes place not only in space, but also in time! The time operator associated with every object becomes classical, by which we mean that it takes the form of a c-number times a unit matrix. Moreover, the statistical fluctuations associated with the operator space–time degrees of freedom also become more and more significant, leading to the emergence of a classical space–time.

We conclude that the localisation of macroscopic objects occurs in conjunction with the emergence of a classical space–time. This is consistent with the Einstein hole argument, namely that classical matter fields and the gravitational metric which they produce are both required in order to give physical meaning to the point structure of space–time. Only when the Universe is dominated by macroscopic objects, as is true for today's Universe, can we assume the existence of a classical space–time. In this approximation, the trace proper time s of the generalised TD can be identified with ordinary classical proper time. Once the Universe achieves such a classical state, it sustains itself therein, because of continual action of stochastic fluctuations on macroscopic objects, thus concurrently achieving the existence of a classical space–time geometry. Because the underlying generalised trace dynamics is Lorentz invariant, the emergent classical space–time is also locally Lorentz invariant. There is however a fundamental difference: in contrast with the underlying theory, in the classical approximation, light-cone structure and causality are approximate emergent features, since the space–time coordinates are now c-numbers.

We may now argue how standard quantum theory for a microscopic system emerges from GQD once a classical background universe with its classical space–time, is given. Regardless of whether there is a classical space–time background, a microscopic system is described at a fundamental level via its non-commutative space–time (16.9), through the associated generalised TD. Subsequent to coarse-graining and construction of the equilibrium statistical thermodynamics, this leads to the system's GQD (16.11) with its own trace time. In the approximation that the stochastic fluctuations can be ignored, this GQD possesses commuting $\hat{t}$ and $\hat{\mathbf{x}}$ operators. As a consequence of their commutativity, these can be mapped to the c-number t and $\mathbf{x}$ coordinates of the pre-existing classical universe, and trace time can then be mapped to ordinary proper time. This is hence a mapping to ordinary space–time, and one recovers standard quantum mechanics in this manner. If this program can be constructed rigorously, it will explain how ordinary quantum theory is recovered from a reformulation which does not depend on classical time. Statistical fluctuations and collapse models then explain the classical limit of quantum mechanics, as before, through the stochastic non-linear Schrödinger equation.

We regard it as a profound inference that the dynamical collapse of the wave function is intimately and directly responsible for the emergence of a classical space–time. Collapse of the wave function is not just an aspect concerning the interaction of a quantum system with a measuring apparatus in the laboratory. Rather, wave-function collapse is responsible for the emergence of classical objects in the universe and is hence also responsible for the emergence of classical space–time. This appears to be a robust and intuitive inference, independent of the mathematical details of trace dynamics. It also appears plausible because it is collapse which localises macroscopic objects, thereby creating space as that is 'between objects.' It does not appear meaningful to talk of space if there were no collapse and macroscopic objects were delocalized and 'everything was everywhere' so to say: what use would the concept of space be in that case? This possibly helps us understand another strange feature of collapse: the wave function, satisfying the Schrödinger equation, lives in a Hilbert space, not in physical space. On the other hand, the so-called jump operator of collapse models maps the wave function to a localized position in physical space: there is somehow a mysterious connection between the complex wave function living in Hilbert space, and the real point in physical space to which the particle being described by the wave function gets localized. This apparent mystery can be removed if we think of the collapse process as taking place in the non-commutative space of the underlying trace dynamics. The process is then described in the Hilbert space of the states in the underlying generalised trace dynamics, which includes the non-commutative space, to which the physical space is an emergent approximation. The state of the system, fundamentally speaking, does not jump from the Hilbert space to physical space; rather it always lives in the non-commutative space, before the jump, as well as after the jump.

As an illustration, let us consider the famous double slit experiment with electrons. Electrons are released one by one from the electron gun; each electron passes through the slits and, upon reaching the screen, collapses to one or the other position, in accordance with the Born probability rule. Once a large number of electrons have collected on the screen, they are distributed as per the Born rule, forming the interference pattern. What happens to an electron after it leaves the gun and before it reaches the screen? There is really no satisfactory description of 'what happens' in standard quantum theory. The electron is supposed to behave like a wave and go through both slits, and the two wavelets (one from each screen) interfere with each other. But what is this wave a wave of? It cannot possibly be the wave function—which is a complex valued quantity, which lives in Hilbert space, not in physical space. Often it is suggested that the wave is a probability wave - that also seems strange; because then we do not seem to attach any physical reality to the wave function. Comprehending the situation becomes easier if we think of this entire process as taking place in the underlying non-commutative space, to which ordinary space is an approximation. The complex-valued description of the state as a 'wave' appears natural if we think of it as a wave in 'non-commutative' space. Problems arise only when we try to describe quantum phenomena on a classical space–time background—the two are in conflict with each other; classical vsg is external to quantum theory and not really a part of a fundamental description of quantum theory.

We conclude that in our study, the problem of time and the problem of measurement are closely related. Suppose we start from a formulation of quantum theory which does not refer to classical time. Then, so as to recover classical time and space–time geometry from this reformulation, we also need to recover the macroscopic limit of matter fields. This is because classical geometry and classical matter fields co-exist. Furthermore, to explain the classical behaviour of macroscopic objects is the same as solving the quantum measurement problem. This is because the latter problem can be stated differently as follows: why are macroscopic objects never observed in superposition of position states? The measurement problem is a small part of a larger problem: how does the classical structure of space–time and matter degrees of freedom arise from an underlying quantum theory of matter and space–time? We have not yet addressed the issue of the gravitational field on non-commutative space–time, nor whether one is constrained to select a specific frame of reference when studying statistical fluctuations and collapse in GQD.

16.4 Quantum Non-locality and Space-time Structure

Non-local quantum correlations cannot be used for superluminal signalling; nonetheless they suggest an acausal influence amongst the entangled particle pairs when a measurement is made on one of them. It is as if the pair behaves like a rigid body, even if the two particles lie outside each other's light cone. To many researchers, this calls for the need for an explanation, even though there is no known experimental conflict between quantum theory and special relativity. Since the influence arises only when a measurement is made and hence wave function collapse takes place, it is natural to expect that an explanation for the non-local influence must be tied up with the explanation for wave-function collapse.

A possible explanation has been suggested by Adler, in the context of trace dynamics and the related explanation for wave-function collapse in terms of the statistical fluctuations of the Adler-Millard charge: these modify the Schrödinger equation to a stochastic non-linear equation. Since the stochastic terms are non-linear, they do result in violation of local causality, as changes in the wave function at one point in space are instantaneously communicated to every other point. However, the average of the stochastic terms over the density matrix obeys a linear equation, thus rendering superluminal signalling impossible. Adler also notes that when the statistical fluctuations are taken into account (i.e., when one goes beyond the linear quantum theory), it becomes necessary to choose a specific inertial frame of reference (so as to meaningfully define the canonical ensemble). This choice of frame, which is possibly the cosmological rest frame of the CMB, breaks Lorentz invariance and collapse is hence not a relativistically invariant process. This could be one possible reason why it is difficult to make a relativistic theory of collapse.

While at some level this maybe an adequate explanation, one might still be left wondering if 'instantaneous communication' across vast expanses of space can be regarded as physically reasonable. There is also the earlier question of the complex wave function living in Hilbert space and how it could be responsible for such communication in physical space. Perhaps the existence of such communication might appear more reasonable if there was a quantum notion of 'zero physical distance' as opposed to the enormous spatial distance that we perceive in ordinary space, over which the entangled pair 'communicates.' In the context of the scheme developed in the present work, we offer such an explanation as follows [42].

We explained earlier how a classical space–time emerges from the GQD in conjunction with the domination of the universe by macroscopic objects. Given this background universe, let us consider the entangled EPR pair of particles. So long as the pair is in flight, its evolution is described by the Schrödinger equation, which is equivalent to the GQD given by Eqn. (16.11). This equivalence is possible because, in the absence of statistical fluctuations (which become significant only at the onset of collapse), the commuting operator coordinates $(\hat{t}, \hat{\mathbf{x}})$ can be mapped to ordinary space–time coordinates $(t, \mathbf{x})$. However, when one of the entangled particles in the pair interacts with the classical apparatus, and the measurement is done, the collapse-inducing stochastic fluctuations of the Adler-Millard charge become significant. By implication, fluctuations in the space–time operators $\hat{t}, \hat{\mathbf{x}}$ associated with the quantum system become important. These operators hence carry information about the arbitrary commutation relations of the generalised TD and therefore they no longer commute with each other. This implies that these operators cannot be mapped to the space–time coordinates $(t, \mathbf{x})$ of special relativity. As a result, one can define simultaneity only with respect to the trace time s, and there is no relativistic theory of wave-function collapse, because it is not possible to describe the collapse process on a classical space–time background.

In this scenario, collapse and the non-local quantum correlation takes place only in the non-commutative space–time (16.9), which is devoid of point structure, light-cone structure, and also devoid of the notion of spatial distance. Therefore we can only say that collapse and the so-called influence from one particle to the other takes place at a specific trace time, which is Lorentz invariant, and it is not physically meaningful to talk of an instantaneous influence travelling across physical space, nor should we call the correlation non-local. In this scenario, the entangled pair does not know distance. The state can be meaningfully described only in the non-commutative space–time, as in the case of the double-slit experiment. We once again conclude that removing classical space–time from quantum theory removes another of its peculiarities, i.e., the so-called spooky action at a distance.

If we attempt to view and describe the quantum measurement on the entangled quantum pair from the viewpoint of the Minkowski space–time of ordinary special relativity, then the process appears to violate local causality. However, such a description cannot be considered valid, because there does not exist a map from the fluctuating and non-commuting $(\hat{t}, \hat{\mathbf{x}})$ operators to the commuting t and $\mathbf{x}$ space–time coordinates of ordinary special relativity. No such map exists even in the non-relativistic case. However, in the non-relativistic case, because there exists an absolute time, it is possible to model the statistical fluctuations as a stochastic field on a given space–time background. This is what is done in collapse models: collapse is instantaneous in this absolute time, but it does not violate local causality.

Conclusions

Dynamical collapse models are phenomenological, and there ought to exist an underlying theory from which they are emergent. The spooky action at a distance in non-local quantum correlations possibly suggests a need for revising our understanding of space–time structure. We have argued that resolving the problem of time in quantum theory suggests that space and space–time are emergent concepts, resulting from dynamical collapse of the wave function of macroscopic objects in the universe. The underlying non-commutative structure of space–time makes it much easier to understand the apparently non-local influence in EPR correlations. We are suggesting that many of the apparently puzzling features of quantum theory arise because we attempt to describe the theory on a classical space–time background. This choice of background is unsatisfactory and inadequate for describing certain quantum phenomena, because it is approximate and external to quantum theory. We hope that further work will put these heuristic ideas on a firm mathematical footing, and also suggest experiments to test these ideas.

Acknowledgement

I have benefitted greatly from useful discussions with my collaborators Angelo Bassi, Hendrik Ulbricht, Saikat Ghosh, Shreya Banerjee, Srimanta Banerjee, Sayantani Bera, Suratna Das, Sandro Donadi, Suman Ghosh, Kinjalk Lochan, Seema Satin, Priyanka Giri, Navya Gupta, Bhawna Motwani, Ravi Mohan, and Anushrut Sharma.

References

[1] A. Bassi, K. Lochan, S. Satin, T. P. Singh, and H. Ulbricht. Models of wave function collapse, underlying theories, and their experimental tests. *Rev. Mod. Phys.*, 85:471, 2013.

[2] Angelo Bassi and GianCarlo Ghirardi. Dynamical reduction models. *Phys. Rep.*, 379: 257–426, 2003. ISSN 0370-1573.

[3] P. Pearle. Combining stochastic dynamical state-vector reduction with spontaneous localization. *Phys. Rev. A*, 39:2277–2289.

[4] GianCarlo Ghirardi, Alberto Rimini, and Tullio Weber. Unified dynamics for microscopic and macroscopic systems. *Phys. Rev. D*, 34:470–491, 1986.

[5] GianCarlo Ghirardi, Philip Pearle, and Alberto Rimini. Markov processes in Hilbert space and continuous spontaneous localization of systems of identical particles. *Phys. Rev. A*, 42:78–89, 1990. ISSN 1050-2947.

[6] P. Pearle and E. Squires. Bound state excitation, nucleon decay experiments and models of wave function collapse. *Phys. Rev. Lett.*, 73:1–5, Jul 1994. doi: 10.1103/PhysRevLett.73.1.

[7] Stephen L. Adler. Lower and upper bounds on CSL parameters from latent image formation and IGM heating. *J. Phys. A*, 40:2935–2957, 2007. ISSN 1751-8113.

[8] M. Arndt and K. Hornberger. Testing the limits of quantum mechanical superpositions. *Nature Physics*, 10:271, 2014.

[9] C. Wan, M. Scala, G. W. Morley, ATM. A. Rahman, H. Ulbricht, J. Bateman, P. F. Barker, S. Bose, and M. S. Kim. Free nano-object Ramsey interferometry for large quantum superpositions. *Phys. Rev. Lett.*, 117:143003, 2016.

[10] M. Aspelmeyer, T. J. Kippenberg, and F. Marquardt. Cavity optomechanics. *Rev. Mod. Phys.*, 86:1391, 2014.

[11] F. Laloe, W. J. Mullin, and P. Pearle. Heating of trapped ultracold atoms by collapse dynamics. *Phys. Rev. A*, 90:052119, 2014.

[12] B. Collett and P. Pearle. Wavefunction collapse and random walk. *Found. Phys.*, 33:1495, 2003.

[13] S. Bera, B. Motwani, T. P. Singh, and H. Ulbricht. A proposal for the experimental detection of csl induced random walk. *Scientific Reports*, 5:7664, 2015.

[14] M. Rashid, T. Tommaso, J. Bateman, J. Vovrosh, D. Hempston, M. S. Kim, and Ulbricht. H. Experimental realisation of a thermal squeezed state of levitated optomechanics. *Phys. Rev. Lett.*, 117:273601, 2016.

[15] M. Bahrami, A. Bassi, and H. Ulbricht. Testing the quantum superposition principle in the frequency domain. *Phys. Rev. A*, 89:032127, 2014.

[16] J. Millen, P. Z. G. Fonesca, T. Mavrogordatos, T. S. Monteiro, and P. F. Barker. Cavity cooling a single charged levitated nanosphere. *Phys. Rev. Lett.*, 114:123602, 2015.

[17] M. Bahrami, M. Paternostro, A. Bassi, and H. Ulbricht. Non-interferometric test of collapse models in optomechanical systems. *Phys. Rev. Lett.*, 112:210404, 2014.

[18] D. Goldwater, M. Paternostro, and P.F. Barker. Testing wavefunction collapse models using parametric heating of a trapped nanosphere. *Phys. Rev. A*, 94:010104, 2015.

[19] M. Billardello, A. Trombettoni, and A. Bassi. Collapse in ultracold Bose Josephson junctions. arXiv:1612.07691 [quant-ph], 2016.

[20] Y. Li, A. M. Steane, D. Bedingham, and G. A. D. Briggs. Detecting continuous spontaneous localisation with charged bodies in a Paul trap. arXiv:1605.01881 [quant-ph], 2016.

[21] M. Carlesso, A. Bassi, P. Falferi, and A. Vinante. Experimental bounds on collapse models from gravitational wave detectors,. *Phys. Rev. D*, 94:124036, 2016.

[22] A. Vinante, M. Bahrami, A. Bassi, O. Usenko, G. Wijts, and T. H. Oosterkamp. Upper bounds on spontaneous wave-function collapse models using millikelvin-cooled nanocantilevers. *Phys. Rev. Lett.*, 116:090402, 2016.

[23] A. Vinante, R. Mezzena, P. Falferi, M. Carlesso, and A. Bassi. Improved noninterferometric test of collapse models using ultracold cantilevers. *Phys. Rev. Lett.* 119, 110401 (2017).

[24] GianCarlo Ghirardi, Renata Grassi, and Philip Pearle. Relativistic dynamical reduction models: general framework and examples. *Found. Phys.*, 20:1271–1316, 1990. ISSN 0015-9018.
[25] P. Pearle. In A. I. Miller, editor, *Sixty-Two Years of Uncertainity: Historical Philosophical, and Physics Inquires into the Foundations of Quantum Physics*. Plenum, New York, 1990.
[26] Oreste Nicrosini and Alberto Rimini. Relativistic spontaneous localization: A proposal. *Foundations of Physics*, 33:1061–1084, 2003.
[27] P. Pearle. How stands collapse II. In Wayne C. Myrvold and Joy Christian, editors, *Quantum Reality, Relativistic Causality, and Closing the Epistemic Circle*, volume 73 of *The Western Ontario Series in Philosophy of Science*, pages 257–292. Springer Netherlands, 2009.
[28] P. Pearle. Relativistic collapse model with tachyonic features. *Phys. Rev. A*, 59: 80–101.
[29] Daniel J. Bedingham. Relativistic state reduction dynamics. *Found. Phys.*, 41:686 [arXiv:1003.2774], 2011.
[30] Daniel J. Bedingham. Relativistic state reduction model. arXiv:1103.3974, 2011.
[31] Roderich Tumulka. A relativistic version of the Ghirardi–Rimini–Weber model. *Journal of Statistical Physics*, 125:821–840, 2006.
[32] Fay Dowker and Joe Henson. Spontaneous collapse models on a lattice. *Journal of Statistical Physics*, 115:1327–1339, 2004.
[33] Fay Dowker and Isabelle Herbauts. Simulating causal wavefunction collapse models. *Classical and Quantum Gravity*, 21:2963, 2004.
[34] D. Bedingham. Collapse models and their symmetries. arXiv:1612.09470 [quant-ph], 2016.
[35] T. P. Singh. Quantum mechanics without spacetime: a case for noncommutative geometry [arxiv:gr-qc/0510042]. *Bulg. J. Phys.*, 33:217, 2006.
[36] Stephen L. Adler. *Quantum theory as an emergent phenomenon*. Cambridge University Press, Cambridge, 2004. ISBN 0-521-83194-6.
[37] S. L. Adler. Generalized quantum dynamics. *Nucl. Phys. B*, 415:195, 1994.
[38] S. L. Adler and A. C. Millard. Generalized quantum dynamics as pre-quantum mechanics. *Nucl. Phys. B*, 473:199, 1996.
[39] T. P. Singh. The problem of time and the problem of quantum measurement. In Thomas Filk and Albrecht von Muller, editors, *Re-thinking Time at the Interface of Physics and Philosophy: The Forgotten Present*, (arXiv:1210.81110). Berlin-Heidelberg: Springer, 2015.
[40] K. Lochan and T. P. Singh. Trace dynamics and a noncommutative special relativity. *Phys. Lett. A*, 375:3747, 2011.
[41] K. Lochan, Seema Satin, and T. P. Singh. Statistical thermodynamics for a noncommutative special relativity: Emergence of a generalized quantum dynamics. *Found. Phys.*, 42:1556, 2012.
[42] S. Banerjee, S. Bera, and T. P. Singh. Quantum nonlocality and the end of classical spacetime. *Int. J. Mod. Phys.*, 25:1644005, 2016.

17

The Weight of Collapse: Dynamical Reduction Models in General Relativistic Contexts

ELIAS OKON AND DANIEL SUDARSKY

Inspired by possible connections between gravity and the foundational question in quantum theory, we consider an approach for the adaptation of objective collapse models to a general relativistic context. We apply these ideas to a list of open problems in cosmology and quantum gravity, such as the emergence of seeds of cosmic structure, the black hole information issue, the problem of time in quantum gravity, and, in a more speculative manner, to the nature of dark energy and the origin of the very special initial state of the universe. We conclude that objective collapse models offer a rather promising path to deal with all of these issues.

17.1 Introduction

The project of constructing a quantum theory of gravity is often regarded as entirely independent of the one devoted to clarifying foundational questions within standard quantum theory—the latter, a task mostly performed in the non-relativistic domain. There are, however, very suggestive indications that the two topics are intimately connected. To begin with, the standard interpretation of quantum mechanics, crucially dependent on the notions of *measurement* or *observer*, seems ill suited to be applied to non-standard contexts, such as radiating black holes or the early universe. With this kind of scenarios in mind, observer-independent versions of quantum mechanics, such as Bohmian mechanics ([1]) or objective collapse models ([2, 3]), appear to be better options.[1] Moreover, throughout the years, people like R. Penrose or L. Diósi, among others, have uncovered intriguing connections between quantum-foundational issues and the quantum-gravity interface that are worth taking seriously (see, e.g., [12, 13, 14, 15, 16, 17, 18, 19, 20]).

---

[1] Everettian theories inspired by [4] and the consistent histories approach developed in [5, 6, 7] also aim at constructing observer-independent formalisms. Unfortunately, at least for now, both of these programs seem to be plagued by insurmountable problems (see e.g., [8, sec. 4] and [9, 10, 11]).

In this chapter we describe a line of research[2] strongly influenced by this type of ideas and show that it has substantial promise in addressing various open issues, confronting not only the quantum-gravity interface in general, but also questions usually thought to belong solidly to the domain of quantum gravity research. The program explores the way in which objective collapse models (also known as dynamical reduction theories) can be used in dealing with a list of open issues in cosmology and quantum gravity. In particular, we have argued that objective collapse theories can help in:

[1] Solving a critical conceptual problem with the inflationary account of the emergence of seeds of cosmic structure.
[2] Revising the expectation that inflation will give rise to, by now detectable, B-modes in the cosmic microwave background (CMB).
[3] Explaining dark energy in terms of energy non-conservation.
[4] Diffusing the black hole information puzzle.
[5] Dealing with the problem of time in quantum gravity.
[6] Shedding light on the origin of the second law of thermodynamics and the very special initial state of the universe.

An obvious initial problem with all this, though, is that most well-developed collapse models, such as GRW of CSL, are non-relativistic; and while at least three relativistic versions have been proposed recently ([38, 39, 40]), none of them, as they stand, seems adequate for a general relativistic context. Therefore, as we will see, part of the project we are describing has focused on developing methods and ideas in order to adapt objective collapse models for scenarios involving gravity.

Before getting into the details of this and related issues, it is convenient to say a few words regarding the philosophy behind the research line we have followed, in which we tackle the exploration of the interface between general relativity and quantum theory in a *top-down* approach. In the usual *bottom-up* quantum gravity programs (e.g., string theory, loop quantum gravity, causal sets, dynamical triangulations, etc.), one starts by assuming one has constructed a fundamental theory of quantum gravity and attempts to connect it to regimes of interest in the "world out there," such as cosmology or black holes. In the *top-down* approach, in contrast, one tries to push existing, well-tested theories to address open issues that seem to lie just beyond their domain of applicability—and in doing so, one considers introducing suitable modifications of the theoretical framework. The hope is for this latter approach to lead to the formulation of effective theories that could provide clues about the nature of more fundamental theories, such as quantum gravity. We see

---

[2] Papers in which this project has been developed include: [21, 22, 23, 24, 25, 26, 27, 28, 29, 30, 31, 32, 33, 34, 35, 36, 37].

this path as emulating the development of quantum theory itself, which started with *ad hoc* additions to the classical theory that were latter employed in the construction of the general formalism. The idea, then, is to push general relativity and quantum field theory in curved space–time into realms often deemed to be beyond their reach and to consider questions usually not explored in such contexts.

This chapter is organized as follows. In section 17.2 we present a brief discussion of possible connections between gravity and foundational questions in quantum theory. Section 17.3 is devoted to the task of adapting objective collapse models to a general relativistic setting. In section 17.4 we consider the application of these ideas to inflationary cosmology, the dark energy issue, the black hole information puzzle, the problem of time in quantum gravity, and the problem of explaining the special nature of the initial state of the universe. We wrap up with some final thoughts in section 17.5.

17.2 Gravity and the Foundations of Quantum Mechanics

As has been recounted many times before, one of the most notable moments during the Einstein-Bohr debates occurred when Bohr surprisingly used considerations based on general relativity in order to counter one of the best challenges Einstein managed to present against the then-new quantum theory (see [41]). That certainly was a sensational knock out by Bohr, which not only showed that he was a serious intellectual contender for Einstein but contributed to the widespread perception that, regarding the general discussion concerning the suitability of quantum mechanics as a basic theory of the micro-world, Bohr carried the day. The resurgence of interest in foundational questions of quantum theory, largely motivated by the work of J. Bell (see [42]) and reflected by the increasing number of conferences and workshops devoted to the subject (and, indeed, the fact that this very book is being written), attest to the fact that Bohr's victory that day was not the end of the story.

As we mentioned in the introduction, even though (in spite of Bohr's move!) work in quantum gravity is often regarded as independent of foundational questions within quantum theory, there are very suggestive indications that the two issues might be intimately connected. Of course, chief among the conceptual problems of quantum theory is the so-called *measurement problem*, which roughly corresponds to the fact that the standard formulation of quantum theory crucially depends on the concept of measurement, even though such notion is never formally defined within the theory. With standard laboratory applications in mind, this might not seem that problematic (one might consider, for instance, adopting an instrumentalist point of view). However, when the system under study is the whole universe, the inadequacies of such an approach become quite evident: in such case, there is nothing outside the system that could play the role of an observer. It is clear, then, that if one wants to apply quantum theory in such scenarios, it is essential to employ an alternative to standard quantum mechanics designed to deal with the measurement problem (see [43, 44] for similar assessments).

Despite the fact that many interesting alternatives to the standard interpretation of quantum mechanics have been proposed, a fully satisfactory solution to the measurement problem, particularly one that is adequate for the relativistic domain, has not yet been found.[3] Such fact has not precluded quantum mechanics of displaying a spectacular predictive power. This, in turn, has led to a widespread pragmatic attitude in physics regarding the conceptual problems of quantum mechanics and, naturally, people working on different quantum gravity programs, such as string theory or loop quantum gravity, have inherited this attitude. Nevertheless, it is quite possible for the solutions to both of these issues to be related: since we already have managed to develop successful quantum theories of all other known interactions, the solution to the measurement problem may well lie in the quantum theory that is still lacking, that is, quantum gravity. On the other hand, the numerous failed attempts to construct a theory of quantum gravity, working under the assumption that the standard interpretation is correct, suggest that it may be necessary to solve the measurement problem in order to achieve the goal of building a quantum theory of gravity.

Intriguing connections between quantum-foundational issues and the quantum-gravity interface have been pointed out throughout the years. For instance, in [13], Penrose considers a large box containing a black hole in thermal equilibrium with a surrounding environment and argues that, since different initial states give rise to the same final one (due to the no-hair theorems), the black hole causes trajectories in phase space to converge. He claims, however, that this loss of phase space volume has to be compensated by some kind of bifurcation in the behavior of the non-black hole region and suggests that such bifurcations could be identified with the multiplicity of possible outcomes resulting from a quantum measurement (where several outputs may follow from the same input). This, together with alleged intrinsic space–time instabilities when macroscopic bodies are placed in quantum superpositions of different locations, lead him to conjecture in [46] a link between quantum collapses and gravity.

Following a similar line of thought, in [14, 47] it is noted that solutions to the Newton–Schrödinger equation that represent quantum particles as self-gravitating objects could, in turn, be natural states that result from a kind of spontaneous self-localization due to the collapse of the wave function. More recently, in [48], an interesting proposal has been put forward that seems to allow for a self-consistent,

---

[3] In fact, as argued in [45], one faces serious problems just by attempting to extend the standard interpretation of quantum theory to the quantum field theory context.

semiclassical treatment in which classical, Newtonian gravity is sourced by the collapsing quantum states of matter (see also [49, 50]). Additionally, in [17] it is suggested for the gravitational curvature scalar to be the field that causes the spontaneous collapses in objective collapse models, and in [18] it is argued that the major conceptual problems of canonical quantum gravity, namely, the problem of time and the problem of diffeomorphism invariant observables, are automatically solved by employing a Bohmian version of quantum mechanics.

Another recent work in which connections between gravity and quantum foundations are allegedly uncovered is [51], where it is argued that a universal, gravity-induced type of decoherence explains the quantum-to-classical transition of systems with internal degrees of freedom. However, in [52, 53], we rebut the claim that gravitation is responsible for the reported effect, we contest the fact that the effect is universal, and, finally, we challenge the ability of decoherence to explain the quantum-to-classical transition.[4] From all this, we conclude that gravity does not help account for the emergence of classicality as claimed.

An important issue within the foundations of quantum mechanics is the ontological question of "what is reality made of according to quantum mechanics?" (see for instance [54]). It is worth noting that answers to such question are often framed in a context where the existence of a classical space–time is taken for granted. That is, things are postulated to populate a classical space–time, which is assumed to exist independently of the rest of the stuff that the theory is supposed to refer to (see, e.g., [55]). Of course, in contexts in which one is considering a quantum description of space–time itself, the issue can no longer be addressed in this standard form (for example, in such a context, one cannot only say that what exists are Bohmian particles or a mass distribution, as both of them require a background space–time to give them their full meaning). The point is that discussions regarding ontology within a quantum gravity context will probably require a substantial reformulation and even a change of language and concepts before they can be fully addressed. It is clear, though, that at the level of the discussion of the present work, immersed in a semi-classical setting, we need to content ourselves with one of the various ontological pictures developed in a non-gravitational context and take it only as a partial and effective answer to the issue.

It is worth noting at this point that, in contrast to what is often assumed, the interface between quantum theory and gravitation need not involve the Planck regime. Consider, for instance, trying to describe the space–time associated with a macroscopic body in a quantum superposition of being in two distinct locations. Clearly, such a scenario implicates both gravity and quantum mechanics, without

---

[4] See [35, sec. 2] for a deeper analysis of why decoherence does not explain the quantum-to-classical transition or help in the resolution of important foundational problems in quantum theory.

ever involving Planck scale physics. In fact, [56] describes an experiment devised to explore precisely such a situation whose results are taken as evidence against semiclassical general relativity. First, they argue that if quantum states undergo standard (Copenhagen-style) collapses, then the semiclassical Einstein equations are inconsistent because during collapses, the energy-momentum tensor is not conserved. That leads them to adopt an Everettian no-collapse interpretation, which, however, they claim is in conflict with the performed experiment. Either way, they conclude that semiclassical general relativity is not viable. Another well-known argument against a semiclassical theory is offered in [57], where a thought experiment is used to defend the claim that a semiclassical scenario leads to violations of either momentum conservation or the uncertainty principle, or to faster-than-light signaling.

These results, like many other in the literature that try to reach the same conclusion, namely, that semiclassical general relativity is not an option, seem rather strong. However, as has been shown repeatedly in, e.g., [58, 59, 60, 61], none of them is really conclusive. For example, regarding [56], it is clear that its argument relies on Copenhagen or Many Worlds being the only alternatives for interpreting the quantum formalism, but neither seems to be viable for the scenario considered. As for [57], it has been shown that the proposed device is impossible to build and, even if built, would not actually lead to a no-go theorem for a semiclassical formalism. In this project we work in a semiclassical setting; however, we do so keeping in mind that such description should not be taken as fundamental, but only as a good approximation, valid under appropriate circumstances (see, e.g., [62]).[5] As we see in what follows, this allows us not only to address a number of open issues is cosmology and quantum gravity but also to search for clues regarding a more fundamental description of space–time.

Most people working on quantum gravity expect general relativity to be replaced by a theory of quantum gravity that will, among other things, "cure the singularities" of the former. However, our current understanding regarding how this replacement is going to work is still filled with serious gaps that offer enough room to accommodate a wide range of ideas. A key unresolved conceptual difficulty in quantum gravity research is the problem of recovering a classical space–time out of a theory that does not contain such a concept at the fundamental level. Many people have suggested that space–time could be an emergent phenomenon ([63, 64, 65]) and many others) and, if so, just as the thermodynamic notion of heat has no clear counterpart at the statistical mechanics level of description, space–time concepts

---

[5] In fact, most attempts at connecting bottom-up approaches to quantum gravity with "real-world" scenarios rely on a semiclassical approach, in which matter is described quantum mechanically but space–time is described classically.

could only become meaningful at the classical level; and even in some of the schemes where the geometric degrees of freedom do appear at the fundamental level (e.g., causal sets and loop quantum gravity), the recovery of standard space–time concepts often is possible only at the level of large aggregated systems.

The important point for us is that, under these scenarios, it would be unreasonable to expect the Einstein equations to provide a description of space–time that goes beyond a phenomenologically useful characterization. Instead, we should think of the relativistic description of space–time in analogy with, say, a hydrodynamic description of a fluid in terms of the Navier-Stokes equations. In such a case, it is clear that the offered description is no more than a good approximation, which does not contain the fundamental degrees of freedom and breaks down under certain circumstances (i.e., turbulence or the break of a wave). Therefore, we should not be unduly surprised if, in association with the quantum gravity interface, we encounter situations in which the Einstein equations fail to be satisfied exactly. In fact, there might be situations in which the metric characterization of space–time provides a good description, which is, however, punctuated by small interruptions associated with the onset of phenomena that cannot be fully accounted for at that level of description. We submit that this kind of situation is precisely what one finds during quantum collapses, where the semiclassical description simply breaks down. However, instead of abandoning the project at that point, and in tune with our top-down approach, we propose to treat such breakdowns phenomenologically, within a trial-and-error frame of mind that seeks to extract clues about a suitable description for the more fundamental degrees of freedom involved. That is, we propose to push forward by adding suitable adjustments, from which we hope to eventually obtain hints for the construction of a more fundamental theory of quantum gravity.[6] In the next section, we explain in detail how we propose to do this.

17.3 Objective Collapse in a General Relativistic Setting

One of the most promising proposals in order to deal with the measurement problem is the dynamical reduction or objective collapse program. Early examples of theories within such a program include GRW ([2]) and CSL ([3]). The main idea of the project is to add non-linear, stochastic terms to the dynamical equation of the standard theory in order to achieve a unified description of both the quantum behavior of microscopic systems and the emergence of classical behavior (e.g., absence of

---

[6] A somehow comparable approach is adopted in [66], where, in order to deal with the recovery of time in quantum gravity, some dynamical variable is designated as a "physical clock," relative to which wave functions for other variables are constructed. By doing so, it is found that one recovers only an approximate Schrödinger equation, with corrections that violate unitarity. The problem is that the precise scheme adopted in this and related works relies heavily on the notion of decoherence and, as such, suffers from rather serious problems discussed on general grounds in [35]. That of course does not mean that taking a slightly more general view of the subject could not lead to something that could lie at the bottom of collapse theories.

superpositions) at the macroscopic level. Of course, the price to pay is a departure from Schrödinger's evolution equation, accompanied by a breakdown of unitarity. In contrast, one of the advantages of these theories is that they leave much of the structure of quantum mechanics intact, so that such things as the EPR correlations are clearly accounted for. Moreover, they are susceptible to experimental testing and, in fact, a robust program for that is currently under way (see [17]).

As we mentioned, most objective collapse models, such as GRW or the original version of CSL, are non-relativistic. However, recently, fully relativistic versions such as [38, 39, 40] have been developed. The problem is that none of them, in their present form, is adequate for a general relativistic context. The objective of this section is to describe one particular way in which this could be achieved. Before doing so, though, we will quickly review the GRW and CSL models (see [2] and [3], respectively).

Non-relativistic GRW postulates each elementary particle to suffer, with mean frequency λ_0, sudden and spontaneous localization processes around judiciously chosen random positions. Such localizations are then shown to mimic the Born rule and to ensure a quick elimination of superpositions of well-localized macroscopic states. In [38], a relativistic version of the GRW model for *non-interacting* Dirac particles is introduced. Such model works in Minkowski space–time as well as in (well-behaved) curved-background space–times.

The non-relativistic Continuous Spontaneous Localization model, or CSL, is defined by a modified Schrödinger equation, whose solutions are given by

$$|\psi, t\rangle_w = \hat{\mathcal{T}} e^{-\int_0^t dt' \left[i\hat{H} + \frac{1}{4\lambda_0}[w(t') - 2\lambda_0 \hat{A}]^2 \right]} |\psi, 0\rangle, \tag{17.1}$$

with $\hat{\mathcal{T}}$ the time-ordering operator and $w(t)$ a random white-noise function, chosen with probability

$$PDw(t) \equiv {}_w\langle \psi, t | \psi, t \rangle_w \prod_{t_i=0}^{t} \frac{dw(t_i)}{\sqrt{2\pi \lambda_0 / dt}}. \tag{17.2}$$

As $t \to \infty$, the CSL dynamics inevitably drives the state of the system into an eigenstate of the operator $\hat{A}$. Therefore, it unifies the standard unitary evolution with a "measurement" of such an observable. In a non-relativistic setting, for the collapse operator $\hat{A}$, one usually chooses a smeared position operator.

In a relativistic setting, the natural thing to do is to assign a quantum state to every Cauchy hypersurface Σ and to postulate a Schwinger–Tomonaga-type equation to implement the evolution from one such hypersurface to another hypersurface Σ'. For example, the evolution equation introduced in [39] is

$$d_x \Psi(\phi; \Sigma') = \left\{ -iJ(x)A(x)d\omega_x - (1/2)\lambda_0^2 N^2(x)d\omega_x + \lambda_0 N(x)dW_x \right\} \Psi(\phi; \Sigma), \tag{17.3}$$

where $d\omega_x$ is the infinitesimal space–time volume separating Σ and Σ', $J(x)$ is an operator constructed out of matter fields, W_x is a Brownian motion field, and $A(x)$ and $N(x)$ are operators that modify the state of an auxiliary quantum field (for more details, we remit the reader to [39]). As for the value of λ_0 in all of these models, it has to be small enough in order to recover normal quantum behavior at a microscopic level but big enough in order to ensure a rapid localization of macroscopic objects. This can be achieved by setting, for example, $\lambda_0 \sim 10^{-16} sec^{-1}$.

At this point it is important to make a few comments regarding the *physical interpretation* of objective collapse theories. As we explained, in order to solve the measurement problem, such theories avoid relying on the standard interpretation of the quantum state in terms of the Born rule. However, by removing such a probabilistic interpretation *without* substituting it by something else, one does not yet arrive at a proper physical theory capable of making empirically verifiable predictions because one lacks a translation rule between the mathematical formalism and the physical world the theory is supposed to describe. Therefore, objective collapse models require an alternative interpretation of the quantum state.

One option in this regard is to interpret the wave function directly as a physical field. However, it seems that by doing so, one is obligated to also treat configuration space as physical (see [68]). Another alternative, proposed in [69], is to take the GRW collapses as the objects out of which physical stuff is made. Yet another option, first presented in [70], is to interpret the theory as describing a physical mass-density field, constructed as the expectation value of the mass-density operator on the state characterizing the system; the relativistic version of this interpretation, with the energy-momentym tensor in place of mass density, is discussed in [71]. Notice that if one adopts the mass density ontology described earlier, then the semiclassical approach becomes much more natural because, within such an interpretation, the expectation value directly represents the energy and momentum distribution predicted by the theory.

Going back to our proposal, as we explained, we will regard semiclassical general relativity as an approximate description with a limited domain of applicability, which must, however, be pushed beyond what is usually expected. In particular, we will incorporate quantum collapses to such a picture. It is clear, though, that during collapses the expectation value of the energy-momentum tensor is not conserved. Therefore, at such events, the semiclassical Einstein equations

$$G_{\mu\nu} = 8\pi G \langle \xi | \hat{T}_{\mu\nu} | \xi \rangle \tag{17.4}$$

are not valid. The proposal, as we mentioned, is to see the issue as analogous to a hydrodynamic description of a fluid, in which the Navier–Stokes equations will not hold in some situations (e.g., when a wave is breaking), but can be taken to hold before and after the fact. Analogously, we will take the semiclassical Einstein equations not to hold during a collapse but to do so before and after.

It is clear, then, that the approach requires some kind of recipe to join the descriptions just before and after a collapse. To do so, we use the notion of a *Semiclassical Self-consistent Configuration* (SSC) introduced in [26], which is defined as follows: the set $\{g_{ab}(x), \hat{\varphi}(x), \hat{\pi}(x), \mathcal{H}, |\xi\rangle \in \mathcal{H}\}$ is a SSC if and only if $\hat{\varphi}(x)$, $\hat{\pi}(x)$ and $\mathcal{H}$ correspond a to quantum field theory for the field $\varphi(x)$, constructed over a space–time with metric $g_{ab}(x)$, and the state $|\xi\rangle$ in $\mathcal{H}$ is such that

$$G_{ab}[g(x)] = 8\pi G \langle \xi | \hat{T}_{ab}[g(x), \hat{\varphi}(x)] | \xi \rangle, \qquad (17.5)$$

where $\langle \xi | \hat{T}_{\mu\nu}[g(x), \hat{\varphi}(x)] | \xi \rangle$ stands for the expectation value in the state $|\xi\rangle$ of the renormalized energy-momentum tensor of the quantum matter field $\hat{\varphi}(x)$, constructed with the space–time metric g_{ab}. In a sense, this construction is the relativistic version of the Newton–Schrödinger system ([14]), and just like it, it involves a kind of circularity in the sense that the metric conditions the quantum field, and the state of the latter conditions the metric.

As we explain in what follows, the SSC formulation allows us to incorporate quantum collapses of the matter sector into the semiclassical picture we are considering. The idea is for different SSCs to describe the situation before and after the collapse and for them to be suitably joined at the collapse hypersurface. In more detail, a spontaneous collapse of the quantum state that represents matter is accompanied by a change in the expectation value of the energy-momentum tensor, which, in turn, leads to a modification of the space–time metric. Note however that this demands a change in the Hilbert space to which the state belongs, so the formalism forces us to consider a transition from one complete SSC to another one rather than simple jumps of states in one Hilbert space. That is, collapses should be characterized as taking the system from one complete SSC construction into another.[7]

The concrete proposal to implement an objective collapse model within the SSC scheme is the following (for simplicity, we consider a model with discrete collapses, akin to GRW). We start with an initial SSC, which we call SSC1, and we give rules to randomly select both a spacelike hypersurface Σ_C of the space–time of SSC1, on which the collapse takes place, and a *tentative* or *target* postcollapse state $|\chi^t\rangle \in \mathcal{H}_1$. Such state is used to determine $|\xi^{(2)}\rangle$, the actual postcollapse state. Then, a new SSC, which we call SSC2, is constructed for the collapsed state to live in and the two SSC's are glued to form a "global space–time." To do so, the SSC2 construction is required to posses an hypersurface isometric to Σ_C, which

---

[7] As noted in [72], this is connected with issues appearing in other approaches, such as the stochastic gravitation program of [73].

serves as the hypersurface where the two space–times are joined (such condition is analogous to the Israel matching conditions for infinitely thin time-like shells). Finally, in order to construct $|\xi^{(2)}\rangle \in \mathcal{H}_2$ out of $|\chi'\rangle \in \mathcal{H}_1$, we demand that, on Σ_C,

$$\langle \chi' | \hat{T}^{(1)}_{ab}[g(x), \hat{\varphi}(x)] | \chi' \rangle = \langle \xi^{(2)} | \hat{T}^{(2)}_{ab}[g(x), \hat{\varphi}(x)] | \xi^{(2)} \rangle, \quad (17.6)$$

where $\hat{T}^{(1)}$ and $\hat{T}^{(2)}$ are the renormalized energy-momentum tensors of SSC1 and SSC2, which depend on the corresponding space–time metrics and the corresponding field theory constructions (see [26] for details).

Think again about the hydrodynamics analogy in which, before and after a wave breaks, the situation is accurately described by Navier–Stokes equations, but the breakdown itself is not susceptible to a fluid description. If we now take the limit in which the duration of the break tends to zero, we have two regimes susceptible to a fluid description, joined at an *instantaneous break*. The spacelike hypersurface Σ_C that joins SSC1 and SSC2 is analogous to that. The intention of all this is to construct an effective description to resolve urgent issues and to eventually explore these matching conditions in search of clues about the underlying theory.

One extra element we have considered, initially in the context of black holes and later in cosmology, is the possibility for the collapse rate λ_0 to depend on the local curvature of space–time. For example, one could have something like

$$\lambda(W) = \lambda_0 \left[1 + \left(\frac{W}{\mu}\right)^\gamma \right] \quad (17.7)$$

with W some scalar constructed out of the Weyl tensor, $\gamma \geq 1$ a constant and with μ providing an appropriate scale. As we will see, in the context of black holes, this dependence brings about all of the information destruction required in order to avoid a paradox and, in cosmology, it helps explain the very special initial state of the universe. An independent motivation for a curvature-dependent collapse parameter comes from the fact that studies of experimental bounds on such a parameter suggest it must depend on the mass of the particle in question ([74, 17]). Having the parameter to depend on curvature seems like an attractive way to implement this in a general relativistic context.

Finally, generalizing all this scheme to multiple collapses is straightforward. However, one must be careful of the fact that, unlike in a unitary quantum field theory, in which $\langle \xi_\Sigma | \hat{T}_{ab}(x) | \xi_\Sigma \rangle$ is independent of the hypersurface Σ, in collapse theories such an expression does depend on the choice of hypersurface. An interesting solution to this problem, spelled out in [71], consists in stipulating that the hypersurface one must use in order to calculate $\langle \xi_\Sigma | \hat{T}_{ab}(x) | \xi_\Sigma \rangle$ has to be the past light cone of the point x. Finally, extending all this to the context of a continuous collapse theory, such as CSL, would involve some type of limiting procedure.

While we do not expect such a process to introduce extra conceptual difficulties, technically, it will surely be extremely demanding.

17.4 Applications

Bellow we consider promising applications of the scheme described in the previous section to several open problems in cosmology and quantum gravity.

17.4.1 Cosmic Inflation and the Seeds of Cosmic Structure

Contemporary cosmology includes *inflation* as one of its most attractive components. Such process was conceived in order to address a number of naturalness issues in the standard big bang cosmological model, such as the horizon problem, the flatness problem and the exotic-relics problem ([75, 76]). However, inflation's biggest success is to correctly predict the spectrum of the cosmic microwave background (CMB) and to account for the emergence of the seeds of cosmic structure.

The staring point of such an analysis is a flat FriedmannRobertsonWalker (FRW) background space–time, which is inflating under the influence of a homogeneous inflaton background field. On top of that, one considers perturbations of the metric and the inflaton, which are treated quantum mechanically and are initially set at an appropriate homogeneous vacuum state (essentially the so-called Bunch–Davies vacuum). Then, from the *quantum fluctuations* of these perturbations, the primordial inhomogeneities and anisotropies are argued to emerge. These primordial inhomogeneities and anisotropies, in turn, are said to constitute the seeds of all the structure in our universe as, according to our current cosmological picture, they later evolved, due to gravitational attraction, into galaxy clusters, galaxies, stars, and planets. The result of all this is a remarkable agreement between predictions of the inflationary model and both observations of the CMB and structure surveys at late cosmological times.

It is indeed remarkable that, with one exception, the theoretical predictions of inflation are in exquisite agreement with observations. The exception has to do with the fact that inflationary models, along with the formation of inhomogeneities and anisotropies, generically also predict the formation of primordial gravity waves, which should become manifest in the so-called B-modes in the CMB polarization ([77, 78, 79]). The problem is that, up to this point, these B-modes have not been observed. We have more to say about this at the end of our analysis, but, in the meanwhile, we return to the central topic of the emergence of the seeds of structure.

According to the standard picture, during inflation, the universe was homogeneous and isotropic (at both the classical and quantum levels of description); in spite of this, the final situation was not homogeneous and isotropic: it contained the primordial inhomogeneities that resulted in the structure that, among other things, permitted our own existence. A natural question arises: how did this transition from a symmetric into a non-symmetric scenario happen, given that the dynamics of the whole system does not break those symmetries? The problem is that the quantum fluctuations from which the seeds of structure are supposed to emerge are not really *physical* fluctuations but only a characterization of the *width* of the quantum state. Therefore, those fluctuations are incapable of breaking the symmetries of the initial state. The standard account, then, implicitly assumes a transition from a symmetric into a non-symmetric state, without an understanding of the process that leads, in the absence of observers or measurements, from one to the other. Such fact renders the standard account unsatisfactory.

A similar issue was considered by N. F. Mott in 1929 concerning the α nuclear decay. In such a scenario, one starts with a $J = 0$ nucleus undergoing a rotationally invariant interaction that leads to a state characterized by a spherical outgoing wave function. The problem is that what one observes in a bubble chamber are the straight paths of the outgoing α particles, which clearly break the spherical symmetry. It is often assumed that the issue was satisfactorily addressed at the time, but a close examination of that work shows this not to be the case. The problem is that the proposed solution was based on the assumption that atoms in the bubble chamber were, on the one hand, highly localized, thus breaking the symmetry of the complete system, and, on the other hand, taken to act as classical detectors with well-defined excitation levels (in contrast with suitable quantum superpositions thereof). Both of these issues are clearly related to the measurement problem we mentioned earlier. However, in the cosmological scenario, the problem presents itself in an aggravated form as, even if we accepted to give observers a fundamental role in the theory (something we do not do), one could not call upon such observers to play the role of inducers of quantum collapse in the very early universe, where no such beings existed.

It is worth noting that we face here a rather unique situation, where quantum theory, general relativity and observations come together. We should contrast the present case with other scenarios one might initially think explore similar realms, such as the COW experiment ([80]) or experiments that study neutron quantum states in earth's gravitational field ([81]). In fact, these experiments are only tests of the *equivalence principle*, in the sense that the situations considered might be described in terms of falling reference frames and, when doing so, gravity simply disappears from the scene. In the inflationary context, the situation is radically different because the quantum state of the inflaton field and, in particular, the spatially dependent fluctuations actually gravitate and affect things like the space–time curvature. Therefore, in order to address the question at hand, namely the emergence of the seeds of cosmic structure, we need to call upon a physical process that occurs in time. After all, emergence actually *means* for something not to be there at a time but to be there at a latter time. In the present case, what we need is a *temporal* explanation of the breakdown of the symmetry of the initial state, and the point we want to stress is that spontaneous objective collapse theories can accomplish this. What we propose, then, is to add to the standard inflationary paradigm an objective, spontaneous quantum collapse of the wave function in a form suitably adapted to the situation at hand (see [21, 22, 23, 24, 25, 26, 27, 28, 29, 36]). In what follows, we develop these ideas in some detail.

Objective Collapse and the Seeds of Cosmic Structure

In the following, we show how the implementation of a CSL dynamics resolves the issues we pointed out previously regarding the standard account for the emergence of the seeds of structure during the inflationary era. In principle, all this should be done using the SSC formalism we described in section 17.3. However, due to its complexity, we use instead a approximated scheme in which we always use the Hilbert space construction corresponding to the unperturbed space–time metric (see what follows), and the collapse thus simply induces a jump into a different vector in the same Hilbert space. In [26] we have checked that, at the lowest order in perturbation theory, this is equivalent to the full-fledged SSC treatment.

As in the standard account, we split the metric and the inflaton into a homogeneous background and a potentially inhomogeneous fluctuation, i.e., $g = g_0 + \delta g, \phi = \phi_0 + \delta\phi$. The metric background corresponds to a flat FRW universe, with line element $ds^2 = a(\eta)^2\{-d\eta^2 + \delta_{ij}dx^i dx^j\}$ (with η a conformal time), and the field background to a homogeneous scalar field $\phi_0(\eta)$. Such $\vec{k} = 0$ mode of the field, which is responsible for the overall inflationary expansion, is treated classically in this effective approximation. However, as shown in [26], it can be treated quantum mechanically within the full SSC formalism. We here then concentrate on the $\vec{k} \neq 0$ modes of the field. Moreover, since our approach calls for quantizing the scalar field but not the metric perturbation, we do not use the so-called Muckhanov–Sassaki variable as is customary in this field.

The evolution equations for the background field and metric are given by

$$\ddot{\phi}_0 + 2\frac{\dot{a}}{a}\dot{\phi}_0 + a^2 V'(\phi_0) = 0 \quad \text{and} \quad 3\frac{\dot{a}^2}{a^2} = 4\pi G\left(\dot{\phi}_0^2 + 2a^2 V(\phi_0)\right), \quad (17.8)$$

with $V(\phi)$ the inflaton potential, " $\dot{}$ " $\equiv \partial/\partial\eta$ and " $'$ " $\equiv \partial/\partial\phi$. The solution for the scale factor corresponding to the inflationary era is $a(\eta) = -\frac{1}{H_I \eta}$, with $H_I^2 \approx (8\pi/3)GV$ and with the scalar field ϕ_0 in the slow roll regime, i.e., $\dot{\phi}_0 = -(a^3/3\dot{a})V'(\phi_0)$. As in the standard inflationary scenario, inflation is followed by a reheating period in which the universe is repopulated with ordinary matter fields. Such stage then evolves toward a standard hot big bang cosmology regime leading up to the present cosmological time. The functional form of a during these latter periods changes, but we can ignore those details because most of the change in the value of a occurs during the inflationary regime. We set $a = 1$ at the present cosmological time and assume that the inflationary regime ends at a value of $\eta = \eta_0$, which is negative and very small in absolute terms.

Next one considers the perturbations. For the metric, we write

$$ds^2 = a^2(\eta)\{-(1+2\Psi)d\eta^2 + [(1-2\Phi)\delta_{ij} + h_{ij}]dx^i dx^j\}, \qquad (17.9)$$

where we are using the so-called longitudinal gauge. The scalar field, on the other hand, is treated using quantum field theory on such background, with its state $|\xi\rangle$ and the metric satisfying the semiclassical Einstein equations (17.4).

As we already explained, at the early stages of inflation (which we characterize by $\eta = -\mathcal{T}$), the state of the scalar field perturbation is described by the Bunch–Davies vacuum. As a result, space–time at the time is totally homogeneous and isotropic and the operator $\delta\phi$ and its conjugate momentum $\pi_{\delta\phi}$ are characterized by Gaussian wave functions, centered on 0, with uncertainties $\Delta\delta\phi$ and $\Delta\pi_{\delta\phi}$ (for ease of notation, we omit the "hats" over the operators $\delta\phi$ and $\pi_{\delta\phi}$). Next enters the collapse, which randomly and spontaneously modifies the quantum state and, generically, changes the expectation values of $\delta\phi$ and $\pi_{\delta\phi}$. We assume that the collapse is controlled by a stochastic function, mode by mode. Such an educated guess will later be contrasted with observations. It is important to keep in mind that, in this picture, our universe corresponds to one specific realization of these stochastic functions (one for each $\vec{k}$). Thus, if we were given that specific realization, the prediction of what we would see in the CMB would be completely transparent and unambiguous. Given, however, that the theory is fundamentally stochastic, we have no way of *a priori* determining the specific realization of such functions, and thus, in order to make concrete predictions, we need to resort to some statistical manipulations. We come to this issue shortly.

From equation (17.4) and what we have said so far, we obtain

$$\nabla^2 \Psi = 4\pi G \dot{\phi}_0 \langle |\delta\phi|, \rangle \qquad (17.10)$$

from which it follows that, as soon as the expectation value of $\delta\phi$ deviates from zero, which it generically will due to the spontaneous collapses, so will the metric perturbation Ψ. In the Fourier representation, the preceding equation reads

$$-k^2 \Psi_{\vec{k}} = 4\pi G \dot{\phi}_0 \langle |\delta\phi_{\vec{k}}| = \rangle \frac{4\pi G \dot{\phi}_0}{a} \langle |\pi_{\vec{k}}|. \rangle \qquad (17.11)$$

In order to derive observational consequences of all this, we note that there is a direct connection between $\Psi_{\vec{k}}$ and the temperature fluctuations of the CMB observed today. In fact, we have

$$\frac{\Delta T(\theta,\varphi)}{\bar{T}} = c \int d^3k e^{i\vec{k}\cdot\vec{x}} \frac{1}{k^2} \langle|\hat{\pi}_{\vec{k}}(\eta_D)|\ \rangle \text{with}\quad c \equiv -\frac{4\pi G \dot{\phi}_0(\eta)}{3a}, \qquad (17.12)$$

where $\vec{x}$ is a point on the intersection of our past light cone with the last scattering surface ($\eta = \eta_D$) corresponding to the direction on the sky specified by θ, φ. Decomposing in spherical harmonics, we obtain

$$\alpha_{lm} = c \int d^2\Omega Y^*_{lm}(\theta,\varphi) \int d^3k e^{i\vec{k}\cdot\vec{x}} \frac{1}{k^2} \langle|\hat{\pi}_{\vec{k}}(\eta_D)|.\rangle \qquad (17.13)$$

It is worthwhile pointing out that, within the standard approach, the expression analogous to the last equation is never shown. That is because the prediction in such approach for this quantity is in fact zero. Of course, what practicing cosmologists do at this point is to bring some loosely worded arguments explaining that quantum theory should not be used to predict what we see directly but only for predicting averages. Therefore, they argue, one should instead consider the quantity

$$C_l \equiv \frac{1}{2l+1} \Sigma_m |\alpha_{lm}|^2, \qquad (17.14)$$

which characterizes the orientation average of the magnitude of the a_{lm}. This line of thought raises several questions: why does the theory only make predictions for the orientation average and not for the a_{lm} directly?, or why should we ignore the fact that we actually observe the temperature at each pixel in the sky and not the average? We do not think the standard approach offers good answers for these questions. Within the approach we are considering, in contrast, the conceptual picture is clear, with a straightforward explanation for where the randomness occurs, what is its source, and how it plays a role in yielding predictions.

In more detail, within our approach, the quantity in equation (17.13) can be thought of as a result of a *random walk* on the complex plane, each step of the walk determined by the random function controlling the CSL evolution of mode $\vec{k}$, and the integration representing the addition of the infinitesimal contributions of each mode to the complete walk. One of course cannot predict the end point of such walk, but one can instead focus on the magnitude of the total displacement to compute $|\alpha_{lm}|^2$ and estimate such value via an ensemble average

$$\overline{\langle\pi|_{\vec{k}}(\eta)|\rangle\langle|\pi_{\vec{k}'}(\eta)|\rangle^*} = f(k)\delta(\vec{k}-\vec{k}'). \qquad (17.15)$$

It can be checked that agreement between predictions and observations requires that $f(k) \sim k$.

In order to compute the average in (17.15) using CSL, we work with a rescaled field $y(\eta,\vec{x}) \equiv a\delta\phi(\eta,\vec{x})$ and its momentum conjugate $\pi_y(\eta,\vec{x}) = a\delta\phi'(\eta,\vec{x})$. We also put everything in a box of size L (to be removed at the end) and focus on a single mode $\vec{k}$, so we write

$$\hat{Y} \equiv (2\pi/L)^{3/2} y(\eta, \vec{k}), \qquad \hat{\Pi} \equiv (2\pi/L)^{3/2} \pi_y(\eta, \vec{k}). \tag{17.16}$$

What we need to do next is to evaluate the ensemble average $\overline{\langle \hat{\Pi} \rangle^2}$ and determine under what circumstances, if any, it behaves as $\sim k$. Before doing so we note that, as shown in the expression given, we must consider $\overline{\langle \hat{\Pi} \rangle^2}$ and not the quantity $\langle \hat{\Pi}^2 \rangle$, which is the focus of attention in traditional approaches. One of the reasons for doing so is made evident by the observation that the latter is generically non-vanishing even for states that are homogeneous and isotropic, such as the vacuum state, while the former will vanish when the state under consideration is the vacuum—thus clearly exhibiting the fact that no anisotropies can be expected in fully isotropic situations.

In order to apply CSL to compute the required average, we need to choose the operator $\hat{A}$ driving the collapse. In this regard, we first assume that the operator driving the collapse of the mode $\vec{k}$ is the corresponding operator $\hat{\Pi}$. That is, we set $\hat{A} = \hat{\Pi}$ in the CSL evolution equation (17.1). The calculation is straightforward although quite long, so we refer the reader to [27] for details. In the end, one obtains

$$\overline{\langle \hat{\Pi} \rangle^2} = \frac{\lambda k^2 \mathcal{T}}{2} + \frac{k}{2} - \frac{k}{\sqrt{2}\sqrt{1 + \sqrt{1 + 4\lambda^2}}}. \tag{17.17}$$

As we mentioned, in order to obtain results consistent with CMB observations, we need $\overline{\langle \hat{\Pi} \rangle^2}$ to be proportional to k. We can achieve this if we assume that the first term is dominant and that

$$\lambda = \tilde{\lambda}/k \tag{17.18}$$

with $\tilde{\lambda}$ a constant independent of k. Note that this replaces the dimensionless collapse rate parameter λ with $\tilde{\lambda}$ having dimensions of time^{-1}. Doing this, we obtain

$$\overline{\langle \hat{\Pi} \rangle^2} = \frac{\tilde{\lambda} k \mathcal{T}}{2} + \frac{k}{2} - \frac{k}{\sqrt{2}\sqrt{1 + \sqrt{1 + 4(\tilde{\lambda}/k)^2}}}. \tag{17.19}$$

Analogously, if instead of $\hat{\Pi}$ as the generator of collapse, we take $\hat{Y}$, we obtain

$$\overline{\langle \hat{\Pi} \rangle^2} = \frac{\lambda \mathcal{T}}{2} + \frac{k}{2} \left\{ 1 - \frac{(1 + 4(\lambda/k^2)^2)}{F(\lambda/k^2) + 2(\lambda/k^2)^2 F^{-1}(\lambda/k^2) - 2(\lambda/k^2)(k\eta)^{-1}} \right\}, \tag{17.20}$$

where $F(x) \equiv \frac{1}{\sqrt{2}} \sqrt{1 + \sqrt{1 + 4x^2}}$. Therefore, in order to obtain results consistent with observations, we need to assume that the first term dominates and that

$$\lambda = \tilde{\lambda} k. \tag{17.21}$$

This time the collapse rate parameter λ of dimension time^{-2} is replaced with the parameter $\tilde{\lambda}$ of dimension time^{-1}. Doing this leads to

$$\overline{\langle\hat{\Pi}\rangle^2} = \frac{\tilde{\lambda}kT}{2} + \frac{k}{2}\left\{1 - \frac{(1+4(\tilde{\lambda}/k)^2)}{F(\tilde{\lambda}/k) + 2(\tilde{\lambda}/k)^2 F^{-1}(\tilde{\lambda}/k) - 2(\tilde{\lambda}/k)(k\eta)^{-1}}\right\}. \quad (17.22)$$

Finally, comparisons with observations, using the GUT scale for the value of the inflation potential and standard values for the slow-roll parameter, leads to an estimate of $\tilde{\lambda} \sim 10^{-5} MpC^{-1} \approx 10^{-19} sec^{-1}$. The fact that this is not very different from the GRW suggested value for the collapse rate is indeed encouraging.

Tensor Modes

As we mentioned, the theoretical predictions of inflation are in exquisite agreement with observations, with one exception. Inflationary models generically predict the production of B-modes in the CMB, but these B-modes have not been observed. Such fact has been used to severely constrain the set of viable inflationary models (see [77, 78, 79]). Here we show that the incorporation of objective collapses into the picture dramatically alters the prediction for the shape and size of the B-mode spectrum, explaining why we have not seen them.

Within the standard approach, both the scalar and tensorial perturbations are treated equally. It is not surprising, then, that standard inflationary models predict a precise relationship between their amplitudes and shapes. In fact, the standard estimates for the power spectra of the scalar and tensor perturbations are given by

$$P_s^2(k) \sim \frac{1}{k^3}\frac{V}{M_p^4 \epsilon} \quad \text{and} \quad P_h^2(k) \sim \frac{1}{k^3}\frac{V}{M_p^4} \quad (17.23)$$

respectively, and direct measurements of ϵ have been used to limit the viability of some of the simplest inflationary models. Previously we showed that adding collapses into the picture solves a grave conceptual problem for the standard account without modifying the prediction for the scalar power spectrum. Next, we show that the story is different with the tensor perturbations.

From (17.4), the equation of motion for the tensor perturbations is

$$(\partial_0^2 - \nabla^2) h_{ij} + 2(\dot{a}/a)\dot{h}_{ij} = 16\pi G \langle (\partial_i \delta\phi)(\partial_j \delta\phi)\rangle^{tr-tr}, \quad (17.24)$$

where $tr - tr$ stands for the transverse traceless part of the expression. In terms of a Fourier decomposition, we need to solve the equation,

$$\ddot{h}_{ij}(\vec{k},\eta) + 2(\dot{a}/a)\dot{h}_{ij}(\vec{k},\eta) + k^2 h_{ij}(\vec{k},\eta) = S_{ij}(\vec{k},\eta), \quad (17.25)$$

with zero initial data, and source term given by

$$S_{ij}(\vec{k},\eta) = 16\pi G \int \frac{d^3 x}{\sqrt{(2\pi)^3}} e^{i\vec{k}\vec{x}} \langle (\partial_i \delta\phi)(\partial_j \delta\phi)\rangle^{tr-tr}(\eta,\vec{x}). \quad (17.26)$$

Formally, this is a divergent expression, but we must introduce a cut-off P_{UV} that one might take as the scale of diffusion dumping. Doing so leads to our prediction for the power spectrum of tensor perturbations

$$\mathcal{P}_h^2(k) \sim \frac{1}{k^3} \left(\frac{V}{M_p^4}\right)^2 \frac{P_{UV}}{k}, \tag{17.27}$$

which is substantially smaller than the standard prediction from before. Therefore, within this approach, we expected not to see tensor modes at the level they are being looked for. In fact, in order to have any possibility of detecting them, we would need to improve sensitivity by various orders of magnitude and to look for them at very large scales.

17.4.2 Dark Energy From Energy and Momentum Non-conservation

We have seen that a key issue to be confronted in incorporating collapse theories into the context of general relativity at the semiclassical level is the generic violation of the condition $\nabla^a \langle T_{ab} \rangle = 0$. In order to deal with such a problem, we have taken semiclassical relativity as an effective description (analogous to that provided by the Navier–Stokes equations for fluid dynamics) and argued that collapses might be incorporated using the SSC formalism to join descriptions of before and after a collapse event. There is, however, a class of scenarios that can be dealt with using a slightly modified theory of gravity known as *unimodular gravity*, a theory based on a traceless version of Einstein's equations

$$R_{ab} - \frac{1}{4} g_{ab} R = 8\pi G \left(T_{ab} - \frac{1}{4} g_{ab} T \right). \tag{17.28}$$

Unimodular gravity arises naturally when one considers reducing the general diffeomorphism invariance of general relativity to invariance under volume-preserving diffeomorphisms only. That, in turn, seems as a natural modification if one contemplates incorporating a "constant rate of collapse events per unit of space–time volume," as would be natural in a relativistic collapse theory. Moreover, in turns out that, in the unimodular version of general relativity, the energy-momentum conservation does not follow from the dynamical equations, so it has to be imposed by hand as an independent assumption. The point we want to make is that, with collapses in mind, we need not do that! It is important to remember, though, that in this case the Bianchi identity leads to $\nabla_{[c} \nabla^a T_{b]a} = 0$, which is now the self-consistency condition for integrability of the equations.

Suppose, then, that we have a situation in which we want to consider a collapse theory in a semiclassical setting, in which energy-momentum is (of course) not conserved, $\nabla^a \langle T_{ab} \rangle \neq 0$ but where $\nabla_{[c} \nabla^a \langle T_{b]a} \rangle = 0$. One such situation is in fact provided by cosmology, where the homogeneity and isotropy of the system guarantees the integrability condition is automatically satisfied. For the case of simply connected space–times, such condition reduces to $J_a = \nabla_a Q$ for some scalar quantity Q (where we defined $J_a \equiv \nabla^a \langle T_{ab}\rangle$) and, as shown in [37], in that case one can recast the equations as

$$R_{ab} - \frac{1}{2}Rg_{ab} + \Lambda_{\text{eff}}g_{ab} = \frac{8\pi G}{c^4}\langle T_{ab}\rangle. \tag{17.29}$$

where the effective cosmological constant is given by $\Lambda_{\text{eff}} = \left(\Lambda_0 + \frac{8\pi G}{c^4}\int J\right)$, with Λ_0 an integration constant.

When energy-momentum is conserved, one of course recovers standard general relativity, with a cosmological constant given just as an integration constant. However, the interesting observation is that if one has some argument that allows one to fix the initial condition Λ_0 at any time, together with an explicit mechanism for violation of energy and momentum, one can estimate the value of the effective cosmological constant at other times. Of particular interest in this regard is the unexpected value for the cosmological constant needed to explain the late accelerated expansion of the universe discovered close to two decades ago ([82, 83]). The observed value of $\Lambda^{\text{obs}} \approx 1.1\,10^{-52}$ m^{-2} is unexpected because the seemingly natural values for Λ are either zero or a value that is 120 orders of magnitude larger than the one favored by observations (see [84]).

In [37] it was noted that collapse theories can offer an attractive possibility in order to solve this puzzle. As is well known, non-relativistic CSL with collapse parameter proportional to mass and with collapse operators given by smeared mass-density operators leads to a spontaneous creation of energy that is proportional to the mass of the collapsing matter quantum state [17, 74]. In the cosmological setting, such an effect generates an energy-momentum violation given by

$$J = -\lambda_{\text{CSL}}\rho^b dt, \tag{17.30}$$

with λ_{CSL} the CSL parameter and ρ^b the energy density of the baryonic contribution to the universe mean density (the contribution of lighter particles can be expected to be sub dominant). Now, taking as initial time the era of hadronization (corresponding to $z_h \approx 7\,10^{11}$), one finds

$$\Lambda_{\text{eff}} - \Lambda_0 \approx -\frac{3\Omega_0^b H_0 \lambda_{\text{CSL}}}{\sqrt{\Omega_0^r}c^2}z_h \approx -\frac{\lambda_{\text{CSL}}}{4.3\,10^{-31}\,\text{s}^{-1}}\Lambda^{\text{obs}}, \tag{17.31}$$

where standard values for the cosmological parameters where used ([85]). Given that the current allowed range for the CSL parameter is $3.3\,10^{-42}\text{s}^{-1} < \lambda_{\text{CSL}} < 2.8\,10^{-29}\text{s}^{-1}$, we arrive at a prediction that, remarkably, yields the correct order of magnitude.

It is very important to notice, though, that this cannot be the whole story, because the quantity estimated has the opposite sign as the observed cosmological constant. On the other hand, such an estimate only contains contributions from baryonic matter in the late cosmological era, where the dominant form of matter is non-relativistic. We should also consider contributions from previous epochs, where the dominant forms of matter were relativistic. Moreover, in order to compare the theoretical estimate with observations, we need a condition fixing the value of Λ_0 at a suitable time. This latter input might possibly emerge from the inflationary scenario, where the spatially flat condition is an attractor of the dynamics, setting $\Omega_{total} = 1$, or from a suitable condition at the Planck scale. At any rate, the lesson we take from all this is that this perspective offers new paths for a possible reconciliation between reasonable theoretical predictions for the cosmological constant and its observed value.

17.4.3 The Black Hole Information Issue

According to general relativity, the end point of the evolution of any sufficiently massive object is a black hole. Furthermore, all black holes are expected to eventually settled into one of the small number of stationary black hole solutions. In fact, within the setting of the Einstein-Maxwell theory, these are fully characterized by just three parameters, namely, mass M, charge Q, and angular momentum J. Work in theories including additional kinds of fields, such as scalar fields, have also led to so-called no-hair theorems that severely limit the possibility of enlarging this class of stationary black holes. Consideration of more general theories, such as those involving non-abelian Yang Mills fields, do lead to a new class of solutions. However, as it turns out, these new solutions represent unstable black holes. Therefore, if we limit ourselves to the stable class of black holes as possible end points of the evolution, we face a situation in which these final states are fully characterized by the three parameters M, Q, and J.

All this brings up the following issue. In general, an initial configuration that leads to the formation of a black hole requires an extremely large amount of data to be fully characterized. However, as we just saw, the final state is fully described by only three parameters. What happens, then, to the information-preserving (i.e., deterministic) character of the laws of physics, which is supposed to allow us to use both initial data to predict the future and final data to retrodict the past? Well, if the black hole in question is eternal, as general relativity holds, then the problem can be avoided by arguing that M, Q and J serve only to characterize the black hole's exterior, with all the remaining information about the initial state encoded in the interior. However, incorporating quantum aspects into the picture radically changes the situation.

As shown in [86], quantum field theory effects cause black holes to radiate. It is true that Hawking's calculation does not include back-reaction, but strong confidence in energy conservation brings people to the conclusion that, as Hawking's radiation takes away energy, the mass of the black hole has to diminish. Such loss of mass further increases the amount of radiation, apparently leading to the complete evaporation of the black hole in a finite time.[8] It seems that all this leads to a problem, since, after the black hole evaporates completely, there is no longer an interior region to contain the initial information. Note, however, that the system contains a singularity "into which information can fall," so still no proper inconsistency arises, because it can be argued that the initial information merely escapes through the singularity. Remember, though, that quantum gravity is widely expected to cure such singularities, and if that is the case, then a conflict finally seems inevitable.

One possible way out of all this is to hold fast to the assumption that information is always conserved, in which case there are two options. First, one can assume that the Hawking radiation carries the information. However, as has been argued in [87], that leads to the formation of a firewall on the horizon. Second, one can assume that the information is preserved in a remnant or in low-energy modes that survive the evaporation. The problem with this second option is that one expects remnants or surviving modes to have an energy of the order of the Plank mass (because one expects Hawking's calculation to hold until the deep quantum gravity regime is reached), so it's really hard to envisage how they could encode an arbitrarily large amount of information.

The other alternative is to assume that information is indeed lost during the evaporation process, from which one could conclude that quantum mechanics needs modifications when black holes are involved. However, it is much more natural (and exciting!) to think of a more fundamental modification of quantum theory involving gravity. In such a scenario, violations of unitarity could be associated with the excitation of certain degrees of freedom characteristic of quantum gravity (which we might want to describe as "virtual black holes"). Such excitations would, in turn, generate modifications of the Schrödinger evolution equation in essentially all situations. Could this be the source of the collapse events in collapse theories? Could there be a unified picture in which loss of information in black holes is accounted for by similar features as those occurring in collapse theories? Later we show how these questions can be answered in the affirmative.[9] Before doing so, though, we need to say a few words about density matrices.

---

[8] Quantum gravity effects could eventually stop the evaporation, leading to a stable remnant. However, given the low mass such objects would have, they are not expected to help in the resolution of the information loss issues (see what follows).

[9] See [31, 33] for extended discussions of all these issues.

It is important to distinguish between two different types of density matrices that are often confused but have very different *physical* interpretations. On the one hand, density matrices are used to describe either *ensembles* or situations in which one only has statistical information regarding the actual pure state of a system;[10] these cases are referred to as *mixed* states or *proper* mixtures. On the other hand, density matrices are used for the description of a subsystem of an isolated quantum system and are constructed by taking the trace over the rest of the system of the pure density matrix of the full system; density matrices of this type are called *reduced* density matrices or *improper* mixtures (see, e.g., [88]). It is therefore important to distinguish between proper and improper thermal states: proper thermal states describe either ensembles with states distributed thermally or a single system for which we lack information regarding its true pure state but know that it is distributed thermally; improper thermal states, in contrast, describe a single subsystem, which is entangled with the rest of the system in such a way that its reduced density matrix is thermal (e.g., the Minkowski vacuum, described in the Rindler wedge, using Rindler modes adapted to the boost Killing field, after tracing over the rest of space–time).

Going back to black holes, what Hawking's calculation shows is that the initially pure state of the quantum field evolves into a final one, which, when tracing over the interior region of the black hole, becomes a reduced or improper thermal state. The important question, then, is how to interpret such a final state when the black hole evaporates completely, so there is no longer an interior region to trace over. If, with us, one assumes that information gets lost during the process, then what one needs to do is to explain how the initial pure state transforms into a proper or mixed thermal state when the black hole evaporates.

Next we explain how this transformation from a pure state into a proper thermal state, during the black hole evaporation process, can be achieved by incorporating objective collapses into the standard picture. As in [30], we do so in a toy model based on the following simplifications:

[1] We assume that the black hole evaporation leaves no remnant.
[2] We work with a simple 2D black hole known as the Callan–Giddings–Harvey–Strominger (CGHS) model, where most calculations can be done explicitly.
[3] We use a non-relativistic toy version of CSL naively adapted to quantum field theory in curved space–time.
[4] We make a few natural simplifying assumptions about what happens when quantum gravity cures the black hole singularity.

---

[10] We are assuming, together with the standard interpretation, that individual isolated systems always possess pure quantum states represented by individual rays in the Hilbert space (or corresponding objects in the algebraic approach).

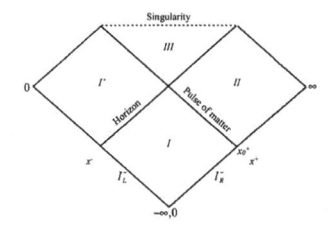

Figure 17.1 Penrose diagram for the CGHS model.

Moreover, as we explain in what follows in detail (see also section 17.3), we work under the assumption that the CSL collapse parameter is not fixed but increases with the local curvature. In the next section, we a give sketch of our proposal (see [30] for more details).

Black Holes and Objective Collapses

We begin with a brief review of the CGHS black hole ([89]), which is a very convenient 2D toy model, in order to study black hole formation and evaporation. The starting point is the action

$$S = \frac{1}{2\pi} \int d^2x \sqrt{-g} \left[e^{-2\phi} \left[R + 4(\nabla \phi)^2 + 4\Lambda^2 \right] - \frac{1}{2}(\nabla f)^2 \right], \quad (17.32)$$

where R is the Ricci scalar for the metric g_{ab}, ϕ is a dilaton field, Λ is a cosmological constant, and f is a scalar field characterizing matter (note that the dilaton is included in the gravity sector). The CGHS solution for the metric (see Figure 17.1) is given by

$$ds^2 = -\frac{dx^+ dx^-}{-\Lambda^2 x^+ x^- - (M/\Lambda x_0^+)(x^+ - x_0^+)\Theta(x^+ - x_0^+)} \quad (17.33)$$

and corresponds to a null shell of matter collapsing gravitationally along the world line $x^+ = x_0^+$, leading to the formation of a black hole. That is, in regions *I* and *I'* the metric is Minkowskian, but in regions *II* and *III* it represents a black hole (with region *II* its exterior and region *III* its interior).

Next we consider quantum aspects of the model. Given our semiclassical approach and the fact that, in the CGHS model, the matter sector is represented only by f, only such an object has to be treated quantum mechanically. We take the null past asymptotic regions I_L^- and I_R^- as the *in* region and both the interior and exterior of the black hole as the *out* region. Regarding the Bogolubov transformations from *in* to *out* modes, it turns out that the only non-trivial ones, and, in fact, those that account for the Hawking radiation, are the transformations from *in* to *exterior* modes. In particular, for the initial state, which is taken to be the vacuum for the right-moving modes and the left-moving pulse that forms the black hole, we have

$$|\Psi_{in}\rangle = |0_{in}\rangle_R \otimes |Pulse\rangle_L = N \sum_\alpha C_{F_\alpha} |F_\alpha\rangle^{ext} \otimes |F_\alpha\rangle^{int} \otimes |Pulse\rangle_L \quad (17.34)$$

where the $|F_\alpha\rangle$ represent states with a *finite* number of particles, N is a normalization constant, and the C_{F_α} are determined by the Bogolubov transformations (see [90]). Note that it is at this stage that one normally takes a trace over the interior degrees of freedom to find an improper thermal state that corresponds to the Hawking radiation. We will *not* do so here because we want to provide a description of the complete system.

We finally consider modifications to the evolution of the system arising from a dynamical reduction theory such as CSL. The idea is to work in the interaction picture, treating the CSL modifications, formally, as an interaction. Therefore, we take the free evolution as that encoded in the quantum field operator, while the state of the field evolves as a result of the modified dynamics. In order to use CSL, we need a foliation. For this we use hypesurfaces with constant Ricci curvature R in the inside, which match suitable hypesurfaces in the outside (see details in [30]). We also introduce a "foliation time" parameter τ to label these hypersurfaces.

The next step is to select a suitable CSL collapse operator, for which we judiciously choose one that drives initial states into states of definite numbers of particles in the inside region (note that the CSL equations can be generalized to drive collapse into a state of a joint eigen-basis of a set of commuting operators). We also incorporate the assumption that the CSL collapse mechanism is amplified by the curvature of space–time. We do so by postulating the rate of collapse λ to depend on the local curvature as in equation (17.7); however, in this toy model, instead of Weyl scalar, we use the Ricci scalar. We thus take

$$\lambda(R) = \lambda_0 \left[1 + \left(\frac{R}{\mu}\right)^\gamma\right] \quad (17.35)$$

where λ_0 is the flat space collapse rate, R is the Ricci scalar of the CGHS space–time, $\gamma > 1$ is a constant, and μ provides an appropriate scale. As a result, in the region of interest (i.e., the black hole interior, where the CSL modifications can become large), we effectively have $\lambda = \lambda(\tau)$. The point of this modification is to ensure for the resulting evolution to drive the initial state into an eigenstate of the collapse operator in a *finite* time (and not an infinite one as in standard CSL).

With all these assumptions, we can characterize the effect of CSL on the initial state in equation (17.34) as simply driving it to one of the eigenstates of the joint number operator. Thus, at any hypersurfaces of constant τ located very close to the singularity, the state of the field is

$$|\Psi_{in,\tau}\rangle = |F_{\tilde\alpha}\rangle^{ext} \otimes |F_{\tilde\alpha}\rangle^{int} \otimes |Pulse\rangle_L, \quad (17.36)$$

where $\tilde{\alpha}$ labels the particular state of definite number of particles randomly chosen by the CSL dynamics to collapse the initial state into.

The next ingredient we need to incorporate in our treatment is the role of quantum gravity. That we do by making a couple of mild assumptions regarding the nature of the underlying quantum gravity theory. First, we assume that it resolves the singularity of the black hole and that it leads, after the complete evaporation, to some reasonable region of space–time that can be described, to a good approximation, by classical notions. Second, we assume that quantum gravity does not lead, at the macro-scale, to large violations of the basic space–time conservation laws. With these assumptions, we can further evolve the interior state to a *postsingularity* (*ps*) stage an write

$$\left|\Psi_{in,ps}\right\rangle = |F_{\tilde{\alpha}}\rangle^{ext} \otimes |0^{ps}\rangle, \qquad (17.37)$$

where $|0^{ps}\rangle$ represents a zero energy and momentum state, corresponding to a trivial region of space–time (recall that we have ignored possible small remnants). We end up, then, with a pure quantum state. However, we do not know which one it is because that is determined randomly by the specific realization of the stochastic functions of the CSL dynamics.

What we have to do at this point, then, is to consider an ensemble of systems, all prepared in the same initial state of equation (17.34). Then we consider the CSL evolution of the ensemble up to the hypersurface just before the singularity, and, finally, we use what was assumed about quantum gravity to further evolve those states to a postsingularity stage. As a result of all this, together with the fact that the CSL evolution at the ensemble level leads to probability distributions that are compatible with Born's rule, one can check that the density matrix characterizing the ensemble in the postsingularity stage is

$$\begin{aligned}\rho^{ps} &= N^2 \sum_{\alpha} e^{-\frac{2\pi}{\hbar} E_{\alpha}} |F_{\alpha}\rangle^{ext} \otimes |0^{ps}\rangle \langle F_{\alpha}|^{ext} \otimes \langle 0^{ps}| \\ &= \rho^{ext}_{Thermal} \otimes |0^{ps}\rangle \langle 0^{ps}|. \end{aligned} \qquad (17.38)$$

Thus, the starting point is a pure state, and at the end the situation is described by a proper thermal state, which expresses the fact that, as a result of the stochastic character of the CSL evolution, we only have statistical information regarding the actual final pure state of the system. Information was lost as a result of the (slightly modified) quantum evolution. Clearly, there is nothing paradoxical in the resulting picture.

One might harbor various concerns about the overall picture we have presented. The first one is the dependence of the results on the choice of foliation and, more generally, the fact that the whole scheme employed is non-relativistic. Of course this is a serious problem, but it can be overcome by employing a relativistic version of the collapse dynamics. In fact, a similar analysis was performed in [32] using a relativistic version of CSL recently developed in [39]. We also note that our choice of collapse operators was *ad hoc*, and one might worry whether the end result crucially depends on that choice. It is important to point out, though, that the no-signaling theorem (also valid within GRW and CSL) ensures that the density matrix characterizing the situation in the region exterior to the horizon is insensitive to the choice of collapse operators relevant for the dynamics in the interior.

There are other aspects in which the stated treatment needs improvement. One is the question of back-reaction, which we have not, at this point, incorporated in any meaningful sense (beyond the simple expectation that as the black hole radiates, its mass decreases). The other major concern is the question of the nature of the universal collapse operators that should appear in the general form of the collapse dynamics. Such a theory has to involve a universal collapse operator that in the non-relativistic context reduces to a smeared position operator and in the situation at hand corresponds to something leading to similar results as those obtained in the treatment described. Finally, we have made some important assumptions about quantum gravity that, of course, could simply turn out to be false.

17.4.4 The Problem of Time in Quantum Gravity

The diffeomorphism invariance of general relativity leads to very problematic implications when one attempts to apply a canonical quantization procedure to the theory. In order to implement such a method, one starts with the Hamiltonian formulation of the classical theory, which corresponds to choosing a foliation Σ_t and taking as canonical data the 3-metrics h_{ab} of the hypersurfaces and their conjugate momenta π^{ab} (for simplicity, we will restrict ourselves to general relativity in the absence of matter fields). The foliation is then characterized in terms of the so-called lapse function N and shift vectors N^a, which determine the points of the Σ_t and $\Sigma_{t+\Delta t}$ hypersurfaces used to describe the evolution of the canonical variables. Moreover, the canonical data is constrained by

$$\mathcal{H}(h_{ab}, \pi^{ab}) = 0 \quad \text{and} \quad \mathcal{H}_a(h_{ab}, \pi^{ab}) = 0 \quad (17.39)$$

where $\mathcal{H}$ and $\mathcal{H}_a$ are specific functions of the canonical variables. These are known as the Hamiltonian and diffeomorphism constraints, respectively. Finally, the Hamiltonian that generates the evolution along the vector field $t^a = n^a N + N^a$ can be expressed as

$$H = \int d^3x \sqrt{h} [N\mathcal{H} + N^a \mathcal{H}_a]. \quad (17.40)$$

The canonical quantization procedure involves replacing the canonical variables h_{ab} and π^{ab} with operators in a Hilbert space $\hat{h}_{ab}$ and $\hat{\pi}^{ab}$, such that the Poisson brackets are suitably replaced by commutation relations. In order to impose the constraints, one starts with an auxiliary Hilbert space $\mathcal{H}_{Aux}$ (usually taken to be the space of wave functionals on the configuration variables), from which the physical Hilbert space $\mathcal{H}_{Phys}$ is constructed as the subset of $\mathcal{H}_{Aux}$ satisfying the operational constraints

$$\hat{\mathcal{H}}\Psi(h_{ab}) = 0 \quad \text{and} \quad \hat{\mathcal{H}}_a\Psi(h_{ab}) = 0. \qquad (17.41)$$

As always, time evolution is controlled by the Schrödinger equation, but since, on physical states, $\hat{H} = \int d^3x\sqrt{h}[N\hat{\mathcal{H}} + N^a\hat{\mathcal{H}}_a] = 0$, it is clear that the state of the system is independent of t. We end up, then, with physical states that may depend on the spatial metric h_{ab} but not on time. Time has then completely disappeared for the physical description provided by the theory; this is the problem of time in quantum gravity.

The problem of time, then, is related to the fact that the Hamiltonian, which is the generator of change, is equal to zero. The situation clearly changes when collapse processes are incorporated into the picture, that is, if the standard evolution equation is replaced by something like

$$id\Psi(h_{ab}) = \left\{\int dt \int d^3x\sqrt{h}[N\hat{\mathcal{H}} + N^a\hat{\mathcal{H}}_a] + \int d^4x\hat{C}(x)\right\}\Psi(h_{ab}), \qquad (17.42)$$

where $\hat{C}(x)$ is an operator characterizing the effects of the collapse dynamics. In such case, the full evolution will not be controlled exclusively by a Hamiltonian constructed out of the theory's constraints, so, even though the constraints annihilate the physical states, they will display a non-trivial evolution. The upshot is that temporal change is brought back into the picture, thus solving the problem of time.

17.4.5 The Weyl Curvature Hypothesis

The second law of thermodynamics has generated intense debates throughout the years. The main issue under discussion is the fact that it is not clear how such a time–asymmetric law can emerge from fundamental laws of nature, like those of general relativity and quantum theory, which are essentially time-symmetric. A popular solution to such a question, the so-called *past hypothesis*, is to postulate that the universe started in a state of extremely low entropy. In [91], Penrose conjectured that the past hypothesis arises from a constraint on the initial value of the Weyl curvature, keeping it very low. Penrose's proposal is very different from standard physical laws, which govern the dynamics rather than the initial conditions, and some have found this odd feature dissatisfying (e.g., lacking explanatory power). Here we show how the adoption of an objective collapse model, particularly one with a curvature-dependent collapse rate as in equation (17.7), allows for a *dynamical* explanation of Penrose's conjecture (see [34] for more details).

We begin by assuming that the very early universe was characterized by wildly varying, generically high values of the Weyl curvature W. As a result, the collapse rate $\lambda(W)$ was very large, and the evolution was dominated by the stochastic component of CSL (i.e., the non-standard term in equation (17.1), with λ_0 substituted by $\lambda(W)$). That implied an extremely stochastic evolution for both matter and geometry. Such type of evolution continued until, by mere chance, a small value of W was obtained. From then on, the evolution settled into the Hamiltonian-dominated regime (i.e., the standard part of equation (17.1)), associated with an almost constant value of the matter density, a value of R with similar characteristics and a very small value of W. Of course, such scenario is precisely what Penrose conjectured. Therefore, a collapse model with a curvature-dependent collapse rate offers a dynamical justification for what Penrose introduced as a constraint on initial conditions in order to account for the very special initial state of the universe.

After a small value of W is randomly achieved, the appropriate conditions for the onset of inflation develop. Inflation further flattens the spatial geometry and leads to the standard story we described in section 17.4.1, with collapses playing a crucial role in the production of seeds of structure. After that, the rate of collapse diminishes, but not to zero, and becomes essential for the job it was originally designed for, i.e., the suppression of superpositions of well-localized ordinary macroscopic objects. The regime where W is large is encountered again within black holes, where, as we explained in section 17.4.3, the collapse mechanism takes care of the information loss required in order to avoid paradoxical situations.

17.5 Conclusions

We have reviewed some known arguments and offered some additional ones, suggesting a deep connection between the conceptual problems at the foundations of quantum theory and topics that are usually considered to belong to cosmology and quantum gravity research. More particularly, we have shown that one of the proposals for dealing with the measurement problem in quantum mechanics, namely the dynamical reduction approach, offers attractive resolutions of various outstanding problems that occur at the interface of the quantum and general relativistic realms. These include the origin of the seeds of cosmic structure from quantum fluctuations during the inflationary era, the black hole information puzzle, and the problem of time in quantum theory.

Moreover, we have seen that, following such a line of thought, one also finds natural explanations for the lack of detection of primordial gravity waves (trough the search for B-modes of polarization in the CMB), for the peculiar value of the cosmological constant, and for some of the special features that, according to Penrose, must have characterized the initial state of the universe. Of course, all these results involve, at this point, several simplifying considerations and some *ad hoc* assumptions, so, in order to be considered as completely satisfactory answers, they require a detailed elucidation of some of the underlying premises. However, we find that the progress achieved so far must be regarded as rather promising, in particular when considering the short time the program has been under development and the scarce number of researchers involved. We are convinced that, by having a well-defined program in which to work, problems and obstacles become visible; and by facing them, one moves forward either by ruling out specific alternatives or by pinpointing crucial characteristics that can make them viable.

Acknowledgments

We acknowledge partial financial support from DGAPA-UNAM project IG100316. DS was further supported by CONACyT project 101712.

References

[1] D. Bohm. A suggested interpretation of quantum theory in terms of 'hidden' variables. *Phys. Rev.*, 85:166–193, 1952.
[2] G. C. Ghirardi, A. Rimini, and T. Weber. Unified dynamics for microscopic and macroscopic systems. *Phys. Rev. D*, 34:470–491, 1986.
[3] P. Pearle. Combining stochastic dynamical state vector reduction with spontaneous localization. *Phys. Rev. A*, 39:2277–2289, 1989.
[4] H. Everett. 'Relative state' formulation of quantum mechanics. *Rev. Mod. Phys.*, 29(3), 1957.
[5] R. B. Griffiths. Consistent histories and the interpretation of quantum mechanics. *J. Stat. Phys.*, 36:219, 1984.
[6] R. Omnes. *Interpretation of Quantum Mechanics*. Princeton University Press, 1994.
[7] M. Gell-Mann and J. B. Hartle. Quantum mechanics in the light of quantum cosmology. In W. Zurek, editor, *Complexity, Entropy, and the Physics of Information, SFI Studies in the Sciences of Complexity, Vol. VIII*. Addison Wesley, 1990.
[8] S. Saunders, J. Barrett, A. Kent, and D. Wallace, editors. *Many Worlds? Everett, Quantum Theory, and Reality*. Oxford University Press, 2010.
[9] E. Okon and D. Sudarsky. On the consistency of the consistent histories approach to quantum mechanics. *Found. Phys.*, 44:19–3, 2014.
[10] E. Okon and D. Sudarsky. Measurements according to consistent histories. *Stud. Hist. Phil. Mod. Phys.*, 48:7–2, 2014.
[11] E. Okon and D. Sudarsky. The consistent histories formalism and the measurement problem. *Stud. Hist. Phil. Mod. Phys.*, 52:217–22, 2015.
[12] B. Mielnik. Generalized quantum mechanics. *Comm. Math. Phys.*, 35:221–256, 1974.
[13] R. Penrose. Time asymmetry and quantum gravity. In C. J. Isham, R. Penrose, and D. W. Sciama, editors, *Quantum Gravity II*, pp. 244–272. Clarendon Press, 1981.

[14] L. Diosi. Gravitation and quantum-mechanical localization of macro-objects. *Phys. Lett. A*, 105:199, 1984.

[15] J. Ellis, S. Mohanty, and D. V. Nanopoulos. Quantum gravity and the collapse of the wavefunction. *Phys. Lett. B*, 221:113–119, 1989.

[16] I. C. Percival. Quantum space-time fluctuations and primary state diffusion. *Proc. Roy. Soc. London Ser. A*, 451:503, 1995.

[17] P. Pearle and E. Squires. Gravity, energy conservation, and parameter values in collapse models. *Found. Phys.*, 26:291–305, 1996.

[18] S. Goldstein and S. Teufel. Quantum spacetime without observers: ontological clarity and the conceptual foundations of quantum gravity. In C. Callender and N. Huggett, editors, *Physics Meets Philosophy at the Planck Scale*, pages 275–289. Cambidge University Press, 2001.

[19] S. Ryu and T. Takayanagi. Holographic derivation of entanglement entropy from the anti-de Sitter space/conformal field theory correspondence. *Phys. Rev. Lett.*, 96: 181602, 2006.

[20] J. Maldacena and L. Susskind. Cool horizons for entangled black holes. *Fortsch. Phys.*, 61:781–811, 2013.

[21] A. Perez, H. Sahlmman, and D. Sudarsky. On the quantum mechanical origin of the seeds of cosmic structure. *Class. Quant. Grav.*, 23:2317, 2006.

[22] G. Len and D. Sudarsky. The slow roll condition and the amplitude of the primordial spectrum of cosmic fluctuations: Contrasts and similarities of standard account and the 'collapse scheme'. *Class. Quant. Grav.*, 27:225017, 2010.

[23] D. Sudarsky. Shortcomings in the understanding of why cosmological perturbations look classical. *IJMPD*, 20:509, 2011.

[24] G. Len, A. De Unanue, and D. Sudarsky. Multiple quantum collapse of the inflaton field and its implications on the birth of cosmic structure. *Class. Quant. Grav.*, 28: 155010, 2011.

[25] S. J. Landau, C. G. Scoccola, and D. Sudarsky. Cosmological constraints on nonstandard inflationary quantum collapse models. *Phys. Rev. D*, 85:123001, 2012.

[26] A. Diez-Tejedor and D. Sudarsky. Towards a formal description of the collapse approach to the inflationary origin of the seeds of cosmic structure. *JCAP*, 045:1207, 2012.

[27] P. Caåate, P. Pearle, and D. Sudarsky. CSL quantum origin of the primordial fluctuation. *Phys. Rev. D*, 87:104024, 2013.

[28] G. Len, S. J. Landau, C. G. Scoccola, and D. Sudarsky. Quantum origin of the primordial fluctuation spectrum and its statistics. *Phys. Rev. D*, 88:023526, 2013.

[29] E. Okon and D. Sudarsky. Benefits of objective collapse models for cosmology and quantum gravity. *Found. Phys.*, 44:114–143, 2014.

[30] S. K. Modak, L. Ortiz, I. Peåa, and D. Sudarsky. Non-paradoxical loss of information in black hole evaporation in a quantum collapse model. *Phys. Rev. D*, 91(12):124009, 2015.

[31] E. Okon and D. Sudarsky. The black hole information paradox and the collapse of the wave function. *Found. Phys.*, 44:461–470, 2015.

[32] D. Bedingham, S. K. Modak, and D. Sudarsky. Relativistic collapse dynamics and black hole information loss. *Phys. Rev. D*, 94(4):045009, 2016.

[33] E. Okon and D. Sudarsky. Black holes, information loss and the measurement problem. *Found. Phys.*, 47:120–131, 2017..

[34] E. Okon and D. Sudarsky. A (not so?) novel explanation for the very special initial state of the universe. *Class. Quant. Grav.*, 33:2016.

[35] E. Okon and D. Sudarsky. Less decoherence and more coherence in quantum gravity, inflationary cosmology and elsewhere. *Found. Phys.*, 46:852–79, 2016.
[36] A. Majhi, E. Okon, and D. Sudarsky. Reassessing the link between B-modes and inflation. *arXiv:1607.03523*, 2016.
[37] T. Josset, A. Perez, and D. Sudarsky. Dark energy from violation of energy conservation. *Phys. Rev. Lett.*, 118:021102, 2017.
[38] R. Tumulka. A relativistic version of the Ghirardi–Rimini–Weber model. *J. Stat. Phys.*, 125(821):10, 2006.
[39] D. J. Bedingham. Relativistic state reduction dynamics. *Found. Phys.*, 41:686, 2011.
[40] P. Pearle. Relativistic dynamical collapse model. *Phys. Rev. D*, 91(10):105012, 2015.
[41] N. Bohr. Discussion with Einstein on epistemological problems in atomic physics. In P. A. Schilpp, editor, *Albert Einstein: Philosopher-scientist*, page 201–41. Open Court, 1949.
[42] J. S. Bell. *Speakable and Unspeakable in Quantum Mechanics*. Cambridge University Press, 2nd edition, 2004.
[43] J. S. Bell. Quantum mechanics for cosmologists. In *Quantum Gravity II*. Oxford University Press, 1981.
[44] J. B. Hartle. Generalizing quantum mechanics for quantum gravity. *Int. J. Theor. Phys.*, 45:1390–1396, 2006.
[45] R. D. Sorkin. Impossible measurements on quantum fields. In B. L. Hu and T. A. Jacobson, editors, *Directions in General Relativity, Vol. II: A Collection of Essays in honor of Dieter Brill's Sixtieth Birthday*. Cambridge University Press, 1993.
[46] R. Penrose. On gravity's role in quantum state reduction. *Gen. Rel. Grav.*, 28:581, 1996.
[47] L. Diosi. A universal master equation for the gravitational violation of quantum mechanics. *Phys. Lett. A*, 120:377, 1987.
[48] A. Tilloy and L. Diosi. Sourcing semiclassical gravity from spontaneously localized quantum matter. *Phys. Lett. D*, 93:024026, 2016.
[49] M. Derakhshani. Newtonian semiclassical gravity in the Ghirardi–Rimini–Weber theory with matter density ontology. *Phys. Lett. A*, 378:14–15, 2014.
[50] S. Nimmrichter and K. Hornberger. Stochastic extensions of the regularized Schrödinger–Newton equation. *Phys. Rev. D*, 91:024016, 2015.
[51] I. Pikovski, M. Zych, F. Costa, and C. Brukner. Universal decoherence due to gravitational time dilation. *Nature Phys.*, 11:668, 2015.
[52] Y. Bonder, E. Okon, and D. Sudarsky. Can gravity account for the emergence of classicality? *Phys. Rev. D*, 92:124050, 2015.
[53] Y. Bonder, E. Okon, and D. Sudarsky. Questioning universal decoherence due to gravitational time dilation. *Nature Phys.*, 12:2, 2016.
[54] A. Ney and D. Z. Albert, editors. *The Wave Function: Essays on the Metaphysics of Quantum Mechanics*. Oxford University Press, 2013.
[55] V. Allori. Primitive ontology in a nutshell. *Int. J. Quant. Found.*, 1(3):107–122, 2015.
[56] D. N. Page and C. D. Geilker. Indirect evidence for quantum gravity. *Phys. Rev. Lett.*, 47:979, 1981.
[57] K. Eppley and E. Hannah. The necessity of quantizing the gravitational field. *Found. Phys.*, 7:51–8, 1977.
[58] N. Huggett and C. Callender. Why quantize gravity (or any other field for that matter)? *Phil. Sci.*, 68(3):S382–S394, 2001.
[59] J. Mattingly. Is quantum gravity necessary? In A. J. Kox and J. Eisenstaedt, editors, *The Universe of General Relativity*, pages 325–338. Birkhuser, 2005.

[60] J. Mattingly. Why Epply and Hannah's thought experiment fails. *Phys. Rev. D*, 73: 064025, 2006.

[61] S. Carlip. Is quantum gravity necessary? *Class. Quant. Grav.*, 25:154010, 2008.

[62] C. I. Kuo and L. Ford. Semiclassical gravity theory and quantum fluctuations. *Phys. Rev. D*, 47:4510, 1993.

[63] T. Jacobson. Thermodynamics of spacetime: The Einstein equation of state. *Phys. Rev. Lett.*, 75:1260, 1995.

[64] N. Seiberg. Emergent spacetime. *arXiv:hep-th/0601234*, 2006.

[65] K. H. Knuth and N. Bahreyni. A potential foundation for emergent space–time. *J. Math. Phys.*, 55:112501, 2014.

[66] R. Gambini, R. A. Porto, and J. Pullin. A relational solution to the problem of time in quantum mechanics and quantum gravity: a fundamental mechanism for quantum decoherence. *New J. Phys.*, 5:45, 2004.

[67] A. Bassi, K. Lochan, S. Satin, T. Singh, and H. Ulbricht. Models of wave-function collapse, underlying theories, and experimental tests. *Rev. Mod. Phys.*, 85:471, 2013.

[68] D. Z. Albert. Elementary quantum metaphysics. In J. T. Cushing, A. Fine, and S. Goldstein, editors, *Bohmian Mechanics and Quantum Theory: An Appraisal*, pages 277–284. Kluwer Academic Publishers, 1996.

[69] J. S. Bell. Are there quantum jumps? In C. W. Kilminster, editor, *Schroödinger: Centenary Celebration of a Polymath*, page 109–23. Cambridge University Press, 1987.

[70] G. C. Ghirardi, R. Grassi, and F. Benatti. Describing the macroscopic world: closing the circle within the dynamical reduction program. *Found. Phys.*, 35:5, 1995.

[71] D. Bedingham, D. Dürr, G. C. Ghirardi, S. Goldstein, R. Tumulka, and N. Zanghi. Matter density and relativistic models of wave function collapse. *J. Stat. Phys.*, 154: 623–31, 2014.

[72] D. Sudarsky. The inflationary origin of the seeds of cosmic structure: Quantum theory and the need for novel physics. *Fund. Theor. Phys.*, 177:349, 2014.

[73] B. L. Hu and E. Verdaguer. Stochastic gravity: Theory and applications. *Liv. Rev. Rel.*, 11:3, 2008.

[74] P. Pearle and E. Squires. Bound state excitation, nucleon decay experiments, and models of wave function collapse. *Phys. Rev. Lett.*, 73:1, 1994.

[75] A. R. Liddle. An introduction to cosmological inflation. In A. Masiero, G. Senjanovic, and A. Smirnov, editors, *High Energy Physics and Cosmology*, World Scientific, 1999.

[76] J. A. Peacock. *Cosmological Physics*. Cambridge University Press, 1999.

[77] M. J. Mortonson and U. Seljak. A joint analysis of Planck and bicep2 B modes including dust polarization uncertainty. *JCAP*, 1410:035, 2014.

[78] Planck Collaboration. E- and B-modes of dust polarization from the magnetized filamentary structure of the interstellar medium. *Astron. Astrophys.*, 586:A141, 2016.

[79] M. Kamionkowski and E. D. Kovetz. The quest for B modes from inflationary gravitational waves. *Annual Rev. Astron. and Astroph.*, 54:227–269, 2016.

[80] R. Colella, A. W. Overhauser, and S. A. Werner. Observation of gravitationally induced quantum interference. *Phys. Rev. Lett.*, 34:1472–474, 1975.

[81] V. V. Nesvizhevsky, H. G. Brner, A. K. Petukhov, H. Abele, S. Baeler, F. J. Rue, Th. Stferle, A. Westphal, A. M. Gagarsky, G. A. Petrov, and A. V. Strelkov. Quantum states of neutrons in the earth's gravitational field. *Nature*, 415:297, 2002.

[82] A. G. Riess et al. Observational evidence from supernovae for an accelerating universe and a cosmological constant. *Astron. J.*, 116:1009–1038, 1998.

[83] S. Perlmutter et al. Measurements of omega and lambda from 42 high redshift supernovae. *Astrophys. J.*, 517:565–586, 1999.

[84] S. Weinberg. The cosmological constant problem. *Rev. Mod. Phys.*, 61:1–23, 1989.

[85] Planck Collaboration. Planck 2015 results. i. Overview of products and scientific results. *Astron. Astrophys*, 594:A1, 2015.

[86] S. Hawking. Particle creation by black holes. *Commun. Math. Phys.*, 43:199–20, 1975.

[87] A. Almheiri, D. Marolf, J. Polchinski, and J. Sully. Black holes: complementarity or firewalls? *JHEP*, 62, 2013.

[88] B. d'Espagnat. *Conceptual Foundations of Quantum Mechanics*. Addison-Wesley, 2nd edition, 1976.

[89] C. G. Callan, S. B. Giddings, J. A. Harvey, and A. Strominger. Evanescent black holes. *Phys. Rev. D*, 45:R1005, 1992.

[90] A. Fabbri and J. Navarro-Salas, editors. *Modeling Black Hole Evaporation*. Imperial College Press, 2005.

[91] R. Penrose. Singularities and time-asymmetry. In S. W. Hawking and W. Israel, editors, *General Relativity: An Einstein Centenary Survey*, page 581–38. Cambridge University Press, 1979.

Index

Adler, Stephen L., 47, 49, 53, 55, 56, 135, 148, 248, 307
   theory of trace dynamics of, 295, 301–308
Aharonov, Yakir, 245
Albert, David Z., 111, 125, 146, 147, 157, 158, 237, 239, 240, 243, 264
Allori, Valia, 76, 104, 105, 117, 159, 160, 162
Anandan, Jeeva, 245

Bacciagaluppi, Guido, 79, 259
Barrett, Jeffrey A., 226, 230, 239, 240
Barrett, Jonathan, 243
Bassi, Angelo, 4, 48, 49, 80, 125–127, 142, 143, 159, 257, 258, 271, 309
Bedingham, Daniel J., 74, 81
Bell's Everett (?) theory, 247
Bell's theorem, 176, 179, 181
Bell, John S., 19, 47, 76, 98, 116, 125, 134–137, 158, 160, 169, 188, 209, 213–215, 219, 260, 271, 314
Bohm's theory, 97, 155, 164, 228, 229, 232–236, 242, 247, 248, 254
   Bohmian particles in, 233, 234, 247
   failure to satisfy Born's rule of, 234
   forms of psychophysical connection in, 232–234
   guiding equation of, 247
   origin of the Born probabilities in, 247
   potential problems of, 234
   result assumptions of, 232–234, 253
   superluminal signaling in, 233, 234
Bohm, David, 179, 233
   result assumption of, 233
Bohmian mechanics
   see also Bohm's theory, 104, 135–139, 143, 144, 169, 171–173, 175, 176, 179–181, 207, 213–215, 221, 312
Bohr, Niels, 108, 314
Born rule, 26, 49, 55, 62, 125, 130, 225, 226, 228, 230, 231, 234–236, 246, 251, 253, 306, 319, 320
   derivation of, 62–64, 246
Bose–Einstein condensates, 187, 201, 202, 299

Brody, Dorje C., 47, 49, 55, 70
Brown, Harvey. R., 233, 234
Brownian motion, 47, 49, 53, 55, 63, 64, 70, 218, 301, 320

causal sets, 91, 313, 318
charge cloud, 109, 244
charge distribution, 109, 245, 246
classical space–time, 300, 301, 304–306, 308, 309
   emergence of, 295, 304, 305
collapse models
   see also collapse theories, 85, 207, 304
   energy non-conservation problem of, 88, 90, 145, 177, 252
   energy-conserved, 252
   energy-driven, 47–49, 52, 54–57, 62, 63, 69
   experimental tests of, 187, 201, 298
   general relativistic, 318
   gravity-induced, 187–189, 200, 227, 248, 316
   in terms of RDM of particles, 248
   relativistic, 74, 76, 85, 86, 296, 299, 307
   general, 312
collapse postulate, 106, 107, 156, 167, 177, 178, 181, 211, 220, 348
collapse theories, 8, 10, 97, 111, 113, 125, 134–137, 143, 154–158, 160, 161, 163, 164, 168, 181, 221, 226–229, 232, 239, 240, 242, 247, 248, 253, 254, 313, 318, 320, 322, 325, 330, 331
   and black hole information puzzle, 313, 314, 340
   and quantum gravity, 312
   and special relativity, 161
   distributional ontology for, 118, 121
   empirically testable dynamics in, 3
   matter density ontology for, 134
   particle ontology for, 248
   quantum state monism for, 97
   structured tails problem of, 146, 178, 240
      new solution to, 242
   tails problem of, 111, 139, 146, 158, 159, 178, 228
      solutions to, 159

composition principle, 124, 127, 128
configuration space, 116, 117, 137, 140, 143, 158–160, 169, 171, 173, 175–179, 181, 234, 236, 243, 244, 320
cosmic microwave background (CMB), 307, 313, 323, 326, 328, 329, 341
CSL theory, 8–10, 23–25, 51, 78–80, 82, 84, 169, 219, 297–300, 303, 318, 319, 325, 327, 328, 331, 334–338, 340
  experimental tests of, 298, 299
  relativistic generalisaton of, 300
  relativistic version of, 24

Dürr, Detlef, 159, 160, 162, 164, 169
de Broglie, Louis, 97, 179
determinate-experience problem, *see* measurement problem, 226, 230
determinism, 211, 212, 214, 215, 217, 220
DeWitt, Bryce S., 229, 237
Diósi, Lajos, 3, 47–49, 69, 135, 187, 189, 219, 312
Dirac, Paul A. M., 105, 124, 175, 275, 319
discontinuous jumps, 219, 220
discrete space–time, 74, 76, 88, 90, 253
dualism, 171, 210, 217, 237
  in orthodox quantum mechanics, 215
  of matter in physical space, 171
  physical, 207, 210, 216

eigenstate–eigenvalue link, 98, 106, 107, 111, 119–121
Einstein hole argument, 300, 305
Einstein's equation, 89, 90
Einstein–Bohr debates, 314
energy density operators, 23, 24, 43
EPR correlations, 309, 319
EPR experiment, 179, 181, 274
Esfeld, Michael, 79, 105, 167
Everett's theory, 226–229, 231, 232, 237–240, 242, 247, 248, 253, 254, 348
  failure of psychophysical supervenience in, 238–240
  origin of the Born probabilities in, 247
  understandings of multiplicity in, 237

Feynman, Richard, 172, 268
field ontology, 244, 245
Frigg, Roman, 124, 125, 127
Fuchs, Christopher A., 210
Fuentes, Ivette, 187

Gao, Shan, 49, 115, 125, 225, 227, 230, 235, 236, 238, 243–246, 248, 249, 252, 258
Ghirardi, GianCarlo, 12, 47–49, 51, 107, 108, 110–115, 125–127, 135, 142, 143, 156, 159, 169, 187, 219, 229, 245, 265
Gisin, Nicolas, 49, 54, 105, 135, 207
Goldstein, Sheldon, 76, 113, 116, 159, 160
Graham, R. Neill, 229, 237

Grassi, Renata, 111, 115, 187
gravitational self-energy, 187, 189, 195
GRW theory, 3–9, 49, 77, 79, 82, 84, 116, 117, 124–127, 134, 135, 137–142, 144, 146, 147, 156, 157, 160, 169, 171, 177, 178, 180, 181, 187, 219, 240, 265, 290, 313, 318–320, 329, 338
  and special relativity, 160
  counting anomaly in, 124, 126, 127, 129, 131, 140, 158–161, 164
  flash ontology of, 76, 116, 117, 134, 137, 146, 155, 160, 163, 164, 180
  fuzzy link in, 124, 126–128, 130, 146
  mass density ontology of, 155, 159, 160, 163, 164, 245, 320
  property structure of, 124
GRWf theory, 117, 137–141, 144–148, 169–171, 173, 175, 180, 181
GRWm theory, 117, 137–142, 144–148, 169, 170, 173–175, 177–179

Hilbert space, 12, 13, 47, 105, 106, 116, 132, 306, 321, 325, 334, 338, 339
  Brownian motion in, 218
  continuous stochastic process in, 12, 13, 20, 21
  discontinuous stochastic process in, 12–14, 19, 20
Hoefer, Carl, 125
Hughston, Lane P., 47, 48, 49, 55, 70

incompatibility of Hilbert spaces, 200
Isham, Christopher J., 50
Ismael, Jenann, 259

Károlyházy, F., 48, 187
Kibble, Thomas, W. B., 187
Kochen, Simon B., 237
Kochen–Specker theorem, 237

Landsman, Nicolaas P., 246
Leggett, Anthony J., 135
Leifer, Matthew S., 243
Lewis, Peter J., 98, 111, 124, 126, 127, 129–131, 140, 142, 154, 158, 161, 233, 234, 243
local beables, 98, 104, 116, 134, 137, 169, 208
Loewer, Barry, 111, 125, 146, 158, 239, 240
Lorentz transformations, 147, 303

macroscopic objects, 21, 108, 110, 118, 126, 127, 136, 149, 157–159, 172, 173, 176, 221, 295, 297, 301, 304–309, 320, 340
many-minds theory, 237, 239
many-worlds theory, 145, 147, 164, 168, 207
mass density operators, 24, 108, 111
  smeared, 111
Maudlin, Tim, 99, 135, 148, 149, 159, 160, 164, 167, 178, 226–228, 230, 232, 264
Maxwell, Nicholas, 257–261, 264, 265, 267, 271
McQueen, Kelvin J., 240

measurement problem, 12, 21, 75, 101, 138, 154, 155, 158, 167, 168, 170, 172, 175, 178, 181, 207, 210–212, 216, 217, 220, 221, 226–232, 236, 237, 240, 246, 247, 253, 254, 297, 307, 315, 318, 324, 340
   and psychophysical connection, 226, 228, 231, 232, 253
   Gao's formulations of, 230
   Maudlin's formulation of, 228
   solutions to, *see* Bohm's theory, Everett's theory, and collapse theories, 231, 232
mirror superposition, 201
Monton, Bradley, 111, 126, 127, 130, 131, 134, 138, 157, 243
Mott, Nevill F. , 324
Myrvold, Wayne C., 97, 164

Newton–Schrödinger equation, 196, 315
Newtonian mechanics, 136, 214, 215
Ney, Alyssa, 157, 164
Nicrosini, Oreste, 12
no-signaling theorem, 234, 338
non-material stuff, 217
nonlinear filtering theory, 47, 55

Okon, Elias, 312
Omnès, Roland, 274

particle number density operator
   smeared, 78
particle ontology, 173–176, 181, 244, 245
passage of time, 221
Pearle, Philip, 47–49, 51, 98, 110, 111, 118–120, 135, 145, 148, 187, 219, 229, 290, 291
Penrose, Roger, 3, 47, 48, 51, 83, 135, 187, 227, 248, 263, 265, 312, 315, 335, 339–341
Percival, Ian C., 48, 49, 187
Popper, Karl, 268
primitive ontology, 88, 97, 104, 105, 116, 134, 135, 137, 138, 155, 159–164, 167, 170–173, 175, 176, 181
   epistemic accessibility of, 170, 171, 181
problem of time, 296, 300, 306, 309, 312–314, 316, 338–340
projective measurements, 4
propensities, 261
propensitons, 261, 265
protective measurements, 245
psychophysical connection, 226–229, 231–233, 238, 253
   in Bohm's theory, 232–234
   in Everett's theory, 237–239
psychophysical supervenience, 226, 228, 229, 237–241, 253
Pusey, Matthew F., 75, 243

QBism, 210
quantum entanglement, 172, 207, 244

quantum field theory, 107, 174, 299, 302, 314, 315, 321, 322, 326, 333, 334
quantum gravity, 228, 248, 249, 253, 254, 313–318, 323, 333, 334, 337–340
   loop, 313
   problem of time in, 312, 313
quantum non-locality, 176, 214, 295, 300, 307

random discontinuous motion of particles, 243, 244, 246–248
   and the origin of the Born probabilities, 246, 247
   arguments for, 244, 245
RDM of particles
   see random discontinuous motion of particles, 247
Rimini, Alberto, 24, 49, 51, 110, 156, 169, 187, 219
Rudolph, Terry, 243

Schrödinger's cat, 138, 139, 142, 146, 167, 275
Schrödinger, Erwin, 245, 274
   charge density hypothesis of, 245
Shimony, Abner, 111
simultaneity, 158, 160, 308
Singh, Tejinder P., 295
special relativity, 158, 160, 161
   non-commutative, 295, 300
Specker, Ernst, 237
Squires, Euan J., 187
standard quantum mechanics, 3, 125, 135, 235, 315
   alternatives to, *see* Bohm's theory, Everett's theory, and collapse theories, 247
Stone, Abraham D., 233
Sudarsky, Daniel, 312
superluminal signaling, 233

Tumulka, Roderich, 81, 98, 134, 160, 164, 169

Vaidman, Lev, 245
Valentini, Antony, 234
von Neumann, John, 4, 5, 50, 51, 106, 155, 156, 275
   collapse postulate of, 156

Wallace, David, 106, 147, 163, 178, 233, 234, 247, 259
wave function
   epistemic view of, 227
   ontic view of, 227
   realism, 243, 244, 246
   reality of, 245
Weber, Tullio, 49, 76, 77, 125, 134, 156, 169, 187, 219
Weinberg, Steven, 47, 135, 279
Weyl curvature hypothesis, 339
Wheeler, John A., 136
   delayed-choice paradox of, 136

Zanghì, Nino, 159, 160, 162, 164
Zeh, H. Dieter, 237